POWER
ECONOMICS

D1342525

**Books of Related Interest from IEEE Press**

Understanding Power Quality Problems: Voltage Sags and Interruptions
Math H. J. Bollen
2000        Hardcover        576pp        0-7803-4713-7

Analysis of Electric Machinery and Drive Systems, Second Edition
Paul C. Krause, Oleg Wasynczuk, and Scott D. Sudhoff
2002        Hardcover        624pp        0-471-14326-X

Electricity Economics: Regulation and Deregulation
Geoffrey Rothwell and Tomas Gomez
2003        Hardcover        300pp        0-471-23437-0

# POWER SYSTEM ECONOMICS

## Designing Markets for Electricity

Steven Stoft

IEEE Press

**WILEY-INTERSCIENCE**

A JOHN WILEY & SONS, INC., PUBLICATION

Cover painting by W. Louis Sonntag, Jr. (1869–1898), *The Bowery at Night,* c. 1895. Early deregulated electricity market with trolleys powered by Westinghouse's AC and shops probably illuminated with Edison's DC. The houses may still be lit by gas. The "Third Avenue Elevated" (1878), whose noise and shadows contributed to the decline of New York's once-elegant theater district, will soon be electrified. (*Uncle Tom's Cabin* was first staged in the Bowery Theater visible at the extreme left.) Arc lights, brought to New York streets in 1880 by Charles Brush, transformed night life. Sonntag frequently depicted the resulting sense of glamour and excitement. The watercolor was a Gift of Mrs. William B. Miles to the Museum of the City of New York.

This text is printed on acid-free paper. ∞

For ordering and customer service, call 1-800-CALL WILEY.

*Library of Congress Cataloging-in-Publication Data is available.*

ISBN 0-471-15040-1

*For my mother, whose writing inspired me to think I could,*
*and my father who taught me to test high voltage with one hand behind my back*

# Contents in Brief

# Contents

**Part 2.   Reliability, Price Spikes, and Investment**

## Part 3.   Market Architecture

Contents

# List of Results and Fallacies

## Part 1.  Power Market Fundamentals

# Part 2. Reliability, Price Spikes, and Investment

## Part 3. Market Architecture

## Part 4. Market Power

# Part 5. Basic Locational Pricing

# Preface

My original purpose in writing this book was to collect and present the basic economics and engineering used to design power markets. My hope was to dispel myths and provide a coherent foundation for policy discussions and market design. In the course of writing, I came to understand there is no received wisdom to present on two key issues: price-spikes and pools. While the majority of the book still holds to my first purpose, Parts 2 and 3 are guided as well by a second. They seek to present the two unresolved issues coherently, answer a few basic questions and highlight some of the gaps in our current understanding.

The price-spike issue is how to design the market to accommodate two demand-side flaws underlying the price-spikes that provide incentives for investment in generation. Part 2 shows that some regulation is required until one flaw has been mitigated. The first regulatory goal should be to ensure the revenue from the aggregate price spike is just sufficient to induce a reliable level of generating capacity. This revenue is determine by (1) the duration of the aggregate price spike which is under the influence of NERC guidelines and (2) the height of the price spike which is regulated by FERC. Currently, neither institution appears aware their policies jointly determine investment.

Part 2 provides a framework for computing the level of investment induced by any combination of NERC and FERC policies. Because many combinations will work, it suggests a second goal. Price volatility should be reduced to levels that might be expected from a mature power market—levels far below those observed in the current markets with their incapacitated demand sides. I hope Part 2 will clarify the regulatory options and the need to fix the market's demand side.

While Part 3 presents the standard principles of bilateral markets, exchanges and pools, it is able to make little progress on the second issue, the power-pool question. An exchange is a widely used form of centralized market—the New York Stock Exchange is an example—while pools are peculiar to power markets. Exchanges trade at one price at any given time and location, while pools pay different prices to different generators according to their costs. The differences in transparency and operation are considerable as may be their performance. Unfortunately, little theoretical or empirical research is to be found, and Part 3 can only raise issues and show the answers are far from obvious.

While three pools operate in the eastern U.S. and PJM has been deregulated for nearly four years, no evaluation of their efficiency has been undertaken. The only national effort, a shoestring operation at the Department of Energy, has been crippled by lack of access to data that FERC could easily obtain from the pools. Pro-pool forces within FERC have, for years, blocked any suggestion to evaluate the performance of pools or their potential benefits. No description of any eastern pool, suitable for economic analysis, can be found within FERC or in the public domain.

Theoretical pool descriptions cover ex-ante pricing while knowledgeable observers indicate the eastern ISOs use ex-post pricing. This is said to be based on a philosophy of controlling quantities in real-time and computing prices after the fact. In practice, it involves proprietary calculations that apparently assume the operator's actions were optimal. I could discover no useful discussion of the theory of this critical issue, so readers of Part 3 must wait for a later edition.

Competitive power markets, like regulated markets, must be designed and designed well. Because of the poor quality of many current designs and the lack of a well-tested standard, this book does not recommend a rush to deregulate. A given deregulation may succeed, but economic theory cannot predict when such a complex political process, once begun, will be out-maneuvered by the forces it seeks to harness. If a market is being designed or redesigned, this book is meant to help; if the decision is to wait, this book is meant to make the wait shorter.

## Acknowledgments

Those who undertook to read, correct and criticize drafts provided an invaluable service and deserve thanks from all of my readers, whom they have protected from many confusions, diversions, and errors. For this difficult undertaking I am especially grateful to Ross Baldick, Joe Bowring, Haru Connally, Rob Gramlich, Doug Hale, Alex Henney, Bill Hogan, Mat Morey, Sabine Schnittger, and Jurgen Weiss.

Many others have made more narrowly focused but still invaluable contributions. They provided an ongoing discussion on many topics and constantly provided fresh views and caught errors. Thanks to Darwin Anwar, Gerry Basten, Richard Benjamin, Severin Borenstein, Jason Christian, Ed Mills, Udi Helman, Mike Rothkopf, Erik Hirst, Ben Hobbs, Mangesh Hoskote, Marcelino Madrigal, Dave Mead, Joshua Miller, Alan Moran, Andrew Klingler, Jim Kritikson, Dan Gustafson, Frank Felder, Carl Fuchshuber, Richard Green, Harry Singh, Alasdair Turner, Hugh Outhred, Gail Panagakis, Alex Papalexopolous, Gregory Werden, and James Wightman.

Without the patient support of the IEEE/Wiley staff, John Griffen, Tony Vengraitis, and Andrew Prince, none of this would have been possible. My copy editor, Susan Ingrao, has been a pleasure to work with and tremendously informative, even answering arcane typesetting questions. Remaining errors are the result of my inaccurate corrections or last minute changes.

For support and guidance on every challenge, I have turned first to my wife Pamela who has been my creative advisor, editor, and legal counsel. I thank her for her abundant patience and unerring judgement.

But of all those who have contributed to this book, I owe the most to my mother, Dorothy, who brought her artistry to the dull world of power economics. Through three complete drafts, she gently but persistently corrected and shaped, guided and polished. While the quality of my writing still falls far short of my mother's, it delights me to have learned, at last, a little of her art.

# Acronyms and Abbreviations

|        |                                                        |
|-------:|--------------------------------------------------------|
| AC     | alternating current                                    |
| ACE    | area control error                                     |
| CA ISO | California Independent System Operator                  |
| CFD    | contract for differences                               |
| CLP    | competitive locational price                           |
| CR     | capacity requirement                                   |
| DA     | day ahead (market)                                     |
| DOJ    | Department of Justice                                  |
| FERC   | Federal Energy Regulatory Commission                   |
| FTR    | financial transmission right                           |
| FTC    | Federal Trade Commission                               |
| GT     | gas turbine generator                                  |
| HHI    | Herfindahl-Hirschman index                             |
| ICap   | installed capacity                                     |
| IPP    | independent power producer                             |
| ISO    | independent system operator                            |
| ISO-NE | ISO New England                                        |
| LBMP   | locational-based marginal price                        |
| LHMC   | left-hand marginal cost                                |
| LMP    | locational marginal price                              |
| LRMC   | long-run marginal cost                                 |
| MC     | marginal cost                                          |
| NERC   | North American Electric Reliability Council            |
| NYISO  | the New York Independent System Operator, Inc.         |
| NYSE   | the New York Stock Exchange                            |
| NE     | Nash equilibrium                                       |
| OpRes  | operating reserve                                      |
| PJM    | Pennsylvania-New-Jersey-Maryland Independent System Operator |
| PTR    | physical transmission right                            |
| RHMC   | right-hand marginal cost                               |
| RT     | real time (market)                                    |
| RTO    | regional transmission organization                     |
| SMC    | system marginal cost                                   |
| SMV    | system marginal value                                  |
| TR     | transmission right                                     |
| UC     | unit commitment                                        |
| VOLL   | value of lost load                                     |

## Units Used to Measure Electricity

| | | |
|---|---|---|
| V | volt | The unit of electrical pressure |
| A | amp | The unit of electrical current |
| W | watt | Power (Energy per hour) |
| h | hour | Time |
| Wh | watt-hour | Energy |
| k | kilo | 1000. Used in kW, kWh and kV. |
| M | mega | 1,000,000. Used in MW and MWh. |
| G | giga | 1,000,000,000. Used in GW and GWh. |
| T | tera | $10^{12}$. Used in TWh. |

# Symbols

| Units | Symbol | Definition |
|---|---|---|
| | $\square^e$, $\square^*$ | equilibrium and optimal superscripts |
| | $\square_{peak}$, $\square_{mid}$, $\square_{base}$ | peakload, intermediate, and baseload generating capacity subscripts |
| | $\square_1$, $\square_0$ | day-ahead, and real-time subscripts |
| | $\square_A$, $\square_B$ | bus-A, bus-B subscripts |
| \$/MWh | $AC_E$ | average cost of energy for a load slice |
| \$/MWh | $AC_K$ | average cost of purchasing and using capacity for a load slice |
| \$/MWyear | $ARR$ | annual revenue requirement of a generator |
| \$/MWh | $CC$ | cost of providing spinning-reserve capacity. |
| none | $cf$ | capacity factor |
| none | $C$ | cost of Production |
| none | $D_{LS}$ | duration of load shedding $\quad (L_g > K)$ |
| none | $D_{PS}$ | duration of price spike $\quad (L_g + OR^R > K)$ |
| none | $D_{peaker}$ | duration of peaker use $\quad (L_g + OR^R > K_{base})$ |
| none | $e$ | price elasticity of demand |
| MWh | $E$ | energy |
| \$/MWh | $FC$ | fixed cost |
| \$/MWh | $F_T$ | current price of a future for delivery at time T |
| MW | $g$ | generation out of service |
| none | $h$ | true probability of needing spinning reserves |
| none | $\hat{h}$ | estimated $h$ |
| amps | $I$ | electrical current |
| MW | $K$ | installed generating capacity (ICap) |
| MW | $\bar{K}^e$ | average $K$ in an equilibrium when $K$ is random. |

| MW | $K^R$ | required installed capacity |
|---|---|---|
| MW | $K_{slice}$ | the capacity required by a load slice. |
| MW | $L$ | power demanded by load (economic demand) |
| MW | $\mathcal{L}$ | control-area losses |
| MW | $\ell$ | losses on one line |
| MW | $L_g$ | augmented load $(L + g)$ |
| MW | $L_{g\text{-}max}$ | maximum value of $L_g$. |
| $/MWh | $MC_{LH}$ | left-hand marginal cost |
| MW | $LL$ | lost load $(L_g - K)$ in the Simple Model of Reliability |
| MW | $\overline{LL}$ | average lost load |
| none | $L_X$ | Lerner index |
| $/MWh | $MC$ | marginal cost |
| $/MWh | $MC_R$ | range of $MC$ when $MC$ is ambiguous. |
| $/MW | $OC$ | overnight cost of capital |
| MW | $OR$ | operating reserves |
| MW | $OR^R$ | required $OR$ |
| $/MWh | $P$ | spot market price of energy (may not be a clearing price). |
| $/MWh | $P^a$ | auction price |
| $/MWh | $P_1, P_0$ | day-ahead energy price, and realtime (spot) energy price. $(P_0 = P)$ |
| $/MWh | $P^S(.), P^D(.)$ | supply bid curve, demand bid curve |
| $/MWh | $P_{cap}$ | price limit on the system operator. |
| $/MWh | $P_A^C$ | price of congestion at bus A |
| $/MWh | $P_A^L$ | price of losses at bus A |
| $/MWh | $P_{max}$ | maximum market-clearing price |
| $/MWh | $P_{spin}$ | spin market price of energy |
| $/MWh | $P_1^T, P_0^T$ | transmission price in DA market and RT market |
| $/MWh | $P_T$ | spot price at future time T |
| MW | $Q$ | market quantity traded |
| volt-amps | $Q$ | reactive power flow |
| MW | $q$ | quantity sold by 1 supplier |
| MW | $Q_1, Q_0$ | energy quantity accepted in DA auction, quantity traded in real time |
| MW | $Q_1^T, Q_0^T$ | transmission quantity accepted in DA auction, quantity traded in real time |
| none | $r$ | discount rate for investment in generation (includes risk premium) |
| $/h | $R$ | revenue |

| ohms | $R$ | electrical resistance |
|---|---|---|
| ohms | $R_T$ | transmission-line resistance |
| \$/h | $R_C$ | system revenue from congestion rent |
| \$/h | $R_{TCC}$ | system revenue from selling TCCs |
| \$/MWh | $MC_{RH}$ | right-hand marginal cost |
| none | $S$ | bid score in an auction |
| none | $s$ | market share |
| none | $ss$ | spot-market share |
| \$/MW | $SC$ | startup cost |
| \$/h \| \$/MW | $SR_\pi$ | short-run profit |
| \$/MWh | $SR_\pi(.)$ | the (short-run) profit function |
| \$/MWh | $SR_{\pi F}$ | final short-run profit including day-ahead and real-time payments |
| years \| date | $T$ | plant life, or a specific future time |
| \$/h | $TVC$ | total variable cost. |
| \$/h | $V$ | total surplus = producer + consumer surplus |
| volts | $V$ | voltage |
| \$/MWh | $VC$ | variable costs |
| \$/MWh | $V_{LL}$ | value of lost load. |
| watts | $W$ | power |
| \$/MWh | $Y$ | penalty for violating installed capacity requirement. |

# *Part 1*
# Power Market Fundamentals

# Prologue

*Is it a fact–or have I dreamt it – that, by means of electricity, the world of matter has become a great nerve, vibrating thousands of miles in a breathless point of time? Rather, the round globe is a vast head, a brain, instinct with intelligence! Or, shall we say, it is itself a thought, nothing but thought, and no longer the substance that we dreamed it?*

Nathaniel Hawthorne
*The House Of Seven Gables*
1851

*P*OWER SYSTEM ECONOMICS **PROVIDES A PRACTICAL INTRODUC-TION TO POWER-MARKET DESIGN.** To assist engineers, lawyers, regulators, and economists in crossing the boundaries between their fields, it provides the necessary background in economics and engineering. While Part 1 covers basics, it provides fresh insights ranging from a streamlined method for calculations, to the adaptation of economics to the quirks of generation models, to the distinction between the market structure and market architecture.

Part 2 focuses on the core structure of power markets which determines the basic character of supply and demand. It encompasses demand-side flaws, short-run reliability policy and the rigidities of supply. Together these determine the notorious price-spikes and unstable investment pattern of power markets. Because of its fundamental nature, this analysis can proceed without reference to locational pricing or unit commitment which gives less-technical readers access to the most important and fundamental economics of power markets.

Part 3 discusses the architecture of the day-ahead and real-time markets. This requires the introduction and analysis of the unit-commitment problem—the problem of starting and stopping generators economically. To avoid unnecessary complexity, the other primary problem of power-system economics, network congestion, is postponed until Part 5. This allows a clearer comparison of the three fundamental types of power trading: bilateral trading, exchange trading and pool-based trading.

Part 4 detours from the drive toward an increasingly detailed view of the market to examine market power. Although best understood in the context of Parts 2 and

3, it does not consider the implications of the network effects described in Part 5. It can be read directly after Chapters 1-5 and 1-6 if desired. Part 5 begins with more than enough power engineering to understand losses and congestion pricing which are its central topics. The theme of Part 5 is that locational loss prices and congestion prices are nothing more than ordinary (bilateral) competitive market prices. The problem of market design is not to invent clever new prices but to design a market that will reliably discover the same prices economics has been suggesting since Adam Smith.

Power markets deviate from standard economics in two ways: the demand sides are largely disconnected from the market and the details of supply costs violate the assumptions of competitive economics. Part 2 focuses on the demand-side flaws which require a regulatory intervention for a few hours per year to ensure reliability and efficient investment. Until these flaws are sufficiently reduced they will remain a great danger to power markets that are poorly designed. Part 2, while basic, contains the most important new material in the book.

While the aggregate market supply function is perfectly normal, the details of generation costs violate a basic competitive assumption. Because the violations are small relative to the size of the market, they may require only a slight adjustment to power exchange bids. Alternatively, a complex and opaque power pool may be needed. The contribution of Part 3 is to present this problem in an accessible manner and to highlight crucial questions that still need answers.

The book tells its story largely through examples. These are highly simplified but designed to capture important phenomena and display their essential natures. The key conclusions drawn from the examples are summarized in "Results" and "Fallacies" which are listed after the table of contents. "Result" is not the best of terms, but Fallacy has no acceptable antonym. Results are not theorems because they are not stated rigorously. They are rarely new; most are standard economic wisdom applied to power markets. They are simply the key points that should be understood in each area. The Results distill much theory that is not presented, but the examples reveal the mechanisms at work behind the Results. Fallacies are treated explicitly to help dispel the handful of popular misconceptions that continually cause confusion.

## READING TO DIFFERENT DEPTHS

Readers who wish to read more deeply on some subjects than on others will find all chapters organized to facilitate this. Chapters consist of the following parts:

1. Introduction (untitled)
2. Chapter Summary
3. Section Summaries
4. Sections
5. Technical Supplement (for some chapters)

Use the chapter summary of 50 to 200 words to determine if a chapter covers a topic of interest and to learn its most essential points. For a complete overview of the chapter's content, read the Section Summaries. For the importance and context of the chapter's focus, read the introduction.

When reading an entire Part, the chapter and section summaries can be skipped, though they may still provide a useful orientation. The technical supplement usually contains more difficult mathematics but never more than easy calculus.

## READING OUT OF ORDER

### For the Nonlinear Reader

A few terms and concepts are crucial and, depending on the reader's background, frequently misunderstood. Other terms are specific to this book. These are defined here for the reader who wishes to use the book more as a reference than a text.

Readers with special interests may wish to read chapters out of order. The book may also serve as a reference, with different parts used as they become relevant. Although each Part of the book has been written for sequential reading, any of Parts 2 through 5 may be read after Part 1. Even within Parts, chapters include cross references to assist those who read only a particular chapter. The glossary, which defines all terms set in **bold**, will be particularly helpful for those who have skipped previous chapters.

Skipping chapters or reading them out of order is encouraged, but the reader will need to understand the concepts discussed below. If these seem difficult, the reader is advised to begin by reading Part 1. Further clarifications and corrections can be found at www.stoft.com.

### Conventions Specific to this Book

The book follows five conventions that are not customary although they deviate from custom only to follow conventional economics more closely or more conveniently.

**Units.** Fixed cost and variable cost, as well as the cost of energy, power and capacity, are all properly measured in $/MWh. Duration, though sometimes expressed in hours per year, is represented in equations by a probability. These units must be understood before reading the book's examples. See page 31.

**Marginal Cost.** The standard definition is used when applicable but is generalized to cover models that use right-angle supply curves. These assume the supply curve changes from flat to vertical with an infinitesimal change in output. In such cases, the left- and right-hand marginal costs are defined to be the cost of the last unit produced and the cost to produce the next unit. The marginal-cost range is defined as the set of values from left-hand to right-hand marginal cost. See page 65.

**Variable Cost.**  Because marginal cost is not defined at full output for a generator with a standard right-angle supply function, the constant marginal cost before full output is called the generator's variable cost. This remains defined even when marginal cost is not, as it is a property of the supply curve and not dependent on the generator's level of output. See page 69.

**Scarcity Rent.**  "Scarcity rent" is not defined in economics texts, yet it is a term commonly used in power economics. Popular usage attempts to distinguish between rents earned when supply of all generation is scarce and those earned when only some types are scarce. This book avoids the ambiguity inherent in such definitions by defining scarcity rent as equal to the more conventional economic term, *inframarginal rent*, which is the area below the market price and above the competitive supply curve. See page 70.

**Aggregate Price Spike.**  The aggregate price spike is defined as the upper portion of the price duration curve, specifically the region in which price is above the variable cost of the most-expensive investment-grade peaker (not an old, inefficient peaker). This cut-off value is well defined only in simple models, but it still provides some intuition about real markets. Revenue associated with this spike is called *the price-spike revenue* and is similar to the popular meaning of scarcity rent. See page 127.

## Conventional Terms That Are Sometimes Misinterpreted

**Profits.**  "Profit" is used by economists to mean long-run economic profits, and these average zero under perfect competition. Short-run profit equals scarcity rent minus startup and no-load costs. Short-run profits cover fixed costs. See page 58.

**Cost Minimization.**  This refers to the minimization of production cost, not to the minimization of consumer cost.

**Market Clearing.**  A market has cleared when there are neither offers to buy output for more than the market-clearing price nor offers to sell output for less than the market-clearing price. When the market has cleared, supply equals demand. The market-clearing price need not be the competitive price, an efficient price, a fair price, or the price set by the system operator.

**Efficiency.**  Efficient production minimizes production cost given the output level. Efficient trade maximizes total surplus, the sum of consumer surplus and producer surplus (short-run profits). See page 53.

**Spot market.**  The spot market is only the real-time market, not the day-ahead market.

## Chapter 1-1
# Why Deregulate?

*The propensity to truck, barter, and exchange one thing for another
. . . is common to all men.*

Adam Smith
*The Wealth of Nations*
1776

**I**N THE BEGINNING THERE WAS COMPETITION—BRUTAL AND INEFFI-
CIENT. Between 1887 and 1893, twenty-four central station power companies
were established within Chicago alone. With overlapping distribution lines, competition for customers was fierce and costs were high. In 1898, the same year he was
elected president of the National Electric Light Association, Samuel Insull solved
these problems by acquiring a monopoly over all central-station production in
Chicago. In his historic presidential address to NELA, Insull explained not only
why the electricity business was a "natural monopoly" but why it should be regulated and why this regulation should be at the state level, not the local level. Insull
argued that

> *exclusive franchises should be coupled with the conditions of
> public control, requiring all charges for services fixed by public
> bodies to be based on cost plus a reasonable profit.*

These ideas shocked his fellow utility executives but led fairly directly to regulatory
laws passed by New York and Wisconsin in 1907 establishing the first two state
utility commissions. Reformers of the Progressive era also lent support to regulation
although they were about equally supportive of municipal power companies.[1] Their
intention, to hold down monopoly profits, was at odds with Insull's desire to keep
profits above the competitive level, but both sides agreed that competition was
inefficient and that providing electricity was a natural monopoly.[2]

---

1. See Platt (1991) for information on central station companies and Samuel Insull. The quotation (p. 86)
is from a contemporary account of Insull's address. Platt describes early competition as follows: "The
Chicago experience of rate wars, distributor duplication, and torn-up streets presented an alternative that
was attractive to virtually no one." For state commissions and progressives, see Rudolph & Ridley (1986).

2. Smith (1995) relies on Gregg Jarrell to conclude "regulation was a response to the utilities' desire to
protect profits, not a consumerist response to monopoly pricing." But Knittel (1999) tests causation by
utilities and consumers and finds no significant correlation between profit change and regulation after
correcting Jarrell's endogeneity problem. This result would be expected from an analysis of profit when
two equal forces have opposite motivations with respect to its level.

On the scale of an isolated city, provision of electricity *is* a natural monopoly and requires regulation or municipal ownership, but as transmission technology developed, it brought new possibilities for trade and competition. The earliest electric companies, for instance Brush's company which lighted New York's Broadway in 1880, integrated generation with distribution, and, in fact, sold light, not electricity. Edison initially did the same, installing the light bulbs in the homes he lit and charging by the number of bulbs installed. Westinghouse introduced high-voltage transmission using alternating current (AC) technology to the United States in 1886, and by 1892 Southern California Edison was operating a 10-kV transmission line 28 miles in length. This, too, was an integrated part of a full-service utility.

Integrated utilities remained natural monopolies for many years while expansion of the high-voltage transmission network continued, mainly for purposes of reliability. Eventually the entire Eastern United States and Eastern Canada were united in a single synchronized AC power system. By operating at extremely high voltages, this system is able to move power over great distances with very little loss, often less than three percent in a thousand miles.

Regulated, vertically integrated utilities were well established by the time the transmission system made substantial long-distance trade possible. As a consequence, trade was slow to develop, but the existence of the grid made the de-integration of the electric industry a possibility.[3] Generation could now be split off to form a separate competitive market, while the remaining parts of the utilities remained behind as regulated monopolies. By 1990, encouraged by a general trend toward deregulation, the de-integration trend in electric markets was underway in a number of countries. Today, more than a dozen semi-deregulated electricity markets are operating in at least ten countries, with several operating in the United States.

In spite of this apparent success, many fundamental problems remain. After ten years of operation, the British market has declared itself a failure and replaced all of its market rules. In a single year, the California market managed to cost its customers more than ten years of hoped-for savings. Alberta (Canada) is worried over the results of its recent auction of generation rights that brought in much less revenue than planned. New York saw prices spike to over $6,000/MWh in 2000, and the New England ISO had to close its installed capacity market due to extreme problems with market power. But initial problems do not prove deregulation is doomed; some markets are functioning well. A closer look at fundamental argu-

---

3. See Joskow (2000b, 16) for a similar view of the unimportance of "the demise of natural monopoly characteristics at the generation level" and the importance of the expansion of the grid, and Ruff (1999) for an alternative view of the role of the grid and transmission pricing.

ments for and against deregulation may help explain why such mixed results might
be expected.

**Chapter Summary 1-1:** Improvements in transmission, rather than changes
in generation technology, have removed the natural monopoly character of the
wholesale power market in most locations. This makes possible the replacement
of regulated generation monopolies with deregulated wholesale power markets.
In principle these can be more efficient than the old-style regulation. In practice,
California has proven bad deregulation to be worse than mediocre regulation, and
England has demonstrated that mediocre deregulation can bring cost-saving
efficiencies to a badly regulated generation monopoly.

In the short run, power-market problems tend to be more dramatic than the
benefits. The problems are primarily the result of two demand-side flaws: the almost
complete failure of customers to respond to relevant price fluctuations, and the
customer's ability to take power from the grid without a contract. As fundamental
as these are, it is possible to design a workable market around them, but it *does*
require design as well as extensive and clever regulation. Recent U.S. history has
shown that there are three impediments to such progress: politics, special interests,
and overconfidence. The last is largely due to a dramatic underestimation of the
problem.

**Section 1:  Conditions for Deregulation.** Deregulation requires the market
not be a strong natural monopoly. One view holds that small efficient gas turbines
have overturned the natural monopoly of large coal plants. Yet today's competitive
suppliers are far larger than any coal plant, so if the size of a large coal plant were
problematic for competition, today's markets would be uncompetitive.

**Section 2:  Problems with Regulation.** Regulation can provide strong cost-
minimizing incentives and can hold prices down, but it must trade off one against
the other. Competition can do both at once. In practice, regulators hold prices down
near long-run average costs but leave cost-minimizing incentives too weak. The
result is high costs and high prices.

**Section 3:  The Benefits of Wholesale Competition.** Competition provides
full strength cost-minimizing incentives and, at the same time, forces average prices
down toward their minimum. It may also encourage efficient retail prices.

**Section 4:  The Benefits of Real-time Pricing.** Competition may induce real-
time pricing, which will reduce consumption during periods of peak demand. This
will reduce the need for installed capacity and, if extensively adopted, should
provide a net savings of about 2% of retail price. Although this could be achieved

easily under regulation, competition will provide some additional incentives, but their consequences are still unclear.

**Section 5: Problems with Deregulating Electricity.**  Contemporary electricity markets have inadequate metering. Consequently it does not make sense for load to respond to price fluctuations, and bilateral contracts cannot be physically enforced in real time. As a result demand can and sometimes does exceed supply, and competitive pricing is impossible at crucial times. These flaws result in high prices that must be limited, and they provide ideal conditions for the exercise of market power. Electricity markets are also extremely complex and prone to problems with local market power due to the inadequacies of the transmission system.

## 1-1.1  CONDITIONS FOR DEREGULATION

Scale economies make it possible for **natural monopolies** to produce their output more cheaply than a competitive market would. A 1-MW power plant is not very efficient, and there is no way to produce power cheaply on this small a scale. A 10-MW power plant can always do better. Efficiency continues to increase significantly to about the 100-MW level but ever more slowly beyond this level. It used to increase to about 800 MW, and it was once assumed that nuclear plants would be the most economical and their most efficient size would be even greater. If these economies of scale continued, the cheapest way to provide California with power would be to build a 25,000-MW power plant and a few smaller ones to handle load fluctuations. But a large single power plant could not support competition. A competitive market necessarily utilizes smaller plants and would therefore have higher production costs. Consumers would have to pay more if a natural monopoly is forced to operate as a competitive industry with small-scale plants.

Efficiency gains from the operation of multiple plants are another possible source of natural monopoly. Even if very large plants are not more efficient, large generating companies may be. A large company can hire specialists and share parts and repair crews. If multiplant efficiencies continue to large enough scales, a competitive market would again be less efficient than a monopolist.[4]

If a monopolist can produce power at significantly lower cost than the best competitive market, then deregulation makes little sense. The lack of a natural monopoly is a prerequisite to successful deregulation, or at least, the condition of natural monopoly should hold only weakly.

### Did Cheap Gas Turbines End the Natural Monopoly?

One popular argument for deregulation claims technical progress has recently nullified the conditions for a natural monopoly in generation. This view assumes generation had previously been a natural monopoly because the most efficient size

---

4. See Joskow and Schmalensee (1983, 54) for a discussion of firm-level economies of scale.

power plant was approaching 1000 MW, and new technologies have made 100-MW plants almost as efficient.

If this argument were correct, then an 1000-MW supplier must in some sense be a monopolist, and the market must need suppliers that have capacities smaller than 1000 MW to be competitive. But in this case, deregulation must certainly have failed in the United States because every market contains suppliers with capacities exceeding 1000 MW. Yet no one who suggests small efficient plants are a necessary condition for competition seems worried by the presence of huge suppliers in the new markets.

The beliefs that the most efficient size power plant must be quite small, and that competitive suppliers can own many such plants are contradictory. Most likely the former is incorrect, at least in markets with peak loads of over 5000 MW. Fortunately, vast transmission grids have made such large markets the norm. When small efficient plants are necessary for competition, suppliers with total generating capacity greater than the most efficient size plant should be prohibited. Greater threats of natural monopoly conditions come from the economies of multiplant companies and weaknesses in the power grid that effectively isolate "load pockets" during peak load conditions.

# 1-1.2   PROBLEMS WITH REGULATION

The most common argument for deregulation is the inefficiency of regulation. There can be no quarrel about its inefficiency, but it does not follow that deregulation will be better. Deregulation is *not* equivalent to perfect competition which is well known to be efficient. Electricity markets have their own inefficiencies that need to be compared with the inefficiencies of regulation.

To date, such comparisons have been largely speculative. The most decisive answers inevitably are based on the least information. One side claims regulation is essential because electricity is a basic need. What about housing? That need is even more basic, and housing is 99% deregulated. The other side claims competition provides incentives to reduce costs, while regulation does not. What about the many regulated utilities that have provided reliable power for many years for less than 6¢/kWh? Do they lack all cost-minimizing incentives?

One argument posits that when a regulated utility makes a bad investment, ratepayers pick up the tab; but, when an unregulated supplier makes a bad investment, stockholders pick up the tab. This analysis is myopic. Particular losses fall on the stockholders of particular companies, but the cost of capital takes into account the probability of such mistakes, and every mistake increases the estimated probability. Like all costs, the cost of capital is paid for by consumers. Not only does the cost of capital average in the cost of mistakes, it also adds a risk premium. To the extent stockholders of regulated utilities are sheltered, they demand less of a risk premium than do stockholders of unregulated suppliers. With competition, there may be fewer mistakes, but the mistakes will be paid for by consumers, and a risk premium will be added.

Regulation has two fundamental problems: (1) it cannot provide a strong incentive to suppliers as cheaply as can a competitive market, and (2) regulatory bodies themselves do not have proper incentives. Well-trained regulators could provide much better regulation. But for government to provide competent regulation, the political process would need to change. The first problem, that of incentives provided by regulators, is more susceptible to analysis.

## The Regulator's Dilemma

Truly competitive markets do two things at once; they provide full-powered incentives (1) to hold price down to marginal cost, and (2) to minimize cost. Regulation can do one or the other but not both. It must always make a trade-off because suppliers always know the market better than the regulators.[5]

This trade-off is the core idea of modern regulatory theory. Perfect **cost-of-service (COS) regulation** is at one extreme of the regulatory spectrum. It assures that, no matter what, suppliers will recover all of their costs but no more. This includes a normal rate of return on their investment. Perfect COS regulation holds prices down to long-run costs but takes away all incentive to minimize cost.[6] If the suppliers make an innovation that saves a dollar of production costs, the regulator takes it away and gives it to the customer.

At the other extreme is perfect price-cap regulation. It sets a cap on the supplier's price according to some formula that takes account of inflation and technical progress, and it never changes the formula. Now every dollar saved is kept by the supplier, so its incentives are just as good as in a competitive market. But it's difficult to pick a price-cap formula that can be fixed for twenty years at a time. A perfect (very-long-term) price cap must always allow prices that are well above long-run cost to avoid accidentally bankrupting suppliers. Consequently, prices will be too high.

The reader unfamiliar with the theory of regulation may be tempted to invent clever ways for the regulator to provide full cost-minimizing incentives while holding prices down to cost, but all will fail. The inevitability of this trade-off has been established repeatedly and with great rigor; however, the trade-off can be improved by improving the regulator's effective knowledge.[7] The main technique for making the trade-off is to adjust the price cap more or less frequently. Constant adjustment produces COS regulation while extremely infrequent adjustment produces pure price-cap regulation. In between, incentives are moderately strong and prices are moderately low. If the regulator has a fair amount of information, this trade-off can be quite satisfactory, but it will never equal perfect competition.

---

5. Suppliers having better information than the regulators is at the root of the regulatory trade-off problem which, though fundamental, is too complex to discuss here.

6. In reality, cost-of-service regulation does provide incentives in two ways: regulatory lag (see Joskow 2000b) which is discussed shortly, and the threat of disallowed costs.

7. For instance, yardstick regulation compares the performance of one regulated firm with other similar firms, thus giving the regulator some of the benefit of the other firms' knowledge without requiring the regulator to know details (see Tirole 1997).

## Regulation in Practice

Competition can hold average prices down to long-run costs while putting full strength pressure on cost minimization. At best, regulation does a decent job of both but does neither quite as well as competition. But how does regulation work in practice?

Regulation tends to err in the direction of driving prices down toward cost. In fact, most regulators believe this is their entire job and would implement pure COS regulation if they could. Fortunately, it's just too much bother to re-adjust rates continuously, so the result is roughly a price cap that gets reset about every three years. This inadvertent "regulatory lag" is a major factor in saving COS regulation from providing no incentive at all. It provides some incentive for cost minimization, but less than would be provided with an optimal trade-off. Even that is too little by the standard of a competitive market. In practice, regulation has typically done a passable job in the United States and could do much better if the effort spent deregulating were spent improving regulation.

## 1-1.3   THE BENEFITS OF COMPETITIVE WHOLESALE MARKETS

Competition provides much stronger *cost-minimizing* incentives than typical "cost-of-service" regulation and results in suppliers making many kinds of cost-saving innovations more quickly. These include labor saving techniques, more efficient repairs, cheaper construction costs on new plants, and wiser investment choices.

Distributed generation is an area in which innovation may be much quicker under competition than under regulation; **cogeneration** is one example. Regulated utilities found such projects extremely awkward at best, so avoided them. A competitive market easily allows the flexibility that such projects require.

The other advantage of competition is its ability to *hold price down* to marginal cost. This is less of an advantage simply because traditional regulation has stressed this side of the regulatory trade-off. Sometimes price minimization can still be a significant advantage. Again cogeneration provides an example. Once regulators decided to encourage it, they needed to price cogenerated power. A formula was designed with the intention of mimicking a market price. Naturally, political forces intervened and the result was long-term contracts signed at very high prices (Joskow 2000b). These gave strong, probably much too strong, incentives for cogeneration. A competitive market can get both incentives and prices right at the same time.

While holding down prices, competition also provides incentives for more accurate pricing. Because it imposes the real-time wholesale spot price on the retailer's marginal purchases, wholesale competition should encourage real-time pricing for retail customers. This can be done easily by a regulated retailer, but a competitive retailer should have an added incentive to provide the option of real-time retail pricing because that would reflect its costs.

## 1-1.4   THE BENEFITS OF REAL-TIME RATES

One view holds that the main benefit of competition will come from the demand side of the market more than the supply side. The price spikes of the wholesale market will be passed on to customers—at least for marginal consumption and will cause customers to curb their demand when the price is highest and generation is most costly. This will allow fewer generators to be built and will reduce the total cost of providing power. A competitive market will pass this savings on to consumers.

Arthur Wright of Brighton, England, invented the real-time meter which Samuel Insull heard of while visiting his homeland in 1894. He sent Louis Ferguson to Brighton to study its use. Insull soon replaced most of his meters with Wright real-time meters and by 1898 was installing them with every new residential hookup (Platt 1991). Although Ferguson invented the notion of charging according to the *time* of peak demand, this was rarely used. Instead customers were charged for peak demand (a **demand charge**) and total energy. This combination became known as a Wright tariff. The purpose of such metering was to improve the system's load factor (average load over peak load), and it did so. Central station load curves from Chicago show that their load factor improved from 30.4% in 1898 to 41.7% eleven years later. The Wright tariff is not real-time pricing but has a similar, though generally weaker, effect on peak demand.[8]

The cost-saving effect of real-time pricing cannot be doubted, but how great is it? Unfortunately it is much smaller today than it was in 1898, and that may explain in part why residential customers no longer have real-time meters. Today's load factors are typically near 60%, although, in a system like Alberta's where regulators have imposed heavy demand charges, the "load factor" can approach 80%. Real-time pricing could do the same. It can never cause load to become completely constant (100% load factor) because this would put an end to real-time price fluctuations, and there would no longer be a reason for customers to shift their consumption off peak. Thus load-shifting caused by real-time pricing is self-limiting. Assume that the shift would proceed halfway if real-time pricing were fully implemented. Changing the load factor from 60% to 80% gives about a 25% reduction in needed generation capacity. This is dramatic, and perhaps a bit optimistic, so for numeric convenience, assume the reduction is only 24%.

With an 80% load factor, most load would be baseload because high real-time prices reduce peak load. Peaking generators cost roughly half of what an average generator costs per installed megawatt. The reduction in generation fixed costs should be in the neighborhood of 12%. Because fuel costs are about as large as the fixed costs of generation, total wholesale power costs will be reduced only about 6%. (Real-time pricing shifts load to off-peak hours but its effect on total energy consumption is minimal and could work in either direction.) Wholesale costs are about $\frac{3}{8}$ the cost of retail power, so retail costs are reduced by about 2.25%.

---

8. For a comparison of demand-charges and real-time rates and an explanation of why real-time rates are crucial for reducing market power, see Borenstein (2001b).

Peak transmission use often occurs during shoulder hours, not peak hours, so there is no easy proof of a transmission cost savings. But for simplicity assume that this savings is $\frac{1}{3}$ as great as the saving in generation. That brings total savings to 3%.

---

**Result     1-1.1          Savings from Real-Time Rates Would Be Small**

Fully implemented real-time pricing would improve load factors roughly from 60% to 80%. This would reduce the cost of supply by approximately 2.25% of retail costs mainly due to savings in peak-load capacity. Additional consumer costs of accomplishing the load shift would likely more than cancel any transmission savings for a net savings of about 2% of retail costs.

---

This 3% reduction in supply cost is offset by a cost increase on the demand side. Shifting load to off-peak hours is costly for customers as it requires the purchase of smarter appliances and changes in consumption that they find undesirable. In fact customers will shift load up to the point where the marginal cost of shifting is just as great as the marginal savings on their electricity bill. A rough estimate puts the total cost to consumers at about $\frac{1}{3}$ of the savings and this does not include the cost of the real-time metering and billing. The *net* savings from real-time metering should be in the neighborhood of 2% of total cost of delivered power.[9]

A second question about real-time pricing as an argument for deregulation is why it cannot be done under regulation. Regulated systems have for years computed real-time system marginal cost. Technically, it would be no problem for a regulated utility to install real-time meters and charge customers marginal-cost prices. California is now testing this possibility (Wolak 2001). It may be that regulators simply lack the will to do this or the ability to carry out the details effectively. On the other hand, it may take more will and cleverness to implement a competitive wholesale market than to implement real-time pricing.

---

## 1-1.5    PROBLEMS WITH DEREGULATING ELECTRICITY

Electricity is a peculiar product. It is the only product that is consumed continuously by essentially all customers. In fact it is consumed within a tenth of a second of its production and less than a tenth of a second of power can be stored as electrical energy in the system.[10] These physical properties result in a product whose marginal cost of production fluctuates rapidly and, thus, whose delivered cost also fluctuates rapidly. No other product has a delivered cost that fluctuates nearly this rapidly.

---

9. Significant savings in transmission and distribution should not be expected because transmission lines are typically used most heavily during off-peak hours, and both have very large fixed-cost components that will be unaffected by deregulation.

10. More kinetic energy is stored in rotating generators, much more potential energy can be stored by pumping water up hill and vastly more energy is stored in local fuel supplies, but all of these stores of energy must be converted to electrical energy by the process of generation before they can be delivered.

## Two Demand-Side Flaws

Although real-time metering began in the late 1800s, it has been discontinued for residential customers, and almost no industrial or commercial customers see real-time prices. Consequently, almost no customers respond to the real-time fluctuations in the delivered cost of power.

Even with this demand-side flaw, the market could operate in reasonably close accord with economic principles if not for the second demand-side flaw, the ability of a load to "take power from the grid without a prior contract with a generator" (Ruff 1999, 28; FERC 2001b, 4). If bilateral contracts could be enforced by physically cutting off customers who exceed their contracts, the market could function almost in alignment with the theory of competitive markets. In no other market is it impossible to physically enforce bilateral contracts on the time scale of price fluctuations.

---

| | |
|---|---|
| **Demand-Side Flaw 1:** | **Lack of Metering and Real-Time Billing** |
| **Demand-Side Flaw 2:** | **Lack of Real-Time Control of Power Flow to Specific Customers** |

---

The **first demand-side flaw** causes a lack of demand responsiveness to price or, technically, a lack of *demand elasticity*. The **second demand-side flaw** prevents physical enforcement of bilateral contracts and results in the system operator being the default supplier in real time.

Because demand responds only minimally to price, the supply and demand curve may fail to intersect, a market flaw so severe it is not contemplated by any text on economics.[11] The system operator, as default supplier, is forced to set the price, at least when supply fails to intersect demand. It can also improve the market by setting price under slightly less dire circumstances. Presently, all power markets operate like this and will continue to do so until very-short-run demand elasticity is significantly improved.

These are not just theoretical problems. While the average cost of production is about $35/MWh, and the maximum cost with new equipment is about $100/MWh, prices in the $1,000 to $10,000 range have occurred in many markets. All four of the U.S. markets have formal price caps, and the Midwest market has informal caps set by system operators who refuse to buy required reserves when the price gets too high. With the extreme demand inelasticity caused by the first flaw, scarcity alone would produce high prices, but the flawed demand side coupled with scarcity also produces ideal conditions for market power which pushes prices still higher (Joskow 2001a).

Dramatic though these flaws are, it should still be possible to design a well-functioning market. But it does require design! Deregulating power markets is called "restructuring" in the United States because the resulting competitive markets have more federal regulations than the regulated markets they replaced (Borenstein and Bushnell 2000). In the long run, the demand side of the market should develop enough price elasticity to clear the market at a finite price. There is a good chance

---

11. See Jaffe and Felder (1996), Kahn et al. (2001), and Green (1998).

that eventually price spikes will be low enough to need no caps. This change in market structure should be encouraged from the start. A responsible deregulation of electricity would first fix the demand-side flaws and then start the market—they are cheaper to fix than the problems they have already caused.

## Complexity and Local Market Power

There are two other fundamental problems with deregulating electricity: complexity and local market power. A power system is a delicate, single machine that can extend over millions of square miles. Every generator in the system must be synchronized to within a hundredth of a second with every other generator in the AC interconnection. Voltage must be maintained within a 5% limit at thousands of separate locations. This must be accomplished on a shared facility, half of which (the grid) must be operated for the common good and half of which (the generators) are operated for hundreds of different private interests.

Complexity can be overcome by a sufficiently well designed set of market rules, but the problem of local market power may need to be solved, at least for the present, by interventionist means. So far, it has been. More than half of the generators in the California ISO were declared "must run," meaning sometime during the year they were crucial to system operation and, therefore, had such extreme market power there was no choice but to regulate their price. San Francisco and New York among other cities are "load pockets;" they require more power than they can import. As a consequence, the two generators in San Francisco would have extreme market power during peak hours every day if they were not regulated. These generators are required for their "real" power production, but most must-run generators are required for their "reactive" power, a concept not well understood by regulators (see Chapter 5-2).

The most difficult and costly problems with new electricity markets are mainly matters of market structure as opposed to market architecture (see Chapters 1-7 and 1-8). When this is understood, and demand-side flaws and the problems with market power and transmission are squarely faced, adequate solutions will probably be found. Then wholesale power markets should prove superior to regulated monopoly generation.

# What to Deregulate

*Nothing is more terrible than activity without insight.*

Thomas Carlyle
(1795-1881)

**D**ELIVERED POWER IS A BUNDLE OF MANY SERVICES. These include transmission, distribution, frequency control, and voltage support, as well as generation. The first two deliver the power while the second two maintain power quality; other services provide reliability.

Each service requires a separate market, and some require several markets. This raises many questions about which services *should* be deregulated and which *should not.* Even within a market for a single service, one side—either demand or supply—may need to be regulated while the other side of the market can be deregulated. For instance, the supply of transmission rights must be determined by the system operator, but the demand side of this market is competitive. In contrast, the demands for ancillary services are determined by the system operator while the supply sides of these markets can be competitive.

The most critical service in a regulated or a deregulated power market is that provided by the system operator. This is a coordination service. For a deregulated market it typically includes operation of the real-time markets and a day-ahead market. These provide scheduling and balancing services, but operating these markets is itself an entirely separate service. While the need for the system operator service is agreed to by all, the proper extent of that service is the subject of the central controversy in power market design.

**Chapter Summary 1-2:** Many services are required to bring high-quality reliable power to end users. Each might be provided by free markets, by the state, by regulated suppliers, or by some hybrid arrangement. Bulk power generation is the source of nearly half the cost of retail power and is one of the services most easily provided by a competitive market. Moreover, it seems to offer several possibilities for significant efficiency improvements. If restructuring gets this much right, it will reap most of the benefits presently available from deregulation of electricity.

Markets for reserves may be the next most sensible targets for deregulation because they are so closely tied to the bulk power market. Reserves are provided by the same generators that provide power. These markets can only be deregulated on the supply side, and they are quite complicated to design. Retail deregulation is easier to design, and retailing can be more fully deregulated, but there is much less to gain. Retail costs are at most 5% of total costs, and the potential savings are less obvious than in generation because most of the traditional retail services are irrelevant. This is also the one area in which competition can be expected to add a significant cost—the cost of marketing.

**Section 1: Ancillary Services and the System Operator.** The system operator must keep the system in balance by keeping supply equal to demand. As many as five markets may be required to accomplish this: one for "regulation" which works minute by minute, and four to handle larger deviations and emergencies. Collectively, these services, together with a few others, are known as *ancillary services*.

Ancillary services benefit the entire market and are either **public goods** or have large external effects. Consequently, all of the markets have a fully regulated demand side, but some can be deregulated on the supply side. The system operator service, which coordinates these markets and provides the regulated demand for ancillary services, is a natural monopoly service and can be provided by a nonprofit or a for-profit entity.

**Section 2: Unit Commitment and Congestion Management.** Traditionally (before restructuring), system operators provided two more coordination services: unit commitment and transmission-congestion management. Unit commitment can be left to the generators themselves, though this may result in a small decrease in efficiency. Transmission congestion must be managed, at least in part, by the system operator.

**Section 3: Risk Management and Forward Markets.** Both generators and their customers are risk averse and wish to avoid the fluctuating prices of the spot market. Not only the hour-to-hour and day-to-day price variations but also the year-to-year variations in the spot market's average price are problematic. Forward energy trading can hedge these risks and needs no more regulation than other commodity markets.

**Section 4: Transmission and Distribution.** Delivery of electric power requires a network of high voltage lines. Because duplicate sets of lines are wasteful, both distribution and transmission appear to be natural monopolies. There has been speculation about granting rights to congestion rents to investors in transmission lines and thereby stimulating a competitive market for such lines, but currently no jurisdiction plans to rely on such a market.[1]

**Section 5: Retail Competition.**  Retail service in electricity does not include distribution of power to the customer. In its present form, it usually does not include metering or even meter reading. Most retailing consists of financial transactions, all of which can be provided from an office building. There are relatively few possibilities for value added. Early on, deregulation advocates claimed that retailers would provide innovative products (new types of meters and new qualities of electricity), but now their claims of innovation focus on billing.

The use of green power and time-of-use billing seem to be the most substantive improvements available in the retail market, but there is little reason to believe these could not be provided by a regulated market. In fact, green power might be more easily provided under regulation. Unfortunately, because customers will want or require the right to switch suppliers, retail competition may make it more difficult for generators to sign long-term contracts. This could make the wholesale market more risky and more susceptible to the exercise of market power.

## 1-2.1  ANCILLARY SERVICES AND THE SYSTEM OPERATOR

The system operator must keep the system in balance, keep the voltage at the right level, and restart the system when it suffers a complete collapse. The system operator carries out these basic functions by purchasing what are called "ancillary services." These include various types of reserves, voltage support, and black-start services. Ancillary services are the subject of Chapter 3-2 where they are defined more carefully and more broadly.

### Regulation and Balancing

Any imbalance between supply and demand causes the system frequency to deviate from the standard (60 Hz in the United States).[2] This is problematic for some appliances; nondigital clocks and phonographs depend on the system frequency, but large generators depend most on a constant frequency. Changing frequency makes these enormous machines speed up or slow down, causing added wear and tear. Consequently, maintaining a precise system frequency is given high priority

---

1. Markets for transmission rights that cover the cost of congestion already exist but not competitive markets for transmission lines. Australia and Argentina are experimenting with for-profit transmission lines. A market for monopoly transco franchises is another possibility (Wilson 1997).

2. Frequency is very precisely maintained as indicated by the following report (NERC 2000, 32). "In late July 1999 . . . system frequency on the Eastern Interconnection dipped to one of its lowest levels in history (59.93 Hertz)."

by the system operator. There seems to have been no cost-benefit analysis, but balancing the system with some accuracy is absolutely necessary.

System frequency is exactly the same for all customers in an interconnection. For example, Quebec and Florida have exactly the same frequency. As long as any system operator maintains its frequency, every other system in the Eastern Interconnection will also have exactly the right frequency. As a result, no one has much interest in trying to maintain frequency. Economics calls this a "free-rider" problem. It is costly to maintain the system balance, and everyone would prefer to let the others take care of it. Consequently, some regulatory body must decide to purchase balancing services and must charge (tax) some market participants to pay for them. There is no alternative; with a pure free market approach, the system frequency would be unacceptably unstable. This means the demand for balancing services must be regulated, but they still can be supplied by a competitive market. Every system operator must meet a balancing standard set by the North American Electric Reliability Council (NERC), so it purchases enough balancing services to meet this standard. Because many different generators can supply the services, there is a good chance the market will be competitive. But it is not automatic; it depends on the details of the market.

**Self-Provision of Reserves.**   There are two approaches to the supply of balancing services. The system operator can purchase the necessary services in a market and assign the cost to either loads or generators on a pro rata basis, or it can assign each supplier a fraction of the physical requirement. The latter approach is called "self-providing" and is often described as a less regulated approach.[3] As an example, consider spinning reserves (spin), a key source of balancing services. Not every generator will want to provide spin, so if physical requirements are imposed on suppliers, a market for spin will develop. Those who find it expensive to provide will buy from those who can provide it more cheaply, and some generators will "self-provide."

What is the practical difference between the physical "self-provision" approach and the financial approach? Imagine that a supplier has been told to supply 100 MWh of spin as part of its contribution to balancing services, and that it "self-supplies." If the system operator had instead purchased spin directly and had charged the supplier for 100 MWh of spin, would the supplier have been worse off? No. It would have sold its 100 MWh of spin to the market at the spin-market price, $P$, for a revenue of $100 \times P$, and then the system operator would have charged it $100 \times P$ for its share of the spin requirement. The result is the same. In both cases the generator provides 100 MWh of spin. In neither case does it have any net cost.

So why the agitation to "self-provide?" There are several possible answers. First, some may not understand the concept. Others may believe they can more easily supply poor quality spin when they self-provide. Still others may not be interested in their own requirement but may wish to provide part of the system-operator service. In other words, they may want to be the market maker for spin and take this business away from the system operator. If they can do it more efficiently and

---

3. See Chao and Wilson (1999a) and Wilson (1999).

with less gaming than the system operator, then they should be allowed to take over this service.

The problem with "self-providing" is that the services are not provided to the suppliers themselves but to the system operator. The term is deceptive. There still must be a physical transaction between the supplier and the system operator. It is this physical transaction, the verification of the provided reserves, that is difficult. "Self-provision" is not a way to avoid the regulated nature of the demand side of the ancillary service market, and it does not make the supply side more competitive.

## Voltage Support and Black-Start Capability

Most generators need to take electric power from the grid in order to start themselves. Consequently, if the system goes down, they cannot help it restart. Special generators have the ability to self-start. Like balancing services, black-start services are a public good and must be purchased by the system operator.

Voltage "sags" when too much "reactive" power is taken out of the system. Reactive power, unlike the "real" power that lights up incandescent lights, does not travel very well over power lines. When too much is used locally, for example by motors, fluorescent ballasts, and transmission lines, the voltage sags locally. To counteract this, more reactive power must be injected locally by capacitors, normal generators, or special generators called synchronous condensers.

Reactive power is less of a public good than the other ancillary services. It is possible, though expensive, to measure its use by individual customers and by the grid. In principle, these could be charged for their use and there could be a spot market in reactive power. This would be a complex and expensive market to run although reactive power is usually very cheap to produce. Also, because of the difficulty with transmitting reactive power over long distance, there would be far less competition in the reactive power market.

Voltage support, like other ancillary services, has a free-rider problem unless all customers are metered for its consumption. Customers lack sufficient incentive to replace the reactive power they use because the voltage drop caused by one customer affects many others. Consequently, at a minimum, there is a need for a regulatory requirement for reactive power purchases. For the present, by far the simplest approach is for the system operator to buy what is needed and, when necessary because of market power, to regulate the purchase price.

## The System Operator Service

The system operator service, which coordinates the ancillary service markets and provides the regulated demand for ancillary services, is a monopoly service (Ruff 1999). This monopoly can be either a nonprofit or a for-profit entity. If nonprofit it is typically called an **independent system operator** (ISO) in the United States and can be minimally regulated. Transco is the current U.S. term for a for-profit system operator, an entity that will probably own the grid and will require extensive regulation.

Nonprofit "monopolies" are quite different from for-profit monopolies. For instance, an ISO will have no motive to maximize the rent from congested transmission lines because it cannot keep that rent as profit. An unregulated for-profit system operator could have an extremely strong motive to maximize rent by withholding transmission service. Because such behavior is both unfair and inefficient, the for-profit system operator will need to be regulated.

One opinion holds that the transmission grid will face competition from distributed generation (small local generators) and is, therefore, no longer a natural monopoly. This was the case in 1890, but now such competition is extremely weak. Imagine a modern city being deprived of its importing transmission lines, and ask if this would raise its cost of power if it were forced to rely on within-city distributed generation. It would be enormously expensive. The trend toward increased reliance on transmission has been apparent for more than a century, and deregulation has only accelerated this trend. New micro-generator technology has increased the chances for distributed generation to slow or reverse this trend, but the transmission system will not lose its natural monopoly character for decades to come. Until then, transcos must be regulated.

Although ISOs have no motive to extract monopoly rents, they do have a weakened motive to act efficiently. Their motivation comes mostly from public scrutiny which is enhanced by the attention of market participants ("stakeholders") who have a lot to lose from inefficient ISO operation.

For the present, the system operator service must be provided by a regulated for-profit monopoly transco or by an lightly regulated nonprofit ISO. Both are poor choices and may prove to have fairly similar problems. The Transmission Administrator in Alberta, Canada, a for-profit regulated monopoly, is regulated in such a way that it behaves like a nonprofit in many respects. For example, it proposed substituting large incentive payments and charges on generators for the building of a large new transmission line. While the line may not have been justified (though now it claims it is), the incentives were structured in such a way that they appeared to make money for customers. The costs of inducing generation investment in more costly locations were both hidden and denied. The monopoly transco's incentive to optimize transmission was extremely weak to nonexistent. Instead, it exhibited the same interest as an ISO in gaining favor with the public and stakeholders.

There is one possible way out of this dilemma. Monopolies may sometimes be forced to compete by granting them a temporary franchise which can be won through a bidding process. Robert Wilson (1997) has proposed doing this and has suggested a scheme for providing them with incentives for efficient system operation. Although this approach shows promise, the technical and political problems with implementation are still formidable.

## 1-2.2   UNIT COMMITMENT AND CONGESTION MANAGEMENT

There is nearly universal agreement that the system operator must run the real-time market, and that the demand for ancillary services must be determined by a central authority. But the services of unit commitment and congestion management have caused a great controversy. This is often referred to as the "Poolco" or "nodal pricing" debate. Professor William Hogan of Harvard has put forward the nodal-pricing model which has now been accepted by PJM and the New York, and New England ISOs. California, Australia, Alberta, and England have rejected it in favor of a less centralized approach.

Hogan (1992, 1995, 1998) specifies a system operator that provides both unit commitment and congestion management services.[4] These two are linked together more because of historical system operator practice or by choice than by logic. In fact, the system operator could easily provide only congestion management and balancing, but not unit commitment.

## The Unit Commitment Service

The unit commitment service need not be centrally provided nor even be provided by a private market. Each generator can provide it for itself. This service simply tells generators when to turn on, how much to produce at each point in time, and when to turn off. These decisions can be made privately. The question is, how efficiently will they be made.

If decisions are private, individual generators will predict the price of energy during the time under consideration and will then incur the cost of starting up if they expect to make a profit. When shutting down appears to be more profitable than running, they will shut down. Economic analysis shows this process works almost perfectly if price predictions are perfect and there are no costs to starting and stopping. In real markets, however, neither assumption applies. This raises two questions: (1) would the system operator predict prices more accurately, and (2) can the system operator work around the problem of startup costs more efficiently?

The system operator can probably predict market price slightly better than individual suppliers because it has more information, but this advantage may not be great. Under nodal pricing, the system operator can predict the day-ahead price when it schedules, but it is the unknown real-time price that matters.

The inefficiency described by economic theory and caused by "nonconvex" startup-costs is almost certainly very small.[5] (This is discussed at length in Chapter 3-8.) Although startup costs probably do little damage to the market equilibrium, they make it more difficult to find. This might be a greater source of inefficiency, particularly if this problem interacts with reliability. Currently, there is little, if any, evidence that the system operator can save much by performing the unit commitment service. But there is little to lose and perhaps a noticeable gain, especially in a new market, so the safe course may be to have the system operator provide the service and make it optional. This is now done by PJM and almost all producers make use of the service. But the implications are unclear because of indirect incentives to accept PJM's service.

---

4. More recently Hogan (2001b) has explicitly suggested designs that make the unit commitment service optional.

5. No-load costs, which are greater than startup costs, also cause the production cost function to be nonconvex, but they apparently cause less trouble.

## Transmission-Congestion Management

Congestion management is one of the toughest problems in electricity market design. Although the costs imposed by congestion in an efficiently run system are quite low, badly designed congestion pricing can make the system unmanageable. This was demonstrated by PJM just before the official opening of its market when it instituted a form of average-cost congestion pricing that resulted in massive gaming of the pricing rules (Hogan 1999).

Any efficient method of congestion management will charge for the use of congested lines. If the price charged is set correctly, the demand for the use of the line will equal its capacity unless the price is zero, and it will be zero only if at zero price the line is still not fully utilized. Pricing ensures that those who value the line most get to use it and that the line's capacity is not wasted. This is the only efficient way to manage congestion.

Although there is only one set of efficient congestion prices (the prices set by the "nodal pricing" approach), there are many other approaches that would in theory give these same prices, or a good approximation of them. Wu and Varaiya proposed the most extreme alternative to nodal pricing. In their "multi-lateral" approach (Wu and Varaiya 1995; Varaiya 1996), the system operator would have no knowledge of the congestion prices. It would simply assign certain parties the right to use lines in a "reasonable but arbitrary way" that would prevent overuse of the lines. These parties would then either exercise their rights by using the lines or would sell their rights to those who valued the lines more. If it worked efficiently as a competitive market should, the market would then produce the exact same prices for the use of congested lines as the system operator would compute under nodal pricing.

Congestion prices make money by charging transmission users for a scarce resource. The revenues collected are called the congestion rent. Because congestion prices are the same under any efficient system, so is the congestion rent. Under nodal pricing, the system operator collects the congestion rent. Under other systems, such as the multi-lateral approach, some private party will collect these rents. This is at the heart of a controversy. When generators ask to "self-manage" their congestion, they are really asking to "self-collect" as much of the transmission rent as possible.

Any system of allocating transmission rights can be thought of as selling or giving away transmission rights. In either case the rights originate with the system operator and are specified in such a way that the transmission lines are protected from overuse. The supply of transmission rights must always be regulated.

There is a fundamental choice to be made: should the system operator sell the transmission rights and use the revenues (congestion rents) to help pay for the cost of the wires, or should the system operator give the rights away and let private parties pocket the congestion rents?[6] Where physical rights have been purchased in the past or been acquired by paying for lines or upgrades, grandfathering these rights, and perhaps converting them to financial rights provides regulatory continuity

---

6. If the system operator sells transmission rights and collects the congestion rent, it can sell the rights on a daily or hourly basis in the form of nodal pricing, or it can sell yearly or monthly transmission rights in periodic auctions, or these can be combined or supplemented with other approaches.

and enhances regulatory credibility. Other schemes for giving away new rights are usually obscure and arbitrary. Except when there is clear ownership of rights to begin with, it is preferable for the system operator to create transmission rights, sell them, and use the resulting revenues to pay for present or future transmission lines.

## 1-2.3  RISK MANAGEMENT AND FORWARD MARKETS

The system operator needs to manage the system in real time to keep it physically secure. It also needs a small lead time for planning (scheduling). This lead time is generally accepted to be approximately a day, but beyond that, the system operator has no need to pay attention to the energy market (transmission and generation capacity is another matter).

Both generators and their customers will want to make long-term arrangements for the supply of power either in decentralized forward markets or in highly centralized futures markets. The futures markets will work best if subject to the normal regulations imposed on commodities markets, but electricity futures need no special treatment. These markets should be just as deregulated as any other commodity market.

Transmission rights also need forward markets. (See Hogan, 1997, for one market design.) As with congestion management, the system operator may play a role in the supply of transmission forwards because ultimately it must decide how much power will be allowed to flow on the various transmission lines. Unfortunately, transmission rights are extremely complex because, in present transmission systems, it is impossible to choose the path over which power will flow. If there is a line from A to B and a trader owns only the right to transmit 100 MW over that line, the trader will not have sufficient rights to make any trade at all. If even 10 MW is injected at A and removed at B, a significant fraction of that power, sometimes more than half, will flow on other lines. Where the power actually flows is determined by the laws of physics.

Fortunately, designing forward transmission rights as purely financial rights can simplify the problem of physical flows. A transmission right for 100 MW can confer on its owner the right to the congestion rent from A to B times 100 MW. If the congestion rent averaged $10/MWh during peak hours, and the right covers 100 MW of flow for 320 peak hours, the owner receives $320,000. With such rights the owners can perfectly hedge a transaction from A to B. The markets for such rights need no special regulation, but they may not be liquid enough to serve the needs of power traders fully. Much work remains to be done in designing better hedges for congestion costs.

## 1-2.4  TRANSMISSION AND DISTRIBUTION

If the congestion rent is paid to the owners of transmission lines, too few lines will be built. The market for transmission lines cannot be left unregulated, and even a regulated market may not be up to the job. One way to get lines built is to have an ISO build them. Another way is to grant one company, a transco, the monopoly right to all lines in some large geographical area. Like any monopoly of an essential product, the transco will need regulation.

Several problems need to be considered. Transmission lines can act as both a substitute for and a complement to generation. New transmission lines increase competition between suppliers who may therefore oppose them, but they are a public good because they reduce market power. The siting of a new transmission line is a highly regulated and contentious process. It is also difficult to assign individual physical or financial rights to the power grid in such a way that investors make the appropriate return on their investment.

These complexities make deregulation of the market for transmission lines impossible. The distribution system seems even more difficult to deregulate, and so far there have been few, if any, proposals to do so.

## 1-2.5   RETAIL COMPETITION

The push to deregulate generation was clearly predicated on reducing the cost of generation which accounts for nearly one half the cost of power. Wholesale competition could save a lot of money; retail competition needs a different rationale. When the costs of the electricity industry are analyzed, they are traditionally divided into three major categories: generation, transmission, and distribution—retail is not mentioned. Retail costs could be cut in half, and no one would notice as they are only a small fraction of distribution costs.[7]

What is retailing? It is not distributing power at the local level or even hooking up individual customers. It is typically only financial transactions and sending out bills; occasionally it involves meter reading. Generally, a retailer buys wholesale power, signs up retail customers, and sends out bills. Although an individual retailer may manage to purchase power cheaply, on average a retailer will pay the average cost of wholesale power. Also, there is no reason to believe that competition on the demand side of this market will reduce the cost on the supply side. There may be room to cut billing costs, but there are other motives at work in the push for retail competition.

The impetus for retail competition comes primarily from two sources: those who believe they can profit by being retailers, and big commercial and industrial customers. Some of them believe they are smarter or more desirable customers and so can cut a better deal on their own. These motivations, though strong, do not translate into politically persuasive arguments, so more theoretical explanations are proffered. These fall into three categories: customer choice, innovative products, and price competition.

### Customer Choice

Although those pushing hardest for retail competition are not particularly "green," their main example of customer choice is that of green power. This could have been

---

7. According to Joskow (2000a, 29), the total cost of retailing services amounts to between 3.3% and 4.7% of total retail revenue. A reduction by half would at most amount to a savings of 2.4%.

provided quite easily under regulation, but it wasn't, so the green choice may be a benefit of retail competition. It allows customers to pay a premium for their energy and then makes sure some of that premium finds its way to generators that use renewable energy sources.

One disadvantage of retail competition is that it makes verifying the reality of this product much more difficult. Physically, green power is identical in all respects to coal power. The consumer cannot verify the product without the help of some third party to audit the suppliers. The source of power is more difficult to monitor when there are many competitive green and semi-green private companies instead of just one regulated utility.

Another choice, often suggested but so far not implemented, is the choice of reliability level. Clearly, some customers have a much greater need for reliability, and this is the one way in which electricity service can differ between customers. It is not possible to provide a more stable frequency or better voltage support on a customer by customer basis, but some reliability (avoidance of some blackouts) can be provided individually. This would require installing remotely controlled individual circuit breakers. Unfortunately this is currently quite impractical. Part of the problem is that those with a strong financial interest in this service are those who want more reliability. This cannot be provided by installing a breaker for them; it must be provided by installing breakers for a great many others who would be sacrificed in an emergency to provide reliability for those who would pay for it.

## Innovative Products

One view holds that competition in retail electricity will spawn innovative new electrical products the way competition in telecommunications has spawned new types of phones and phone services.[8] But so far AC power has shown no prospects of becoming wireless. Power engineering went through a period of rapid innovation between 1870 and 1910 that was similar to the current innovation in telecommunications, and someday it may again. Technology has a life of its own. Competition may spur some development of new generating technologies, but the basic AC outlet will be around for a while.

Those who promised dramatic new physical innovations in retail power five years ago have now shifted their focus from new kinds of electricity meters and smart appliances to innovative billing systems. Would they have held telecommunications up as an example of the benefits of deregulation if the last fifteen years had produced only more complex phone bills while telephones remained rotary dial?

The argument for the value of billing innovations is a curious one. In England, electricity deregulation has brought some of the same aggravation to electricity shopping that U.S. customers experience in shopping for phone service. Littlechild argues that the fact that customers spend so much effort comparing rates proves they must benefit from doing so.[9] Perhaps customers draw a different conclusion

---

8. Though competition has spurred innovation in telecom, if competition had arrived forty years earlier, we would not have had vacuum-tube cell phones in the 1950s. Much of what is taken to be the miracle of telecom competition is just normal technological progress.

9. See Littlechild (2000, 11).

than economists because they understand their efforts are as much to avoid bad rates that were previously not a danger as to find good rates that were previously not available.[10] In any case, other than real-time pricing, it is hard to imagine how better electricity bills can bring much improvement to the life of the average customer.

## Price Competition

The latest rationale for retail competition is the benefit of price competition. Even if customers don't think fancy electric bills are as useful as cell phones, cheaper bills would certainly prove popular. Littlechild (2000, p. 9) pushes price competition as the major benefit of retail competition and lists two ways this can happen, first, by "reducing the costs of retailing." Unfortunately, as already noted, the costs of retailing electricity under regulation are exceptionally small. Even with massive improvements in efficiency, customers will probably not notice a difference, and it is quite possible that marketing costs will more than offset any efficiency gains.

Littlechild also mentions a reduction in wholesale costs caused by "improving wholesale power procurement." Where does this savings come from: better generators, better operation, or a reduction in market power? With or without retail competition, generators still keep every dollar they save, no more and no less, so retail competition cannot possibly improve the cost-minimizing incentives of generators. With costs the same, the only room for savings is a reduction in profit. Because the industry cannot survive on below normal profits, this can only come about if there is some excess profit at the outset.

So retail competition can only lower wholesale costs by reducing the market power of wholesalers. But market power on the supply side of the wholesale market would normally be reduced by an increase in competition on the supply side, not by increased competition on the buyer's (retailer's) side. In fact more competition on the buyers' side means less monopsony power to counteract the monopoly power of the suppliers.

There appears to be only one small chance for retail competition to inhibit wholesale market power. If retailers sign long-term contracts with wholesalers, the contracts will inhibit the wholesalers' market power in the spot market. Unfortunately, competitive retailers have a more, not less, difficult time signing long-term contracts than the monopoly utility-distribution companies (UDCs) because their customer base is less stable. The customers of the UDC can't leave if there is no retail competition, but customers of competitive retailers generally can leave on fairly short notice and will leave if they do not like the current price. This makes it difficult for competitive retailers to sign long-term power contracts, especially for the supply of residential customers. If the retailer buys its power with a long-term contract and sells it to customers under short-term contracts, it takes an enormous risk. If the market price goes down in the future, the retailer must either

---

10.  Recently, and long after this section was written, this phenomenon was documented in the PJM market by Michael Rothkopf, who was the recipient of an offer to pay 38% above the going rate with only extra cancellation penalties as compensation. In California deceptive practices similar to those found in phone-rate competition were quickly spotted in the retail power market.

sell at a loss to its residential customers or lose them and resell its power at a loss to someone else.

The net result seems to be that retail competition offers no benefits in reducing wholesale market power. As it will not bring down the costs of generation, it seems to hold little promise of improving wholesale performance. The slim hope that price competition will save more on billing costs than it spends on marketing is a flimsy basis for such a large experiment.

*Chapter 1-3*
# Pricing Power, Energy, and Capacity

*It is not too much to expect that our children will enjoy in their homes
electricity too cheap to meter.*

Lewis L. Strauss
Chairman, Atomic Energy Commission
1954

**P**OWER IS THE RATE OF FLOW OF ENERGY. Similarly, generating capacity, the ability to produce power is itself a flow. A megawatt (MW) of capacity is worth little if it lasts only a minute just as a MW of power delivered for only a minute is worth little. But a MW of power or capacity that flows for a year is quite valuable.

The price of both power and energy can be measured in $/MWh, and since capacity is a flow like power and measured in MW, like power, it is priced like power, in $/MWh. Many find this confusing, but an examination of screening curves shows that this is traditional (as well as necessary). Since fixed costs are mainly the cost of capacity they are measured in $/MWh and can be added to variable costs to find total cost in $/MWh.

When generation cost data are presented, capacity cost is usually stated in $/kW. This is the cost of the flow of capacity produced by a generator over its lifetime, so the true (but unstated) units are $/kW-lifetime. This cost provides useful information but only for the purpose of finding fixed costs that can be expressed in $/MWh. No other useful economic computation can be performed with the "overnight" cost of capacity given in $/kW because they cannot be compared with other costs until "levelized." While the U.S. Department of Energy sometimes computes these economically useful (levelized) fixed costs, it never publishes them. Instead it combines them with variable costs and reports total levelized energy costs.[1] This is the result of a widespread lack of understanding of the nature of capacity costs.

Confusion over units causes too many different units to be used, and this requires unnecessary and sometimes impossible conversions. This chapter shows how to make almost all relevant economic calculations by expressing almost all prices and

---

1. In Tables 14 through 17 of one such report (DOE 1998a) the useful (amortized) fixed costs are not reported, and the fixed O&M costs are reported in $/kW which may be an amortized value reported with the wrong units or, if the units are correct, may represent a misguided conversion of an amortized cost to an "overnight" cost.

costs in dollars per megawatt-hour ($/MWh). The remainder of the book confirms this by working every example in these units.

**Chapter Summary 1-3:** Energy is measured in MWh, while power and capacity are measured in MW. All three are priced in $/MWh, as are fixed and variable costs. Other units with the same dimensions (money divided by energy) may be used, but this book will use only $/MWh. Screening curves plot average cost as a function of capacity factor. The slope of the curve is **variable cost**, and the intercept is **fixed cost**. The average cost $(AC_K)$ plotted in these graphs is not the average cost of using a megawatt-hour of energy produced at a certain capacity factor but rather the average cost of a megawatt-hour of generating capacity. Because the equation for a screening curve is used through the book, understanding this distinction is crucial.

### Working Summary

Readers wishing to gain only a working knowledge of measurement units for use in later chapters should understand the following.

| Quantity | Quantity units | Price Units |
|---|---|---|
| Energy | MWh | $/MWh |
| Power | MW | $/MWh |
| Capacity | MW | $/MWh |

| Cost | Symbol | Cost Units |
|---|---|---|
| Fixed | $FC$ | $/MWh |
| Variable | $VC$ | $/MWh |
| Average | $AC_K = FC + cf \times VC$ | $/MWh |
| Average | $AC_E = FC/cf + VC$ | $/MWh |

| Ratio | Symbol | Units |
|---|---|---|
| Capacity factor | $cf$ | none |
| Duration | $D$ | none |

**Notes:** Energy is a static amount while power and capacity are rates of flow. The average cost of using capacity, $AC_K$, depends on the capacity factor, $cf$, which is the fraction of time the capacity is used. The average cost of energy, $AC_E$, produced by a specific generator also depends on $cf$.

**Section 1: Measuring Power and Energy.** Power is the flow of energy and is measured in watts (W), kilowatts (kW), megawatts (MW), or gigawatts (GW). Energy is an accumulation of power over a period of time. For instance, a kilowatt flowing for one hour delivers a kilowatt-hour (kWh) of energy. The price of both energy and power is expressed in $/MWh. It can also be expressed in "mills," short for "milli-dollars per kilowatt hour," with 1 mill equal to $1/MWh.

**Section 2: Measuring Capacity.** Capacity is the *potential* to deliver power and is measured in megawatts. Like power, it is a flow.

**Section 3: Pricing Capacity.** "Overnight" capacity costs are measured in $/kW and so cannot be added to or averaged with variable costs to find which generator could more cheaply serve load of a specific duration. Screening curves plot the annual revenue requirement $(ARR)$ of a generator as a function of the generation's capacity factor. Fixed cost $(FC)$ is the value of $ARR$ for a capacity factor of zero. Since $ARR$ is measured in $/kWy, the same must be true of fixed cost. Dividing $FC$ by 8.76 converts it to $/MWh, a more convenient set of units. Considering the rental cost of capacity makes these units seem more natural.

To avoid confusion when using screening curves and their associated algebra, the distinction between the average cost of capacity $(AC_K)$ and the average cost of energy $(AC_E)$ should be kept in mind. Traditional screening curves graph $AC_K$.

## 1-3.1   MEASURING POWER AND ENERGY

### Power Versus Energy

**Power** is the rate of flow of energy. This is true for any form of energy, not just electricity. If you wish to boil a cup of water you need a quantity of **energy** to get the job done, about 30 watt-hours. Any specific power level, say a thousand watts (kilowatt, or kW), may or may not make you a cup of tea depending on how long the power continues to flow. A typical microwave oven delivers power at a rate of about 1 kW (not 1 kW per hour). If it heats your water for one second, the water will receive *power* at the *rate* of one thousand watts, but it will gain very little *energy* and it will not make tea. Two minutes in the micro-wave will deliver the necessary energy, $\frac{1}{30}$ of a kWh.

Confusion arises because it is more common to have the time unit in the measurement of a flow than in the measurement of a quantity. Thus if you want to fill your gas tank, you buy a quantity of 15 gallons of gasoline, and that flows into your tank at the rate of 5 gallons per minute. But if you need a quantity of electric energy, that would be 30 watt-hours, and it would be delivered at the rate of 1000 watts.[2] Because a watt-hour is a unit of energy, it would make sense to speak of delivering 1000 watt-hours per hour, but that just boils down to a rate of 1000 watts (1 kW) because a watt-hour per hour means watts times hours divided by hours, and the hours cancel out.

---

**Unit Arithmetic**

Units—kilowatts, hours, and dollars—follow the normal laws of arithmetic. But it must be understood that a kWh means a (kW × h) and a $ per hour means a ($/h).

Also note that "8760 hours per year" has the value of 1, because it equals (8760 h)/(1 year), and (8760 h) = (1 year).

As an example, $100/kWy =

$$\frac{\$100}{kW \times year} \times \frac{1000\,kW}{1\,MW} \times \frac{1\,year}{8760\,h}$$

which reduces to $11.42/MWh.

---

### The Price of Power and Energy

Because power is a flow, its total cost is measured in dollars per hour, not dollars. The total cost of a certain quantity of energy is measured in dollars. Consequently the *price* (per unit cost) of power is measured in dollars per hour per MW of power flow, while the price of energy is measured in dollars per MWh. But these units are the same:

$$\text{(dollars per hour) per MW} = (\$/h)/MW = \$/MWh$$

so the units for the price of power are the same as for the price of energy.

Typically the price of retail energy is about 8¢/kWh.[3] At that price, the price of power would be 8 cents/hour for a kilowatt of power flow, which is the same. These units are convenient for home use but are inconveniently small for bulk power systems. Consequently this book will use megawatts (millions of watts) instead

---

2. Watts per hour has units of watts divided by hours and has no use in the present context.

3. Average revenue per kilowatt hour to ultimate residential consumers was 8.06¢/kWh, according to DOE (2001c, Table 53).

of kilowatts. The same energy price can be re-expressed as $80/MWh. When discussing large markets and annual energy use, power may be measured in gigawatts (GW, or billions of watts) and energy in terawatt hours (TWh, or trillions of watt hours).

Another commonly used unit is the **mill**, short for "milli-dollar," or $\frac{1}{1000}$ of a dollar. This unit might seem particularly inappropriate for wholesale markets, but it is commonly used to compensate for using the kW which is also inappropriately small. Together these give rise to "milli-dollars per kilowatt-hour," often incorrectly shortened to "mills." Scaling both the numerator and denominator up by 1000 has no effect on the numeric value and converts milli-dollars to dollars and kilowatts to megawatts. So 80 mills/kWh is identical to $80/MWh.

# 1-3.2   MEASURING GENERATION CAPACITY

The size of a generator is measured by the maximum flow of power it can produce and therefore is measured in MW. The capacity to produce a flow of power is best conceptualized as a flow just as a MW of power is a flow of energy.[4]

In principle one could define an amount of capacity related to the flow of capacity as energy is related to power, but this is not necessary. Moreover, it is likely to cause confusion because when applied to a generator, it would aggregate a flow of capacity over many years without any discounting. For these reasons, the idea of a capacity amount, different from a capacity flow, will not be introduced or utilized.

Having found that capacity, like power, is a flow measured in MW, it is natural to ask if it is priced in $/MWh as is power. Most would say no, but it is best to look to its use in solving real economic problems before drawing this conclusion. Consider the problem of choosing which generator can most cheaply serve a load of a particular duration. The long tradition of solving this problem by using "screening curves" will provide the key to this puzzle.

# 1-3.3   PRICING GENERATION CAPACITY

## The "Overnight" Cost of Capacity

A generator has an "overnight cost" which is typically given in $/kW. For example, the overnight cost of a coal plant might be $1,050/kW, so a 1000 MW plant would cost $1,050 million. In economic terms, this is the present-value cost of the plant; it would have to be paid as a lump sum up front to pay completely for its construction.

A conventional gas-turbine generator (GT) would have an overnight cost closer to $350/kW. Although the GT is three times cheaper than the coal plant, for some

---

4. The flow of available capacity is interrupted during generator outages, but the flow of installed capacity is continuous. This chapter ignores the difference and assumes that the flow of capacity from a generator is continuous and constant.

purposes the coal plant is the more inexpensive choice. Fuel costs must always be taken into account when evaluating the choice of generators. Coal plants are built because their cost of fuel per unit of energy output is less. Assume coal costs only $10/MWh of energy produced, while the cost of fuel for a GT comes to $35/MWh. Now which plant is cheaper?

More information is needed. The comparison depends on how much the plant will be used, and that depends on the load it will serve. For concreteness, assume that the load has a duration of 25% (2190 hours/year) so the plant serving it will have a capacity factor of 25%. Now, which plant is cheaper?

Focusing on only the basics, the problem seems workable. The overnight cost captures the *fixed cost* of generation, and the fuel cost per unit of output captures the *variable cost*. Duration gives a sufficient description of the load. But the problem is still impossible to solve because the fixed cost of capacity has been measured in the wrong units. Overnight costs measured in $/kW cannot be added to fuel costs measured in $/kWh. This would produce nonsense.

---

*Fallacy*  1-3.1         **Fixed and Variable Costs Are Measured in Different Units**
Because capacity is usually paid all at once, while fuel is paid for over time, variable costs but not fixed costs should include a time dimension.

---

When units have the same "dimensions," they differ only by a *scale factor* (a pure number). Different quantities having units of the same dimension can be added. For example, 1 MWh can be added to 100 kWh to get 1100 kWh (or 1.1 MWh). But quantities whose units have different dimensions cannot be added. This is the meaning of the famous saying, "you can't add apples and oranges." For example, 1 MW cannot be added to 1 MWh. Engineers and physicists pay close attention to mismatched units because they always signal deeper trouble. Any calculation that involves adding MW and MWh simply does not make sense.

## Identifying Fixed Costs on Screening Curves

Screening curves, shown in Figure 1-3.1, are used to compare generation costs by taking account of the three factors of our present problem: fixed cost, variable costs and load duration (which determines the generator's capacity factor). Necessarily, they provide guidance on the proper units for fixed costs. Traditionally, these curves plot "**annual revenue requirement** per kW" (*ARR*) as a function of capacity factor (*cf*). The generator's **capacity factor** is its percentage utilization which is determined by the load's duration.[5]

Traditionally, the **variable cost** component of *ARR* is computed by taking the fuel cost expressed in $/MWh and converting to $/kWy.[6] The result is $87.60/kWy for a coal plant and $306.60/kWy for a GT. This assumes full-time operation, so

---

5. Duration is measured as a percentage (see Chapter 1-4), so if all load served has the same duration, the capacity factor equals load duration. If the load has a range of durations, these must be averaged.

6. For simplicity, this assumes that fuel is the only variable cost. Operation and maintenance include an additional variable cost component which should be expressed in $/kWy.

**Figure 1-3.1**

Use of screening curves
to select a generator.

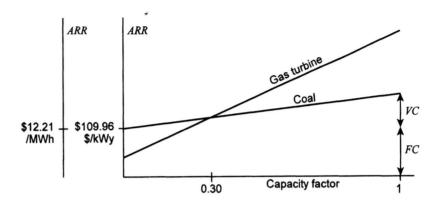

to find the variable component for any particular *cf*, these must be multiplied by
*cf*, 25% in the case of the present example.

The overnight cost of capacity is more problematic. A coal plant with an
overnight cost of $1,000/kW does not cost $1,000/kWy. This would imply a plant
life-time of one year and a discount rate of zero. The correct fixed-cost component
of *ARR* is the overnight cost amortized ("levelized") over the life of the plant. This
is equivalent to computing home mortgage payments based on a mortgage that lasts
the life of the house. Obviously a discount rate (interest rate) is involved. The
formula for amortization is

$$FC = \frac{r \cdot OC}{1 - e^{-rT}} \approx \frac{r \cdot OC}{1 - 1/(1+r)^T} \qquad (1\text{-}3.1)$$

Notice that **fixed cost** (*FC*) depends only on overnight cost (*OC*), the discount rate
(*r*, in % per year) and the life of the plant (*T*, in years).[7]

**Table 1-3.1**  Technology Costs

| Technology | VC ( /MWh) | VC ( /kWy) | OC ( /kW) | FC ( /kWy) | FC ( /MWh) |
|---|---|---|---|---|---|
| Gas turbine | $35 | $306.60 | $350 | $40.48 | $4.62 |
| Coal | $10 | $87.60 | $1050 | $106.96 | $12.21 |

Fixed-costs are based on *r* = 0.1 and on *T* = 20 for gas turbines and 40 for coal plants. Equation 1-3.1
gives fixed costs in $/kWh which are then converted to $/MWh by dividing by 8.76.

*FC* is a constant flow of cost that when added to *VC* gives *ARR*, the annual
revenue requirement per kW of generation capacity. Of course this assumes a
capacity factor of 1. If *cf* is less, *VC* will be reduced proportionally, but *FC* is
unaffected because capacity must be paid for whether used or not. That is why *FC*
is termed the *fixed* cost. The formula for *ARR* is

Screening Curve: *ARR* = *FC* + *cf* × *VC*

---

7. Using *monthly* instead of *annual* compounding in the second formula greatly improves its accuracy as
an approximation. To do this, change *r* to *r*/12 in the denominator and *T* to 12 *T*.

for *ARR* to be valued in $/kWy, both *FC* and $cf \times VC$ must also be valued in $/kWy. As these are the traditional units for *ARR*, the traditional units for fixed cost must also be $/kWy. These units have the same dimension as $/MWh and any quantity expressed in $/kWy can be converted to $/MWh by dividing by 8.76.

Variable cost is naturally expressed in $/MWh, so capacity factor, *cf*, must be a pure number (dimensionless), otherwise, $cf \times VC$ would not have the same units as *ARR*. This is correct; a capacity factor is just the fraction of a generator's potential output that is actually produced. It is actual energy output divided by potential energy output, so the energy units cancel.

---

**Result  1-3.2**   **Energy, Power, and Capacity Are Priced in $/MWh**
Although power is measured in MW and energy in MWh, both are priced in $/MWh. Like power, generating capacity is a flow measured in MW and consequently is also priced in $/MWh.

---

## The Rental Cost of Capacity

Fixed costs are the costs of generation capacity. It may be argued that buying a generator is buying capacity and that generators are measured in MW, not in MWh. This is only partially true. If a 1 MW gas-turbine generator is worth $350,000, does this mean 1 MW of capacity is worth $350,000? No, the gas turbine is worth that only because it has a certain expected lifetime. An identical but older gas turbine is worth less, even though it has the same 1 MW capacity. Thus the price of capacity always involves a time dimension, either explicitly or implicitly.

Measuring capacity in MW indicates that capacity is being considered a flow. A 100-MW generator delivers a 100-MW flow of capacity for some unspecified period of time. That flow must be paid for by a flow of money—so many dollars per hour. This corresponds to a *rental cost*. If a generator is rented, the cost of renting will be so much per hour, or per day, or per year. If this is scaled by the generator's capacity, for easy comparison with the rental rate of other generators, then it is natural to express the rental cost of a generator in $/h per MW, or equivalently in $/MWh.

The above screening curve analysis can be summarized as saying that generation capacity costs should be expressed as a rental rate and not as a one-time (overnight) purchase price. Rental rates naturally have the same units as variable costs and so make total and average cost calculations convenient.

## Two Kinds of Average Cost: Avoiding Confusion

The cost of operating a generator with a specific capacity factor can be read from a screening curve. Traditionally this cost is expressed in $/kWy and called an annual cost. Although a kWy has the dimensions of energy, this cost is **not** the annual cost of *energy* produced by the plant! A screening curve shows the average cost of using the coal plant's *capacity*.

**Figure 1-3.2**

Capacity-cost based and energy-cost based screening curves.

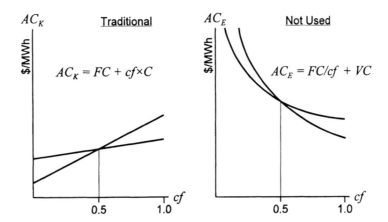

With the price of energy always expressed as an hourly cost, it is more convenient to divide the annual cost, in $/kWy, by 8.76 and arrive at an average cost per hour for the year expressed in $/MWh. Like the annual cost, this hourly average cost is still not a cost of energy produced but the cost of using capacity. This book will always report capacity, energy and power costs in $/MWh for ease of comparison and to make addition of costs and cost averaging possible.

Although screening curves plot the average cost of capacity use, the average cost of energy produced is also interesting and could be used to construct hyperbolic screening curves. A pair of these is shown on the right side of Figure 1-3.2; they are nonlinear (hyperbolic), but they still intersect at exactly the capacity factor at which one plant becomes more economical than the other. The equations for the linear and hyperbolic screening curves are closely related and are shown in Figure 1-3.2.

The average cost of capacity ($AC_K$) used with capacity factor $cf$ is the fixed cost of using that capacity, plus $cf$ times the variable cost of producing energy. If $cf = \frac{1}{3}$, then one third of the variable cost of maximum potential energy output must be added to the constant fixed cost which increases the average cost per unit of capacity by $cf \times VC$.

### Screening Curves

A traditional screening curve plots average cost as a function of capacity factor. When using the screening curve equation, this average cost can be confused with the average cost of the *energy* produced by a certain type of generator. Instead, it is the average cost of using a unit of capacity.

Screening curves could have been defined using the average cost of energy. Then as the capacity factor approached zero the average cost would approach infinity. The average-energy-cost equation is used to analyze market equilibria in later chapters, but nonstandard screening curves are never used.

The average cost of energy ($AC_E$) when the generator runs with capacity factor $cf$ is the variable cost of producing that energy, plus the fixed cost of the capacity divided by $cf$. If $cf = \frac{1}{3}$, then fixed costs must be spread over only $\frac{1}{3}$ of the total possible energy output, so they are multiplied by 3 (divided by $cf$) before being added to $VC$.

A load slice is a horizontal strip cut from a load-duration curve (see Chapter 1-4.1). Depending on its average duration it will be served by some particular technology, baseload, midload, or peaking. Any given load slice is defined by a capacity, $K_{slice}$, which is the height of the slice, and an average duration, $D$. It also has a total energy requirement, $E = D \times K_{slice}$. To serve this load, generation capacity of $K_{slice}$ must be installed and must run with a capacity factor of $cf = D$. Having selected a technology, one can compute the **average cost** per MW-of-*capacity* of serving the load, $AC_K$, and the average cost per MWh-of-*energy* of serving the load, $AC_E$. The total cost of serving load is then given by both $K_{slice} \times AC_K$ and $E \times AC_E$. Because $K_{slice}$ and $E$ are both fixed, choosing the technology that minimizes either $AC_K$ or $AC_E$ will minimize the total cost of energy. This is why either the traditional or the hyperbolic screening curves can be used.

These relationships can be summarized as follows. For a particular load slice served by generators with fixed cost, $FC$, and variable cost, $VC$, the average cost of capacity and energy can be found as follows:

$$\text{Capacity: } AC_K = FC + cf \times VC = FC + D \times VC \tag{1-3.1}$$

$$\text{Energy: } AC_E = FC/cf + VC = FC/D + VC \tag{1-3.2}$$

The capacity factor of the generator, $cf$, equals the average duration of the load, $D$.

No one uses hyperbolic screening curves, but when an average cost is computed for a specific technology, say by DOE, a value for $AC_E$ (not $AC_K$) is always computed. Typically, DOE might report the overnight cost ($OC$), some information about fuel costs, and a value for $AC_E$ based on technical capacity factor ($cf$).[8] In other words, DOE reports some technologically determined value on the technology's *hyperbolic* screening curve.

$AC_K$ is used to determine the optimal durations of various generation technologies, and from these durations the optimal investment in these technologies. Since competitive markets optimize technology, $AC_K$ is also used to determine competitive outcomes. Either $AC_E$ or $AC_K$ may be used to compare the cost of peak energy with the value of lost load, depending on whether **peaker** costs are equated to the value of lost load or the average hourly cost of lost load (see Chapter 2-2 and 2-3). $AC_E$ is also well suited to DOE's interest in alternative technologies—nuclear, wind, solar, and so on. These have in common capacity factors which, even in a market environment, are not affected by normal variations in market structure but are instead technologically determined because their variable costs are almost always below the market price. They run whenever they are physically able, so their capacity factor is determined by their technical capability. The economics of an alternative technology can be assessed by comparing its $AC_E$ with the market's average price.

Standard technology generators have capacity factors determined by the market and not just by their technology. In this case, the duration of the load they serve, which determines their capacity factor, needs to be determined from their fixed and variable costs along with those of other technologies. This is done with screening curve analysis, or with algebra based on screening curves. Traditional linear screening curves prove simplest. These curves and their associated algebra will be used throughout the book, as is the formula for $AC_E$.

---

8. See DOE (1998a) Tables 14–17. In a table of cost characteristics of new generating technologies (DOE 2001a, Table 43) $OC$ is given, but not $FC$. In a slide labeled "Electricity Generation Costs," DOE (2001b) reports the capital costs in mills/kWh. As the title indicates, these are $FC/cf$, for $cf$ determined by technology-based capacity factors, and so are components of $AC_K$, as advertised in the title and thus points on the hyperbolic screening function.

**Table 1-3.2** Fixed and Variable Cost of Generation

| Type of Generator | Overnight Capacity Cost $/kW | Fixed Cost $/MWh | Fuel Cost $/MBtu | Heat rate Btu/kWh | Variable Cost $/MWh |
|---|---|---|---|---|---|
| Advanced nuclear | 1729 | **23.88** | 0.40 | 10,400 | **4.16** |
| Coal | 1021 | **14.10** | 1.25 | 9,419 | **11.77** |
| Wind | 919 | **13.85** | — | — | **0** |
| Advanced combined cycle | 533 | **7.36** | 3.00 | 6,927 | **20.78** |
| Combustion turbine | 315 | **4.75** | 3.00 | 11,467 | **34.40** |

\* Overnight capacity cost and heat rates are from DOE (2001a), Table 43. Plants not labeled "advanced" are "conventional." Rental capacity costs are computed from overnight costs, a discount rate of 12% and assumed plant lifetimes of 40 years except for wind and gas turbines which are assumed to be 20 years. For simplicity, operation and maintenance costs are ignored.

## 1-3.4  TECHNICAL SUPPLEMENT

### Checking Fixed-Cost Units with the Amortization Formula

As a final check on the units of fixed costs, the amortization formula can be analyzed. Interest rates (e.g., 10% per year) has the dimension of "per unit time," and $T$ has the dimension of time, so $rT$ is dimensionless, that is, a pure number. This is necessary for compatibility with "1" in the denominator. In the numerator, $OC$ has the traditional units of $/kW and "r" again has the dimension of $1/$time, so $r \times OC$ has the dimensions of $OC$ per unit time, for example, $/kW per year. If overnight cost is measured in $/kW and interest is given in percent per year, fixed cost must be measured in $/kWy.

### Fixed and Variable Costs for Different Technologies

Table 1-3.1 computes fixed and variable costs for five types of generators as an example of converting overnight cost to fixed costs. The listed values of $FC$ and $VC$ are exactly the values needed to draw screening curves and choose the most efficient plant to serve loads of any duration. For example, the cost of serving load of duration $D$ with an advanced combined-cycle plant is

$$AC_K = (7.36 + 20.78\,D)\ \$/\text{MWh}.$$

To convert this to the more traditional units of $/kWy, both values should be multiplied by $(1\ \text{M}/1000\ \text{k})(8760\ \text{h/y})$ or 8.76. (Note that, including units, this is just multiplication by 1 since 1 M = 1000 k and 8760 h = 1 y.)

To avoid having $AC_K$ appear to have the same units as variable costs, its units are often stated as "$/kW per year" which translates to $/kW/year. But just as $x/y/z = x/(y \times z)$, so $/kW/year equals $/kW-year which is denoted by $/kWy. The phrase "$/kW per year" is correct, but it means no more and no less than $/kWy, which has the dimensions of dollars per energy.

# Power Supply and Demand

*And when the Rain has wet the Kite and Twine, so that it can conduct the Electric Fire freely, you will find it stream out plentifully from the Key on the Approach of your Knuckle.*

Benjamin Franklin
1752

**T**HE PHYSICAL ASPECTS OF SUPPLY AND DEMAND PLAY A PROMINENT ROLE IN POWER MARKETS. Shifts in demand, not associated with price, play a role in all markets, but in power markets they often receive attention to the exclusion of price. This is not simply the result of regulatory pricing; even with market prices, demand shifts will play a key role in determining the mix of production technologies. In this way hourly demand fluctuations determine key long-run characteristics of supply.

Because electric power cannot be stored, production always equals consumption, so the difference between supply and demand cannot be indicated by flows of power. Neither is the instantaneous difference indicated by contracts since real-time demand is determined by customers physically taking power. The short-run supply–demand balance is indicated by voltage and, especially, frequency. This unusual market structure requires some elementary background in system physics. More detail is provided in Chapters 5-1 and 5-2.

**Chapter Summary 1-4:** Load duration curves are still relevant in unregulated markets, but their role in analysis is more subtle because their shape is affected by price and its correlation with load. They can still be used with screening curves to check an equilibrium, but to predict an equilibrium they must be used in combination with price elasticity.

Power production always equals consumption (counting losses as part of consumption) which makes it impossible to assess the supply–demand balance by observing quantities or quantity flows. Instead, frequency is the proper indicator of system-wide balance, and net unscheduled flows between regions are used to share the responsibility of maintaining this balance.

**Section 1: Describing the Demand for Power.** A year's worth of hourly fluctuations can be usefully summarized by a load-duration curve that plots demand against duration, the fraction of the year during which demand is at or above a certain level. It can also be thought of as the probability of finding load above a certain level.

If customers are charged real-time prices, peak demand will be reduced, allowing a reduction in generating capacity. The result will be a load-duration curve with its peak cut off horizontally.

**Section 2: Screening Curves and Long-Run Equilibrium.** If the screening curves of the available technologies are drawn on the same graph, their intersections determine capacity factors that mark technology boundaries. By mapping these capacity factors to durations and then to the load-duration curve, the optimal capacities for these technologies can be read off the vertical axis.

This technique can be used to partially confirm a market equilibrium but not to find one. In a market, price affects the shape of the load-duration curve, so it cannot be taken as given until the equilibrium is known.

**Section 3: Frequency, Voltage, and Clearing the Market.** When consumers turn on ten 100-W light bulbs, they are demanding 1000 W of power, and if generators supply only 900 W, the system will not be "in balance." In spite of this, the power supplied will exactly equal the power consumed (ignoring losses). This equality of power flows is caused by a decrease in voltage sufficient to cause the 100 W bulbs to use only 90 W of power. For motors the same effect is caused by a drop in frequency. Because voltage is automatically adjusted at substations, frequency is the main balancer of power inflows and outflows.

The United States is divided into three AC interconnections: the Western, Eastern, and Texas. The system frequency is constant throughout each AC interconnection, which means that a change in frequency cannot be used to locate a supply–demand imbalance. Instead, net power flows are tracked out of each control area and compared with scheduled power flows. This allows the imbalance to be located.

## 1-4.1   DESCRIBING THE DEMAND FOR POWER

Traditionally the demand for power has been described by a **load-duration curve** that measures the number of hours per year the total **load** is at or above any given level of demand. An example is shown in Figure 1-4.1. Total demand (load) is a demand for a flow of power and is measured in MW. Although the load-duration curve describes completely the total time spent at each load level, it does not include information about the sequence of these levels. The same load-duration curve can

**Figure 1-4.1**

A load-duration curve.

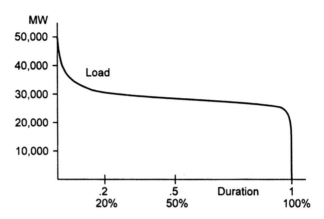

be produced by wide daily swings in demand and little seasonal variation or by wide seasonal variation and limited daily swings.

The introduction of a market adds the dimension of *price*. Economists often represent demand by a *demand curve* which expresses demand solely as a function of price. Nonprice fluctuations of the type captured by a load-duration curve are referred to as shifts in the demand curve and are generally not described in detail. But electricity is different because it is not storable, so peak demand must be satisfied by production from generators that are used as little as 1% of the time. Such generators, **peakers**, are built with technology that differs markedly from that used for **baseload generators** which run most of the time and are stopped only rarely. As a result, power markets face the problem of determining how much generation capacity should be built using each type of technology, for example, coal-fire steam turbines or gas-fired combustion turbines (**gas turbines**). This explains the unusual importance of demand shifts and consequently of load-duration curves in power markets.[1]

## Load-Duration Curves

A load-duration curve can be constructed for a given region (or for any collection of loads) by measuring the total load at hourly intervals for each of the 8760 hours in a year, sorting them, and graphing them starting with the highest load. The result is a curve that slopes downward from the maximum load in the peak hour, hour 1, to the minimum load, **baseload**, in the most off-peak hour, hour 8760 (see Figure 1-4.1).

**Duration** is traditionally measured in hours per year, but both hours and years are measures of time, so duration is **dimensionless**, which means it can be expressed as a pure number, a ratio, or percentage. To convert from units of hours per year (h/year) to a pure number, simply multiply by 1 in the form (1 year)/(8760 h). Duration has a natural interpretation as the probability that load will be at or above a certain level. To use this interpretation pick a load level, say 35 GW, and using the load-duration curve, find the corresponding duration, 20% in this case. This

---

1.  Service industries such as restaurants and airlines often have demand fluctuations which cause similar problem because their output is not storable, but they tend to use the same technology on and off peak.

indicates that load is 35 GW or greater 20% of the time. Put another way, the probability of load being 35 GW or greater in a randomly selected hour is 20%. This interpretation is most convenient.

## The Price-Elasticity of Demand

Presently, demand is almost completely unresponsive to price in most power markets because wholesale price fluctuations are not usually passed on to retail customers. Often retail prices remain under some form of price regulation, but competitive retailers have also been slow to implement real-time pricing. In the longer run, retail prices do change, sometimes seasonally. In the long run a 10% increase in the price of power will cause approximately a 10% reduction in the use of power.[2] This is not a very accurate approximation, but the long-run response to a 10% increase in price is likely to be found between 5% and 15% and is certainly not zero. Economists term this price sensitivity a **price elasticity of demand**, which is often shortened to *demand elasticity*. If a 10% change in price causes a 5%, 10%, or 15% change in demand, the elasticity is said to be 0.5, 1.0, or 1.5, respectively. (Technically, demand elasticities are negative, but this book will follow the common convention of re-defining them to be positive.)

## Real-Time Pricing and the Load-Duration Curve

Under regulation, residential load usually faces a price that fluctuates very little while commercial and industrial load often face time-of-use (TOU) pricing or demand charges. Time-of-use prices are designed to be high when demand is high, but the approximation is crude as they are set years in advance. Consequently they miss the crucial weather-driven demand fluctuations that cause most problematic supply shortages. Demand charges are no more accurate as they are based on individual demand peaks, not system peaks. Coincident-peak charges improve on this by charging customers for their use at the time of the system peak, but these are less common.

Because supply is fairly constant, the market is tightest when demand is highest. Consequently, high wholesale prices correspond well with high demand. If these real-time prices are passed through to customers, then retail prices will track load fairly well. Although real-time prices work best, all four pricing techniques, TOU, demand, coincident-peak, and real-time, tend to raise prices when demand is highest and reduce prices when demand is lowest. This results in lowering the peak of the load-duration curve and raising the low end of the curve.

If load faced real-time prices, the need for generation capacity might be reduced to the point where the load-duration curve under regulation had a duration of, say, 10%. Then, between 0% and 10% duration, supply and demand would be balanced by price. Instead of having generation follow load, load would be held constant by price at the level of installed capacity. In the lowest duration hours, price would

---

2. There is no natural definition of short- and long-run demand elasticity, which can be defined usefully over any time horizon from five minutes to twenty years. This text will use short-run elasticity to mean something on the order of one day and long-run to mean about five years.

**Figure 1-4.2**

The effect of price elasticity on load duration.

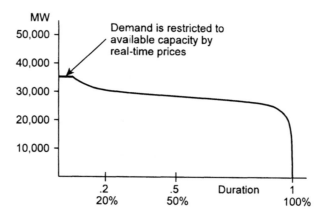

need to be very high to reduce demand to this level. By fluctuating sufficiently, price would control demand and produce a flat-topped load-duration curve with a maximum load just equal to generating capacity as shown in Figure 1-4.2.

## 1-4.2   SCREENING CURVES AND LONG-RUN EQUILIBRIUM

When demand is inelastic or when it faces a fixed price so that the load-duration curve is fixed, this curve can be used to find the optimal mix of generation technologies. The technique was developed for a regulated power system in which price and the load-duration curve are often fixed, but it is still useful for understanding certain aspects of competitive markets.

It assumes that fixed and variable costs adequately describe generators. These are used to draw screening curves for each technology on a single graph as shown in Figure 1-4.3. The intersections of these curves determine capacity factors that separate the regions in which the different technologies are optimal. These capacity factors equal the load durations that determine the boundaries between load that is served by one technology and the next. The screening curves in the figure are taken from Figure 1-3.1 and intersect at a capacity factor, $cf$, of approximately 30%. Consequently all load with a duration greater than 30%, or about 2600 hours, should be served by coal plants, while load of lesser duration should be served by gas turbines. The arrow in the figure shows how the needed capacity of coal plants can be read from the load-duration curve. The optimal GT capacity is found by subtracting this from maximum load which is the total necessary capacity. (Forced outages and operating reserve margins are considered in Parts 2 and 3.)

If customers face the wholesale market price through real-time pricing, this technique cannot be used because the load-duration curve depends on price, and price depends on the choice of technology, and the choice of technology depends, as just described, on the load-duration curve. This circularity is in no way contradictory, but it makes it difficult to find the competitive market equilibrium. Not only is calculation more difficult, but, also, the elasticity of demand must be known.

In spite of this circularity, the traditional technique can be used to partially confirm a long-run equilibrium. The load-duration curve observed in a market

**Figure 1-4.3**

Using screening curves
to find the optimal mix of
technologies.

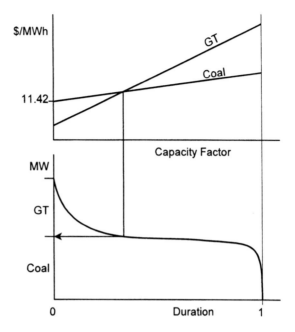

includes the effect of price on demand. When it is used along with the screening
curves of the available technologies, the traditional method should predict the mix
of technologies observed in the market if the market is in long-run equilibrium.
In practice many complications must be overcome.

## 1-4.3    FREQUENCY, VOLTAGE, AND CLEARING THE MARKET

So far, this chapter has considered how to describe demand and how to find the
optimal mix of technology to supply it. This section considers the physical details
of the supply–demand balance in real time. At any instant, customers are using
power, generators are producing it, and the amount produced is exactly equal to
the amount consumed. Some may object to the word "exactly," but the discrepancy
is at least a thousand times smaller than anyone's ability to measure it and is entirely
irrelevant. The determination of the supply–demand balance depends on electrical
phenomena more subtle than the concepts of quantity and quantity flow.

### Losses

In real networks, a few percent of the power consumed is consumed by the network.
This consumption should be considered part of demand even though it serves no
end used. With this convention, the system can be viewed as maintaining a perfect
balance between supply and consumption (including losses) at all times and between
supply and demand whenever customers are getting the power they want.

| Convention | **Loss Provision is Not Considered Part of Supply or Demand** |
| --- | --- |
|  | Losses will be considered as a service paid for by traders and provided separately from the trading arrangement. Consequently, from a trading point of view, power flows can be viewed as lossless. |

Supply precisely equals consumption because there is no storage of power in transmission systems.[3] But if supply equals consumption regardless of price, what signal should be used for price adjustment? How can demand be observed to be either greater than or less than supply? A mismatch between supply and demand is signaled not by power flows but by frequency and voltage. When they are below their target values, demand exceeds supply and vice versa.

## Frequency and Voltage

Power systems attempt to maintain a constant **frequency**, the rate at which alternating current alternates. In the United States, alternating current (AC) reverses direction twice, thus completing one *cycle* and returning to its original direction, 60 times per second. The frequency of AC in the United States is therefore said to be 60 cycles per second, also known as 60 Hertz or 60 Hz. In many countries the target frequency is 50 Hz.

**Voltage** is the amount of electrical pressure that pushes current through electrical appliances such as lights and motors. As with frequency, power systems have a certain target voltage that they attempt to maintain. In the United States the target residential voltage is about 120 volts. In some countries the target voltage is about twice as high. When an unprotected 120-V appliance is plugged into a 240-V outlet, the extra electrical pressure (voltage) causes twice as much current to flow through the appliance. This causes the appliance to use four times as much power (power is voltage times current) and the appliance typically burns out. The important point for this section is that as voltage increases, most appliances use more power, and as voltage decreases most appliances use less power.

Imagine a system with ten generators operating at full throttle supplying ten thousand homes with lights burning and motors running. If one generator goes off line, two things happen. The system voltage and frequence both decrease. Both cause electrical appliances to use less power. This effect has been described for voltage, but for more complex reasons most motors use less power when the system frequency declines. The decline in voltage and frequency is produced automatically by the physics of the entire system including all loads and generators. It happens to the exact extent necessary to balance inflow (supply) and outflow (consumption). If this did not happen, a law of physics, just as fundamental as the law of gravity, would be violated.

Although nothing can prevent the combined drop in frequency and voltage when generation is reduced and load maintained, it is possible to influence the relative extent to which each decreases. In fact, the system has automatic controls that do

---

3. More precisely, the amount stored is minuscule and cannot be utilized for trade.

just that. At substations, where very high transmission voltages are reduced to the lower, but still high, distribution voltage, there are automatically controlled transformers. These adjust so that the distribution voltage remains relatively constant even when the transmission voltage drops. Because of these devices, more of the power flow adjustment that accompanies the loss of a generator is accomplished through frequency reduction than through voltage reduction. Nonetheless both can and do happen.

## Frequency and Interconnections

An interconnection is a portion of the power grid connected by AC power lines. The three interconnections in the United States—the Eastern **Interconnection**, the Western Interconnection, and most of the Great State of Texas—each maintain a uniform frequency. Frequencies in Maine and Texas bear no particular relationship to each other, but the AC voltage in Maine stays right in step with the AC voltage in Florida, night and day, year in and year out. The frequency in every utility in an entire interconnection is exactly the same.[4] If one utility has a problem, they all have a problem.

These three interconnections are connected to each other by a number of small lines, but they are separate interconnections because the connecting lines are all DC lines. No AC power flows between them. On DC lines, the electrical current flows in only one direction; it does not alternate directions. Thus DC lines have no frequency and as a consequence need not (and cannot) be synchronized with the power flow of an AC interconnection. This allows trade between two different interconnections that are not synchronized with each other.

## The Signal for Price Adjustment

When a generator breaks down unexpectedly (a forced outage) and supply decreases, demand is then greater than supply, even though consumption still precisely equals supply. Consumption is less than demand because of rationing. A consumer with a 100-W light turned on is demanding 100 watts of power. During a brown out, however, 100 watts are not supplied to the bulb as power to the bulb is rationed by the suppliers. This rationing is not due to deliberate action but is a consequence of system physics which automatically lowers the voltage and frequency. For simplicity, in the remainder of this section, rationing will be discussed as if it happened solely through frequency reduction, as this is generally considered to be the predominant effect.

A drop in frequency below the target level of 60 Hz is a clear and accurate indication that demand exceeds supply for the interconnection as a whole. Similarly, any frequency above 60 Hz indicates that supply exceeds demand. In other words, more than 100 watts are being delivered to 100-W motors. This extra power is

---

4. If the frequency difference between Maine and Florida were 0.001 Hz, for one minute, it would cause an accumulated phase change between the two states of 22 degrees. This would lead to dramatic changes in power flow. Thus while the frequency lock between utilities is not exact, it is extremely tight, and there can be no persistent frequency difference. One utility cannot have a problem unless they all do.

generally unwanted because appliances are built and operated on the assumption that power will be delivered at a frequency of 60 Hz.

Because frequency indicates the discrepancies between supply and demand, frequency is the right guide for interconnection-wide price adjustment. When frequency is high, price should be reduced; when frequency is low, price should be raised. This is the classic adjustment process for keeping supply equal to demand.

---

| | |
|---|---|
| **Definition** | **Demand** |

The demand for power is the amount of power that would be consumed if system frequency and voltage were equal to their target values for all consumers. Note that shed load is included as part of demand. This is an economic definition and contradicts the engineering definition provided by North American Electric Reliability Council (NERC). (Often "load" is used to mean demand.)

---

| | | |
|---|---|---|
| **Result** | **1-4.1** | **Supply Equals Consumption but May Not Equal Demand** |

As in all markets, demand is the amount customers would buy at the market price were supply available. If voltage or frequency is low, customers consume less power than they would like so supply is less than demand.

---

As always, the real world adds one more layer of complexity. The frequency in every power market in an interconnection is exactly the same. Thus frequency reveals nothing about the supply and demand conditions in any particular market but only about the aggregate supply and demand conditions of the entire interconnection. Consequently individual markets cannot rely on the frequency alone to determine their price adjustments.

NERC defines another control variable that takes account of both frequency and the net excess flow out of a trading region (the net interchange). The net excess outflow is the actual outflow minus the scheduled outflow. An excess outflow is a strong signal that supply is greater than demand in the trading region. If the frequency is high in the interconnection this is a weak signal of excess supply in any particular market. These two signals are combined to form a single indicator of excess supply for each market. The indicator is called the area control error, or **ACE**. Control areas are required to keep their ACE near zero, and they do. ACE is the main indicator of the supply–demand balance in every control area in the United States and when there is a market, it is the signal that determines whether the price will be increased or decreased by the system operator.

# What Is Competition?

*The rich, ... in spite of their natural selfishness and rapacity, ... though the sole end which they propose ... be the gratification of their own vain and insatiable desires, they divide with the poor the produce of all their improvements. They are led by an invisible hand to make nearly the same distribution of the necessaries of life, which would have been made, had the earth been divided into equal portions among all its inhabitants, and thus without intending it, without knowing it, advance the interest of the society.*

Adam Smith
*The Theory of Moral Sentiments*
1759

COMPETITION IS LEAST POPULAR WITH THE COMPETITORS. Every supplier wants to raise the market price, just as every buyer wants to lower it. Perfect competition frustrates both intentions.

Some commodity markets provide almost perfect competition; eventually power markets may work almost as well. But designing such markets is difficult. Economic competition is not like competition in sports, which may be considered perfect when there are just two powerful and equal competitors. Economists consider competition to be **perfect** when every competitor is small enough (*atomistic* is the term used) to have no discernable influence against the "invisible hand" of the market.

Adam Smith guessed intuitively that a perfectly competitive market, in the economic sense, would produce an outcome that is in some way ideal. Many difficulties can cause a market to fall short of this ideal, but even a market that is only "workably competitive" can provide a powerful force for efficiency and innovation.

Power markets should be designed to be as competitive as possible but that requires an understanding of how competition works and what interferes with it. On its surface, competition is a simple process driven, as Adam Smith noted, by selfishness and rapacity; but the invisible hand works in subtle ways that are often misunderstood. Those unfamiliar with these subtleties often conclude that suppliers are either going broke or making a fortune. This chapter explains the mechanisms that keep supply and demand in balance while coordinating production and consumption to produce the promised efficient outcome.

**Chapter Summary 1-5:** The plan of deregulation is to achieve efficiency through competition. Economics guarantees this result provided the market reaches a classic competitive equilibrium. This requires at least three conditions to be met: price taking suppliers, public knowledge of the market price, and well-behaved production costs. Although production costs seem problematic to many, they cause little trouble, and deregulation will probably succeed if markets are designed for maximum competition and transparent prices.

**Section 1:  Competition Means More than Struggle.**  The dictionary defines competition as "a struggle with others for victory or supremacy," but economics does not. Designing markets to be "competitive" in the dictionary's sports-oriented sense produces poor designs about which little can be predicted. Economic competition requires many competitors on each side of the market and results in a lack of market power and "price taking" behavior.

**Section 2:  Efficient Markets and the Invisible Hand.**  The central result of economics states that competition leads to efficiency. But to achieve short-run efficiency, competitive behavior must be supplemented with well-behaved costs and good information. Long-run efficiency requires free-entry of new competitors as well. Efficiency means that total surplus, the sum of profit and consumer surplus, is maximized.

**Section 3:  Short- and Long-Run Equilibrium Dynamics.**  Price and quantity adjustments, usually by suppliers, lead the market to equilibrium. In a competitive market, suppliers adjust output until marginal cost equals the market price and adjust price until the market clears (supply equals demand). They are price takers because when considering what quantity to produce they *take* the market price as given; that is, they assume it will remain unchanged if they change their output.

A long-run competitive equilibrium is brought about by investment in productive capacity. Profit (which means long-run economic profit) is revenue minus costs, and cost includes a normal return on capital (investment). Thus, zero economic profit provides a normal return on investment. If economic profit is positive and the market competitive, new suppliers will enter. In this way profit is brought down to zero under competition, but this is enough to cover all fixed costs and a normal risk premium.

**Section 4:  Why is Competition Good for Consumers?**  Competition minimizes long-run costs and pays suppliers only enough to cover these minimum costs. Although it is possible to depress price in the short run, it is not possible to pay less on average than minimum long-run average cost.

## 1-5.1   COMPETITION MEANS MORE THAN "STRUGGLE"

The dictionary defines **competition** as "a struggle with others for victory or supremacy." This definition is based on sports, not economics, but is quite influential with regulators and politicians. Consequently, when economics says "competition is desirable," this is often interpreted to mean that struggle among market players is desirable. There is a grain of truth to this interpretation, but it misses the main point.

The popular view judges competition mainly on fairness, so market power on the supply side is not a problem provided that demand is similarly endowed. Competition is now in vogue with many regulators, and many who have spent a lifetime passing judgment on the fairness of prices have taken up the call to "let the market do it." They see their new job as making sure the new markets are fair, that "the playing field is level." They believe it is only necessary to ensure the struggle between market players is fair. Because economics promises that competitive markets will be efficient, a good outcome is thought to be assured.

The economic promise of efficiency is not predicated on a fair struggle. Two fairly matched "competitors" do not approximate what economics means by competition. For example, economics makes no guarantee that pitting a monopoly transco against an equally powerful load aggregator will produce an even moderately acceptable outcome. Economics cannot predict the outcome of this kind of "competition" and would view this as a very poorly structured market.

The economist's notion of competition refers to competition among suppliers or among demanders but not between suppliers and demanders. Competition is not a struggle between those who want a higher price and those who want a lower price. The process of economic competition between many small suppliers works by suppliers undercutting each other's price in order to take away the others' customers. This drives the price down to the marginal cost of production but no lower because at lower prices suppliers would lose money. If supply-side competition is stiff enough, the market price will be pinned to the marginal-cost floor. This is the meaning of perfect competition.

When suppliers face such stiff competition that they cannot affect the market price and must simply accept it and sell all they can sell profitably at this price, they are said to be *price takers*. This is the principle requirement for a market to be perfectly competitive and is the primary assumption on which economic claims of market efficiency rest.

Generally it takes many competitors, none of which have a large market share, to produce perfect competition in the economic sense.[1] If there are any large suppliers they are likely to have the ability to profitably raise price. In this case they are not price takers and are said to have market power. They know they can affect the supply–demand balance by reducing their output and thereby drive up the price enough to increase their profit.

---

1. Under the special and uncommon conditions of Bertrand competition, two competitors are enough.

## 1-5.2   THE EFFICIENCY OF PERFECT COMPETITION

One economic result is, without doubt, the most prominent in all of economics. It is the point made by Adam Smith in the *Wealth of Nations:*

> *... he intends only his own gain, and he is in this, as in many other cases, led by an invisible hand to promote an end which was no part of his intention.*[2]

Vague as this may be, Adam Smith, and later Leon Walras, are correctly credited with developing the notion that competitive markets harness the profit motive to produce an efficient and socially useful outcome. This Efficient-Competition Result has been re-examined many times and modern proofs have resulted in "Nobel Prizes" for Kenneth Arrow (1972) and Gerard Debreu in (1983).[3]

### Short-Run Competition

The Efficient-Competition Result has limitations. It does not mean that every free market is efficient, or even that every free market in which suppliers are price takers is efficient. Because of these limitations, economics has carefully defined both competition and efficiency and has added two more concepts: well-behaved cost functions and good information. The modern Efficient-Competition Result can be summarized as follows:

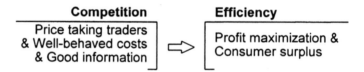

| Competition | Efficiency |
|---|---|
| Price taking traders & Well-behaved costs & Good information | ⇒ Profit maximization & Consumer surplus |

This result can also be summarized as "a competitive equilibrium is efficient."[4] The three conditions listed above under "competition," are necessary to guarantee that the market will reach a **competitive equilibrium**. If suppliers have small enough market shares, they will not have the power to change the market price and profit from doing so, and they will take price as given. This is called acting competitively, but it does not guarantee a competitive equilibrium. First, such an equilibrium must exist and second, traders must be able to find it. If costs are not well behaved—and startup and no-load costs are not—there will be no equilibrium. If traders lack adequate information, including publicly known prices, they may not

---

2. Smith is often quoted as saying that a market is "guided as if by an invisible hand." But a full text search of *The Wealth of Nations* reveals only this one use of "invisible hand." In the same year Smith's book was published, George Washington observed, in his first inaugural address that the "Invisible Hand" (of God) had guided the United States to victory. In fact, both Smith and Washington viewed the invisible hand, God, and the forces of nature as being nearly synonymous.

3. Economists do not get authentic Nobel prizes. The prize in economics is given by the Bank of Sweden, not by the Nobel Foundation.

4. This discussion is necessarily far from rigorous and is meant only to convey a general understanding of the most important concepts. See Mas-Colell et al. (1995) starting on p. 308.

find the competitive equilibrium which consists of an optimal set of trades. (Problems with ill-behaved, **nonconvex**, costs and lack of information are discussed in Chapter 3-8.)

| | |
|---|---|
| **Definitions** | **Perfect Competition** |
| | Agents act competitively, have well-behaved costs and good information, and free entry is brings the economic profit level to zero. |
| | **Act Competitively** |
| | To take the market price as given (be a price taker). |
| | **Well-Behaved Costs** |
| | Short-run marginal cost increases with output and the average cost of production stops decreasing when a supplier's size reaches a moderate level. |
| | **Good Information** |
| | Market prices are publicly known. |

## Long-Run Competition

A short-run competitive equilibrium is (short-run) efficient; it makes the best use of presently available productive resources. A long-run competitive equilibrium guarantees that the right investments in productive capacity have been made but requires that the three short-run conditions be met and adds two new ones. Production costs must not possess the conditions for a natural monopoly (see Section 1-1.1), and competitors must be able to enter the market freely.[5] With **free entry**, if there are above-normal profits to be made, new suppliers will enter which will reduce the level of profits. In this way free entry ensures that profits will not be above normal. A normal profit level is the key characteristic of a long-run competitive equilibrium. **Barriers to entry** is the term used to describe market characteristics that prevent free entry.

## Efficiency and Total Surplus

Almost every proposed market design is declared efficient, but in economics the term has a specific meaning. The simplest meaning applies to productive efficiency which means that what is being produced is being produced at the least possible cost. Minimizing cost is often the most difficult part of the market designer's problem, so this meaning is generally sufficient.

When not qualified as productive efficiency, efficiency includes both the supply and demand sides of the market.[6] **Efficiency** means (1) the output is produced by the cheapest suppliers, (2) it is consumed by those most willing to pay for it, and (3) the right amount is produced. These three can be combined into a single criterion by using the concept of **consumer surplus**.

---

5. See Mas-Colell et al. (1995, 334).

6. The term "allocative efficiency" is almost universally used to mean demand-side efficiency. But this is not the meaning found in economics dictionaries; which include both sides of the market and do not distinguish it from "efficiency."

**Figure 1-5.2**

Total surplus equals the
area between the demand
curve and the marginal
cost curve.

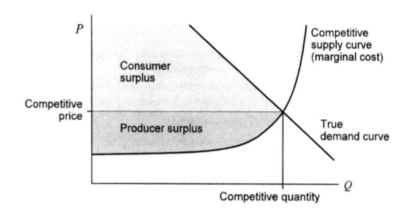

| Definitions | **Productive Efficiency** |
|---|---|
| | Production costs have been minimized given total production. |
| | **Efficiency** |
| | Total surplus has been maximized. This automatically includes minimizing the cost of what is produced and maximizing the value of what is consumed, as well as producing and consuming the right amount. |

A consumer's demand curve measures how much the consumer would pay for the first kilowatt-hour consumed, and the second, and so on. Generally the more consumed, the less would be paid for the next kilowatt-hour. Because the initial kilowatt-hours are so valuable, the total value of consumption is generally much greater than the amount paid. The difference between the maximum a consumer would pay as revealed by the consumer's demand curve and what the consumer actually does pay is the consumer's surplus.

Profit is analogous to consumer surplus and is often called **producer surplus**. It is total revenue minus total cost, while consumer surplus is total value to consumers, $V$, minus total consumer cost, $CC$. ($V$ is sometimes called gross consumer surplus, and is the area under the demand curve for all consumption.) Both $V$ and $CC$ are measured in dollars. If the sum of profit and consumer surplus is maximized, the market is **efficient**, and all three of the above criteria follow.

When profit ($R - C$) and consumer surplus ($V - CC$) are added, the consumer costs ($CC$) and producer revenue ($R$) cancel because they are the same. The result is **total surplus** ($V - C$), consumer value minus producer cost. Consequently efficiency is the same as maximizing total surplus.

### The Efficient-Competition Result

| Result | 1-5.1 | **Competitive Prices Are Short- and Long-Run Efficient** |
|---|---|---|
| | | If productions costs are well behaved so competitive prices exist, these prices will induce short-run (dispatch) efficiency and long-run (investment) efficiency. |

Note that the Efficient-Competition Result does not say that a market's prices will be competitive if costs are well behaved. That conclusion requires the lack of market power and good information. The Efficient-Competition Result says that if market prices are competitive then supply and demand will be efficient. Assuring competitive prices is the main problem addressed by this book.

## Problems Caused by Production Costs

The Efficient-Competition Result depends on well-behaved production costs and these cannot be designed. Either efficient generators have well-behaved costs or they do not. If they do not, and costs are sufficiently problematic, then a standard competitive market design cannot be depended on to provide an efficient outcome.

Cost problems present the most fundamental challenge. Three different problems receive attention: (1) nonconvex operating costs, (2) the fixed costs of investment, and (3) total production costs that decrease up to very large scales of production. The third problem is the problem of natural monopoly and was discussed in Chapter 1-1.

The second problem is the subject of the most frequent misconception, an extremely pessimistic one. It holds that ordinary fixed costs are sufficient to disrupt a competitive market. This pessimism about competition is often accompanied by optimism about less-competitive free markets. The belief is that if the market can avoid the problems of a competitive equilibrium free-market forces will produce a good outcome. Perhaps, it is argued, if the market is not monitored too closely, generators will exercise market power and thereby earn enough to cover fixed costs and keep themselves in business. Chapter 2-1 shows that fixed costs are not a problem for competition and the proposed noncompetitive remedy is unnecessary and detrimental.

The problem of **nonconvex** operating costs is the most difficult. The cost of starting a generator makes generation costs nonconvex because it makes it cheaper per kWh to produce 2 kWh than to produce 1kWh. This causes the market to lack a competitive equilibrium and could easily cause inefficiency in the dispatch of an otherwise competitive market. To circumvent this problem, some markets use a unit-commitment auction that attempts to replace a standard "classic" competitive equilibrium with a different equilibrium which is still efficient. Chapter 3-9 discusses this problem in detail and suggests it may be of minor importance and that a standard competitive market might still provide a very high level of efficiency.

These conclusions are part of a larger pattern. Competitive markets are difficult to design, and none of the three basic requirements of efficiency can be achieved to perfection. But if the two controllable ones, price taking behavior and good information, are well approximated, a very efficient market will result. The quirks of a competitive equilibrium are not much of a problem, but designing a competitive market requires a great deal of care.

## 1-5.3   SHORT- AND LONG-RUN EQUILIBRIUM DYNAMICS

Markets are never in equilibrium, but economics focuses primarily on their equilibrium behavior. The ocean is never in equilibrium, yet it is always found at the lowest elevations where physics predicts its equilibrium to be. In equilibrium the ocean would have no waves. Although markets, like oceans, have "waves," they too usually stay near their equilibrium. An equilibrium may change over time as the globe warms and the ice caps melt, but this does not prove the equilibrium uninteresting. A market's equilibrium is a useful guide to its behavior, even though the market is never exactly in equilibrium.

Both the supply and the demand side of the market adjust their behavior in order to produce a market equilibrium, but competitive economics is primarily concerned with the supply side.[7] This section explores the forces that push the supply side of a market toward a competitive equilibrium.[8]

### The Meaning of Short Run and Long Run

These concepts do not, as is often supposed, refer to specific periods of time but instead refer to the completion of particular market adjustment processes. "In the short run" indicates that adjustments in the capital stock (the collection of power plants) are being ignored, but adjustments in the output of existing plants are being considered.

The phrase "in the long run" indicates adjustments in the capital stock are not only being considered but are assumed to have come to completion. This is a useful abstraction. If the market has not recently suffered an unexpected shock, it should be near a state of long-run equilibrium because business spends a great deal of effort attempting to discern future conditions, and for the last five years, today was the future.

Of course mistakes are made and markets are never in exact long-run equilibrium. But mistakes are as often optimistic as pessimistic, and consequently a long-run analysis is about right on average. However, a newly-created market is more likely than most to be far from its equilibrium.

### The Short-Run Equilibrium

*Marginal cost* is the cost of producing one more unit of output, one more kilowatt-hour. It is also approximately the savings from producing one less kWh. In this section and the next, these are assumed to be so close together that no distinction is necessary, which is typically the case. Chapter 1-6 pays a great deal of attention to the special case where these are different.

In a competitive market suppliers are **price takers**. They cannot change the market price profitably, so they consider it fixed. Price taking also means they can sell all they want at the market price, but they cannot sell anything at a higher price. Most markets are not perfectly competitive, and suppliers find that at a higher price they sell less but more than nothing. This will be ignored as the present purpose is to analyze how a market would work if it *were* perfectly competitive.

A short-run competitive equilibrium determines a market price and a market quantity traded. To bring the market into equilibrium, two dynamic adjustment mechanisms are needed: (1) a price adjustment and (2) a quantity adjustment. In most markets suppliers adjust both, although in some, buyers set the price.

---

7. Demand-side management concerns itself primarily with information problems on the demand side of the market. These problems also deserve attention.

8. For a more complete treatment of the microeconomics of competition presented with power markets in mind, see Rothwell and Gomez (2002).

**Quantity Adjustment.**  A price taking supplier will *increase* output if its marginal cost, *MC*, is less than the market price, *P*, and will *decrease* its output if $MC > P$. Its profit increases by $(P - MC)$ for every unit produced when *P* is higher than *MC* and decreases by $(MC - P)$ when *P* is lower.

**Price Adjustment.**  Whenever demand exceeds supply, suppliers raise their prices, and whenever supply exceeds demand, they lower prices.

**Equilibrium.**  The quantity adjustment dynamic causes the marginal cost to equal the market price in a competitive market. The price adjustment dynamic causes the quantity supplied to equal the quantity demanded. When supply equals demand, the market is said to have *cleared*, and the price that accomplishes this is called the **market-clearing price**, or the equilibrium price, or, for a **competitive market**, the **competitive price**. Together the two adjustment mechanisms bring a competitive market to a **competitive equilibrium**.

---

### Price Taking and Price Adjusting

Suppliers typically name their price. For example, most retailers put price tags on their wares, and customers pay those prices. So how can suppliers be "price takers?"

Because "price-taking" has a specialized meaning, suppliers can be both price takers and price adjusters at the same time. Suppliers take price as given when deciding how much to produce and adjust their price if they notice excess supply or demand in the market.

---

**Price Taking vs. Price Adjustment.**  Notice that "price taking" suppliers adjust their prices in order to clear the market. This is not a contradiction. "Price taking" is something that happens in the quantity-adjustment dynamic but not in the price-adjustment dynamic. Price takers "take the price as given when computing their profit-maximizing output quantity." This means they assume that their choice of output will not affect the price they receive for it.

The quantity dynamic, which causes *MC* to equal the market price, acts as a coordinating mechanism among suppliers because there is only one market price. This is why public knowledge of the market price is a key assumption of the Efficient-Competition Result. Because all suppliers have the same marginal cost in the competitive equilibrium, no money can be saved by having one produce more and another less. This is what makes production efficient.

Some will object to this result on the grounds that coal plants have lower marginal costs than gas turbines even in a competitive equilibrium. This objection is based on a misunderstanding of the definition of "marginal cost," which will be explained in the following chapter.

## The Marginal-Cost Pricing Result

---

**Result     1-5.2**     **Competitive Suppliers Set Output So That *MC* = *P***

A competitive producer sets output to the level at which marginal cost equals the market price, whether or not that is the competitive price. This maximizes profit. ($MC = P$ for all suppliers.)

---

## The Long-Run Equilibrium

The process of long-run competition involves investing in plant and equipment, not simply changing the output of existing plants. This dynamic requires a definition of profit. Profit is of course revenue minus cost, but economics defines costs more broadly than does business. Economics, and this book, define cost to include a normal rate of return on all investment. This rate of return is defined to include a **risk premium**. If a supplier covers its costs, it automatically earns a normal rate of return, including an appropriate risk premium, on its entire investment. Under this definition of "normal," a business that earns more is considered to be worth investing in, and a business that earns less is not. A normal business investment, therefore, has revenues that exactly cover all its costs in the economist's sense. Because **profit** equals revenue minus cost, a normally profitable supplier earns zero profit.

---

**Definition**          **(Economic) Profit**

Revenue minus total cost, where total cost includes a normal, risk-adjusted, return on investment. The normal (economic) profit level is zero. (Business defines a normal return on equity to be profit, while economics defines it as covering the cost of equity.)

**Short-Run Profit**

Revenues minus short-run costs which include variable, startup and no-load costs. The "profit function," defined in Chapter 2-7, computes short-run profits.

---

As defined, profit is synonymous with long-run profit which is different from short-run profit which does not include the cost of capital; that is, it does not include any return on investment. Consequently, short-run profit is expected to be positive on average so these profits can cover the fixed cost of capital.

---

**Result      1-5.3a**          **Under Competition, Average Economic Profit Is Zero**
In a long-run competitive equilibrium, the possibility of entry and exit guarantees that profits will be normal, which is to say zero.

**Result      1-5.3b**          **Under Competition, Fixed Costs Are Covered**

When profit is zero, all costs are covered including fixed costs, so in the long run, competition guarantees that fixed costs will be covered.

**Result      1-5.3c**          **A Supplier with a Unique Advantage Can Do Better**

If a supplier has access to limited cheaper inputs (hydro-power or geothermal energy), it will have greater profits. If the advantage is unlimited, it has a natural monopoly.

---

If the expected market price is so low that a supplier cannot enter the market and cover all costs, no supplier will enter. More specifically, if a new generation unit cannot cover all costs, no new units will be built. The result will be a gradually diminishing supply of generation (due to retirements of old plants) in the face of gradually increasing demand. This tightening of the market will cause the price to rise, and eventually price will be high enough to cover all costs.

Similarly, if price is so high that costs are more than covered, suppliers will build new generating units. This will increase supply and cause the price to fall. The result of this long-run dynamic is that the profit in any competitive market returns to the normal level of profit (zero) in the long-run competitive equilibrium.

## 1-5.4   WHY IS COMPETITION GOOD FOR CONSUMERS?

In the long-run producers cover their fixed costs, and in the short run total surplus is maximized, but what consumers want is a low price. Does competition provide the lowest possible price?

Not in the short run. In the short run, it is possible to design market rules which lower the market price without reducing supply. This is difficult but possible. But at a lower price producers will not cover their fixed costs. This will make future investors think twice. The result will be a risk-premium added to the cost of capital and future production will be more costly than it would have been had cost been left at the competitive level.

Competition does not guarantee the lowest possible price at any point in time. Instead it guarantees that suppliers will just cover the long-run total costs and no more. It also guarantees that the cheapest suppliers will be the ones producing. Together these mean production costs (including the long-run cost of invested capital) are minimized and producers are paid only enough to cover their cost. This implies that the long-run average cost to consumers is also minimized. No market design regulated or unregulated can induce suppliers to sell below cost on average. Competition minimizes long-run average costs of production and long-run average costs to consumers.

*Chapter* **1-6**

# Marginal Cost in a Power Market

*The trouble with the world is not that people know too little,*
*but that they know so many things that ain't so.*

Mark Twain
(1835–1910)

SIMPLIFIED DIAGRAMS OF GENERATION SUPPLY CURVES HAVE CON-
FUSED THE DISCUSSION OF MARGINAL COST. Typically, these supply curves
are diagrammed to show a constant marginal cost up to the point of maximum
generation. Then marginal cost becomes infinite without taking on intermediate
values. Typically it jumps from about $30 to infinity with only an infinitesimal
increase in output. Mathematics calls such a jump a discontinuity. In fact, the curve
would be discontinuous if it jumped only from $30 to $40.

The definition of marginal cost does not apply only to the points of discontinuity.
Hence it does not apply to a right-angle supply curve at the point of full output,
neither does it apply to the points of a market supply curve at which it jumps from
one generators marginal cost to the next. Unfortunately market equilibria sometimes
occur at such points, and concerns over market power often focus on them. Attempts
to apply the standard definition at these points can produce confusing and erroneous
results.

Fortunately, the definition is based on mathematics that generalizes naturally
to discontinuous curves. Applying this generalization to the textbook definition
clears up the confusion and restores the economic results that otherwise appear
to fail in power markets. For example, in power markets, as in all other markets,
the competitive price is never greater than the marginal cost of production.

**Chapter Summary 1-6:** Individual supply curves are often constructed with
an abrupt end that causes the market supply curve to have abrupt steps. The standard
marginal-cost definition does not apply at such points. Instead, left- and right-hand
marginal costs should be used to define the marginal-cost range. Then the competitive price, which remains well defined, will always lie within that range. A market
price exceeding the marginal-cost range indicates market power.

**Section 1:  The Role of Marginal Cost.**  Marginal costs play a key role in cost-based power auctions because they help determine the competitive price. They also play a key role in analyzing market power and gain their importance by defining the competitive supply curve for individual generators. To find the market (aggregate) supply curve, individual supply curves are summed horizontally.

**Section 2:  Marginal-Cost Fallacies.**  In power-market analysis, marginal cost is often defined as the cost of the last unit produced, but this definition is found in no economics text. A second fallacy asserts that when marginal cost is ambiguous, the competitive price is ambiguous. Together these lead to a variety of erroneous conclusions, such as "the competitive price is above marginal cost," and "the competitive price is ambiguous."

**Section 3:  The Definition of Marginal Cost.**  When a marginal-cost curve is discontinuous (has a sudden jump), marginal cost can be specified only within a range at the points of discontinuity. This range extends from the left-hand to the right-hand marginal cost at the point under consideration. For all points where the curve is continuous, the range is a single point equal to the standard marginal cost.

**Section 4:  Marginal Cost Results.**  The competitive price is within the marginal-cost range of every competitive generator and within the marginal-cost range of the market. If even one supplier has a supply curve that is continuous at the market price, the market supply curve is continuous at that price and the competitive price is equal to the standard marginal cost which is well defined. In any case, the competitive price is the price at which the supply and demand curves intersect.

**Section 5:  Working with Marginal Costs.**  This book assumes that supply curves have extremely large but finite slopes rather than the infinite slopes frequently assumed. This is a more realistic assumption and has no practical consequences, but it has the simplifying property of making marginal cost well defined and the marginal cost of all operating competitive generators equal to the market price.

**Section 6:  Scarcity Rent.**  Scarcity rent is revenue less variable cost and is needed to cover startup and fixed costs. Economics refers to this as "inframarginal rent," and has no separate definition of a scarcity rent. A folk-definition defines scarcity rent as actual revenue minus the maximum revenue that is collected just before the system runs completely out of capacity. Used with a stylized model, this definition has some appeal, but when applied to real systems it is highly ambiguous and misleading.

**Figure 1-6.1**

Adding individual supply
curves horizontally to
find the market supply
curve. If B is continuous,
A + B is also.

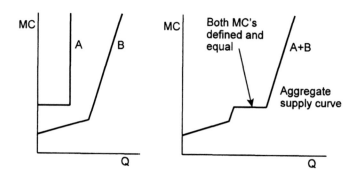

## 1-6.1   THE ROLE OF MARGINAL COST

Marginal cost plays a key role in the economic theory that proves a competitive
market is efficient, but there are also two practical uses of marginal cost that
increase its importance in a power market. First, many power markets rely on a
central day-ahead auction in which generators submit individual supply curves and
the system operator uses these to determine the market price. Because price should
equal marginal cost in an efficient market, the auction rules should be informed
by a coherent theory of marginal cost. Second, many power markets suffer from
potential market-power problems which cause the market price to diverge from
marginal cost. Market monitors need to understand this divergence.

Although the competitive market price usually equals the marginal cost of
production, it is not determined by that alone. At times marginal cost is ambiguous,
yet the competitive price is not. Then, marginal value (to customers) plays the
decisive role. The competitive price is determined by the intersection of the market's
supply and demand curves. Marginal cost determines only the supply curve.

A supply curve can be thought of as answering the question, How much would
a generator produce if the market price were $P$/MWh? As explained in Section
1-5.3, price-taking suppliers adjust output until marginal cost equals the market
price. As a consequence, if $Q$ is the quantity supplied at a given price $P$, then $P$
must equal the marginal cost. Thus a price-taker's supply curve and marginal cost
curve are the same.

The market's supply curve, also called the **aggregate supply curve**, is found
by summing horizontally all of the individual generators' supply curves. For a given
price, the quantity supplied by each generator is read horizontally from each
individual supply curve and these quantities are summed to find the market supply.
This quantity is plotted at the given price, as shown in Figure 1-6.1.

Notice that because one generator has a continuous supply curve (no vertical
section) the market has a continuous supply curve. Notice also that when both
generators are operating and have defined marginal costs, they have the same
marginal cost. Section 1-6.3 generalizes this by showing that every operating
generator either has a marginal cost equal to the market price or has a marginal-cost
range that includes the market price.

## 1-6.2   MARGINAL-COST FALLACIES

### Discontinuous Supply Curves

Individual supply curves are almost always drawn as "hockey sticks." That is, they are drawn with a slight upward slope (or as flat) until they reach the capacity limit of the generator and then they are drawn as perfectly vertical (see curve A, Figure 1-6.1). Textbook supply curves usually have a slope that increases gradually (See curve B, Figure 1-6.1). Curves without a vertical segment are called continuous. Unfortunately, a generator's supply curve, as typically drawn, takes an infinite upward leap when it reaches full output (which is the most common output level for an operating generator). At this point, marginal cost is not smooth but jumps from say $30/MWh to infinity with only an infinitesimal change in output.

The smoothness of textbook supply curves plays a crucial role in keeping the textbook definition of marginal cost simple, and this has led to mistakes and confusion. Eliminating the confusion requires the introduction of a carefully constructed definition which applies to the discontinuous supply curves used in power-market analysis. With this definition of marginal cost, all standard economic results are found to apply to power markets. Once this is understood, the problematic supply curves can be analyzed correctly with a simple rule of thumb. This provides guidance when setting the market price in a cost-base auction and when determining whether market power has been exercised.

### Fallacies

Two basic fallacies underlie a series of misconceptions surrounding competitive pricing and market power. These are (1) the Marginal-Cost Fallacy and (2) the Ambiguous-Price Fallacy. Both of these will be illustrated using Figure 1-6.2, which shows a normal demand curve and a supply curve that is constant at $30/MWh up to an output of 10 GW, the capacity limit of all available generation.

The Marginal-Cost Fallacy takes two forms. The simple form asserts that marginal cost at $Q = 10,000$ MW is $30/MWh in Figure 1-6.2. The subtle form asserts that nothing can be said about the marginal cost at this output level. Some of the conclusions drawn from these assertions are as follows:

1. The competitive price is $30/MWh, and the market should be designed to hold prices down to this level.
2. The competitive price is $30/MWh, and this is too low to cover fixed costs, so marginal-cost prices are inappropriate for power markets.
3. Scarcity rents are needed to raise prices above marginal-cost-based prices.
4. Market power is necessary to raise prices to an appropriate level.
5. The competitive price cannot be determined.

**Figure 1-6.2**

A normal market equilibrium for an abnormal supply curve.

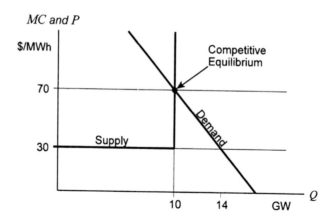

All of these conclusions assume that there is some problem with standard economics caused by the supply curve coming to an abrupt end instead of turning up smoothly as it does in undergraduate texts. In fact, economic theory has no difficulty with this example, and all of the above conclusions are false. Consider a competitive market, with many suppliers and many customers, described by the curves in Figure 1-6.2. What if the price in this market were $30/MWh? At this price, the demand curve shows an excess demand of about 4 GW. Some customers trying to buy more power are willing to pay up to $70/MWh for another MW of supply. They will find a supplier and offer to pay considerably more than $30, and the supplier will accept. This shows that the competitive price is above $30/MWh. The story will be repeated many times, with different values, until the market price reaches $70/MWh. At that price every supplier will produce at full output, so the supply will be 10 GW, and demand will be 10 GW. At any higher price demand would fall short of supply, so the price would fall, and at any lower price, demand would exceed supply, so the price would rise. There is nothing unusual about this equilibrium; it is the classic story of how price clears a market by equating supply and demand.

## The Marginal-Cost Fallacy

*Fallacy* 1-6.1

**Marginal Cost Equals the Cost of the Last Unit Produced**
Marginal cost equals the savings from producing less even when this is different from the cost of producing more.

**(Subtle Version)**

Nothing can be said about marginal cost at the point where a supply curve ends or jumps from one level to another.

But shouldn't price equal marginal cost? In this example, all that can be said is that marginal cost is greater than $30/MWh. So there is no contradiction between price and marginal cost, but they cannot be proven to be equal. The desire to pin down marginal cost precisely seems to arise from a belief that competitive suppliers should set price equal to marginal cost and thereby determine the market price. But this logic is backwards. As explained in Section 1-5.3, suppliers set price to clear the market and set quantity to bring marginal cost in line with price. In this example, the market-clearing forces of supply and demand determine price unambiguously, and although marginal cost is ambiguous, it is greater than $30/MWh which is enough to determine supply unambiguously. Everything of practical importance is precisely determined.

### The Ambiguous-Price Fallacy

*Fallacy*  1-6.2 | **When Marginal Cost Is Ambiguous, so Is the Competitive Price**
Competitive suppliers set price equal to marginal cost; thus when marginal cost is hard to determine, the competitive price is hard to determine.

Having analyzed the example, the preceding list of incorrect conclusions can be restated in their.

1. The competitive price is not $30/MWh, and the market design should not hold price to this level.
2. The competitive price is high enough to contribute significantly to fixed cost recovery.
3. No mysterious "scarcity rent" need be added to the marginal cost of physical production.
4. Market power is not needed if the market is allowed to clear.
5. The competitive price is $70/MWh.

## 1-6.3  THE DEFINITION OF MARGINAL COST

The above discussion is accurate but informal. Because of the controversy in this area, it is helpful to formalize the concepts used in analyzing supply curves with discontinuities or abrupt terminations.

The *MIT Dictionary of Modern Economics* (1992) defines marginal cost as "the extra cost of producing an extra unit of output." Paul Samuelson (1973, 451) defines marginal cost more cautiously as the "cost of producing one extra unit more (or less)." The "or less" is important. The assumption behind this definition is that producing one more unit of output would cost exactly as much as producing one less unit would save. This is true for the continuous marginal-cost curves of textbook economics but not for the discontinuous curves used by power-market analysts. To discuss the marginal cost of a discontinuous supply curve, the definition must be extended to include the points of discontinuity where the cost to produce an extra unit is distinctly greater than the savings from producing one less.

### Left- and Right-Hand Marginal Costs

In the example of Figure 1-6.2, the marginal cost of production goes from $30 on the left of 10 GW to infinity on the right of 10 GW. This is a double complication. Not only does marginal cost change abruptly, it becomes infinite. The present

**Figure 1-6.3**

Right- and left-hand
marginal costs.

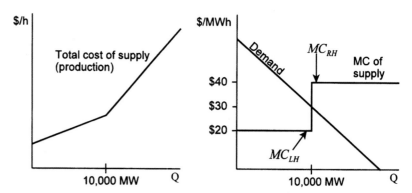

definitions can be illustrated more clearly with a less pathological marginal-cost
curve.

Figure 1-6.3 shows the total cost curve and the marginal cost curve of a simple
market. The discontinuity is the jump in marginal cost at the 10 GW output level.
To the left of 10 GW the marginal cost is $20/MWh, while to the right it is
$40/MWh. But what is the marginal cost precisely at 10 GW? It is undefined, but,
as every textbook would confirm, the answer is *not MC* = $20/MWh.

To formalize this definition, it is useful to consider the mathematics of the total
cost curve shown at the left of Figure 1-6.3. To the left of 10 GW, its **derivative**
(slope) is $20/MWh, while to the right its slope is $40/MWh. But the mathematical
definition of a derivative breaks down at 10 GW, and since marginal cost is just
the derivative of total cost, the definition of marginal cost also breaks down at this
point. Mathematics *does* define two very useful quantities at the 10-GW point, the
left-hand derivative (slope) and the right-hand derivative (Courant 1937, 199–201).
These are, of course, $20 and $40/MWh, respectively. Because marginal cost is
just the derivative, it is natural to define **left-hand marginal cost** $(MC_{LH})$ as the
left-hand derivative, and **right-hand marginal cost** $(MC_{RH})$ as the right-hand
derivative. Other points along the total cost curve also have left and right-hand
derivatives, and these are just equal to the normal derivative. Similarly, $MC_{LH}$ and
$MC_{RH}$ are normally equal to each other and equal to standard marginal cost, $MC$.

The **marginal-cost range**, $MC_R$, is defined as the range of values between and
including $MC_{LH}$ and $MC_{RH}$. This definition is motivated by the idea that marginal
cost cannot be pinned down at a point of discontinuity but can reasonably be said
to lie somewhere between the savings from producing one less and the cost of
producing one more unit of output.

| Definitions | **Left-hand marginal cost ($MC_{LH}$)** |
| --- | --- |
| | The savings from producing one less unit of output. |
| | **Right-hand marginal cost ($MC_{RH}$)** |
| | The cost of producing one more unit of output. When this is impossible, $MC_{RH}$ equals infinity. |
| | **The marginal-cost range ($MC_R$)** |
| | The set of values between and including $MC_{LH}$ and $MC_{RH}$. |

## 1-6.4 MARGINAL COST RESULTS

### Refining the Marginal-Cost Pricing Result

In Figure 1-6.2, the $MC_{LH}$ at 10 GW is $30/MWh, but what is the $MC_{RH}$? It is tempting to say it is undefined, but again mathematics provides a more useful answer. The $MC_{RH}$ at 10 GW is infinite. This definition is both mathematically sound and useful because it allows a simple rewriting of the standard economic results concerning marginal costs.

| Result | 1-6.1 | **Competitive Suppliers Set Output so $MC_{LH} \leq P \leq MC_{RH}$** |
| --- | --- | --- |
| | | A competitive producer sets output to a level at which its marginal-cost range, $MC_R$, contains the market price, $P$, whether or not that is the competitive price. |

First, a price-taking supplier will decrease output as long as $P < MC_{LH}$ because producing one less unit will save $MC_{LH}$ and cost only $P$ in lost revenues. Thus, the savings is greater than the cost. Similarly, if $MC_{RH} < P$, the supplier will increase output. Thus whenever $P$ lies outside the range between left- and right-hand marginal costs, the supplier will adjust output. When the range is below $P$, output is increased, which raises the range and vice versa when $MC_R$ is above $P$. As a result, the marginal-cost range will end up encompassing $P$.

This means that in a competitive market, price will never exceed marginal cost; this would violate basic economics. Technically, $P > MC$ can never be proven true in a competitive market.[1] Competitive price will always be less than or equal to left-hand marginal cost, and there is no need for it exceed this value for fixed cost recovery.[2]

---

1. Even those who best understand these concepts sometimes add to the confusion. *"Thus in the absence of market power by any seller in the market, price may still exceed the marginal production costs of all facilities producing output in the market at that time."* (Borenstein 1999, 3) *". . . the price of electricity has to rise above its short-run marginal cost from time to time, or peaking capacity would never cover its fixed costs."* (Green 1998, 4).

2. Part 3 discusses "nonconvex costs," complexities of the production cost function that require deviations from marginal cost. Essentially this means that startup costs and other short-run, avoidable costs must be covered by price.

**Figure 1-6.4**

The smallest possible change in the supply curve of Figure 1-6.2 restores all normal economic properties.

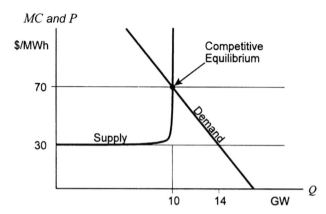

This result can be extended from a single producer to the whole market. The $MC_{RH}$ of the market is the least cost of producing one more unit, so it is the minimum of the individual marginal-cost ranges. Similarly, the market $MC_{LH}$ is the maximum individual $MC_{LH}$. In a competitive market, every supplier is a price taker and adjusts its output until $P$ is within its marginal-cost range. Thus $P$ is less than or equal to every individual $MC_{RH}$, so it is less than or equal to the $MC_{RH}$ of the supplier with the lowest $MC_{RH}$, which is the $MC_{RH}$ of the market. Similarly, $P$ is greater than or equal to the $MC_{LH}$ of the market.

The range from the market $MC_{LH}$ to the market $MC_{RH}$ is contained within the marginal-cost ranges of each individual supplier. If even one supplier in the market has $MC_{LH} = MC_{RH}$, the market will also have this property. In other words, if even one supplier has a well defined-marginal cost at the market price, then the market itself has a well-defined marginal cost.

## The System-Marginal-Cost Pricing Result

| | | |
|---|---|---|
| **Result** | **1-6.2** | **Competitive Price Equals System Marginal Cost** |
| | | In a competitive market, price is within the marginal-cost range of every generator supplying power. It is thus within the market's marginal-cost range. If even one operating supplier has a continuous marginal cost curve, the competitive price actually equals marginal cost as defined by the aggregate supply curve. |

## Finding the Competitive Price

Fortunately the above results are needed only for untangling the current confusions over marginal cost. They demonstrate, among other things, that price does not exceed marginal cost in a competitive power market.

Fortunately, these results are not needed to find the competitive equilibrium, which is determined, as in any other market, by the intersection of the supply and demand curves. This is most easily seen by smoothing out one of the problematic supply curves very slightly.

Standard economic theory applies once the vertical segments have been removed from the cost curves. This can be done with an arbitrarily small change in its shape.

As shown in Figure 1-6.4, giving the marginal cost curve a nearly, but not perfectly, vertical slope makes no noticeable difference to any economic result. And this is how it should be. Economics should not and does not depend on splitting hairs. Notice that in the finitely-sloped model, price really does equal marginal cost at the intersection of the two curves. The price and quantity dynamics of a market with the vertical supply curve will be essentially the same as those of the continuous market. In this example, at a price different from $70 and a quantity different from 10 GW, the markets have essentially identical gaps between supply and demand and between price and marginal cost. So they adjust price and quantity in the same way.

---

**Result      1-6.3**       **Supply Intersects Demand at the Competitive Price**
To find the competitive price and the marginal cost, draw the supply and demand curves, including the vertical parts of the supply, curve if any. The intersection of supply and demand determines "*MC*," *P*, and *Q*.

---

This demonstrates that the standard method of finding the competitive equilibrium works even when the marginal cost curves have infinite slopes. Of course if the slope is infinite at the intersection of supply and demand, marginal cost will be technically undefined. Yet pretending that the true marginal cost is determined by this simple short cut will never give the wrong answer to any real-world question.

---

## 1-6.5   WORKING WITH MARGINAL COSTS

Discussing left- and right-hand marginal costs and the marginal-cost range is cumbersome and unnecessary. If every vertical segment of a marginal cost curve is replaced with a nonvertical but extremely steep segment, the new curve will be continuous and will not jump from one value to another. Such a change may or may not improve its accuracy, but in either case it will make no detectable difference to any economic prediction of consequence.

This book will tacitly assume all supply curves and marginal cost curves that are depicted as having vertical segments actually have extremely steep but finite slopes. In other words, all marginal-cost curves are assumed to be continuous. Consequently, marginal cost is always a well-defined single value.

For example, a supply curve that is constant at $30/MWh up to a maximum output of 500 MW can be replaced with one that is identical up to 500 MW and then slopes upward linearly reaching a value of $30,000/MWh at an output of 500.001 MW. No measurement, however careful, could discern the difference. Yet this supply curve, being continuous, has a well-defined value (marginal cost) at every level of output.

In fact most, if not all generators, have continuous marginal cost curves. Typically, they have an **emergency operating range** above their nominal maximum output level and are willing to produce in this region if well paid or coerced. Most generators in PJM include such

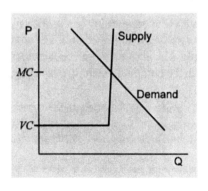

an emergency operating range in their bids and the total capacity available in this range is 1,900 MW out of a total installed capacity of about 60,000 MW.[3] As long as there is one such generator in a market, the market's marginal cost curve is continuous. Real markets always have well-defined marginal costs and the competitive price equals that marginal cost. The difficulties resolved in this chapter only matter for the "simplified" diagrams used by power-market analysts.

This book also will use the same simplified diagrams but without taking the vertical segments literally. Such supply curves will have constant marginal costs up to the nominal "maximum" output level, but above that marginal costs will increase rapidly. If the supply curve is flat at $30 but the market price is $50, the generator's marginal cost will be $50 and it will produce on the steeply sloped segment. When referring to such a generator, it is both wrong and confusing to say its marginal cost is $30 as is the custom. To avoid this confusion, the marginal cost of a generator's supply curve to the left of the "maximum" output level will be termed its **variable cost**. This is not entirely standard, but it is in keeping with the term's normal usage which refers to all costs that vary with the output level.

## 1-6.6   SCARCITY RENT

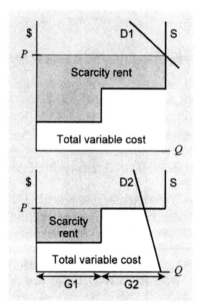

"Scarcity rent" has no formal economic definition but many popular meanings.[4] Although several are useful, most do not lend themselves to careful analysis. However, one essential economic concept comes close to the popular meaning. **Scarcity rent** will be defined as revenue minus variable cost.[5] Economics refers to scarcity rent as **inframarginal rent**.

In the figure at the left, when demand is described by D1, both generators are producing at full output, and load would be willing to pay either generator more than its variable cost of production if it would produce more. In this sense they are both scarce and both earn scarcity rents.

With demand reduced to D2, as shown in the lower half of the figure, generators of type G2 have excess capacity and are no longer scarce and earn no scarcity rent; their variable costs equal the market price. Generators of type G1 are still scarce because load would be more than willing to pay their variable cost if they would produce more. If G2 had a variable cost of $1,000/MWh so that G1 were earning a rent of, say, $950/MWh because G1 could not satisfy the entire load, G1 would commonly be seen as in scarce supply. The above definition coincides with an important concept of economics and with the common meaning of scarcity.

---

3. Personal communication from Joe Bowring, head of PJM's Market Monitoring Unit, January 7, 2002.

4. Samuelson (1973, 623) comes close to using the term when he says "Competitively determined rents are the results of a natural scarcity." His definition of such rents is the long-run analog of the short-run definition of scarcity rent given here.

5. This is greater than short-run profit by the amount of startup costs and no-load costs which will be ignored until Part 3.

**Figure 1-6.5**
Folk-definition of
scarcity rent.

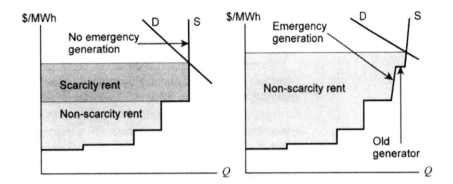

## A Folk Definition of Scarcity Rent

Sometimes, in the field of power system economics, scarcity rent is defined as actual revenue less the highest revenue earned before total generation becomes scarce. This might be called a "folk definition." The notion is that until the system runs out of capacity, price increases are due to increases in marginal cost, but after that point they are driven up by ever increasing scarcity. In an *idealized model*, this definition has some appeal.

---

**Definition**          **Scarcity Rent**
Revenue minus variable operating cost (which do not include startup costs and no-load costs).

---

Say there are only ten types of generators on the market, and call the one with the highest variable cost the **peaker**. Next assume that there are no out-of-date generators with higher variable costs installed in the system. Finally assume that no installed generator has an **emergency operating range** in which its marginal costs increase dramatically as it increases its output beyond its normal rating. With these assumptions, peakers will earn enough to cover more than variable cost only when the system runs out of capacity. In other words, peakers can cover their fixed costs only from scarcity rents but not from any nonscarcity inframarginal rents. All other generators cover their fixed costs from a combination of scarcity and nonscarcity rents. The left half of Figure 1-6.5 illustrates this property of an idealized supply curve.

The folk definition has the advantage of allowing the following types of statements which seem designed to segregate scarcity conditions from the normal operating conditions of the market.

1. Scarcity rents pay capital costs of units that run infrequently.
2. In the long-run competitive equilibrium, scarcity rents are just high enough to cover the fixed costs of peakers.
3. Scarcity rents are paid only infrequently.

This appears to ratify the view that power markets are qualitatively different in their cost structure and consequently cannot be analyzed with the standard marginal-cost apparatus.

In the idealized model, these statements are true, although they give the impression that scarcity rents are mainly or wholly associated with peakers. In fact, under the folk definition, every type of generator receives the same amount of scarcity rent per MWh. In addition, the average scarcity rent in $/MWh does not equal the fixed cost of peakers but is greater by a factor of one over the duration of the peaker's use, something that is not easily determined.

Two problems with this definition make it unworkable in a real market. First, there are likely to be old generators on the system with variable costs greater than the most expensive new generator that would be built (the peaker). In this case scarcity will not set in until the old generator is at full output. This will expand the nonscarcity rents and shrink the scarcity rents to the point where they no longer cover the fixed costs of a peaker. Second, there will be some (probably many) generators with marginal cost curves that continue on up to some very high but ill-defined value. This will reduce scarcity rents to some negligible and indeterminable value. Proving scarcity rents exist requires proving price is above the point where the supply curve becomes absolutely vertical; absolutes are notoriously hard to prove.

Because of these shortcomings and the limited usefulness of the folk definition, this book will use only the definition given above that coincides with "inframarginal rents," a term that has proven itself useful in economics. This is in keeping with the chapter's general view that generation cost functions present no new problems of consequence and require only a minimal expansion of the definition of marginal cost and then only to deal with the stylized mathematics of discontinuous cost functions.

## A Marginal-Cost Example

- Four suppliers can each produce 100 MW but no more.
- Each supplier has constant marginal cost (*MC*) up to this limit.
- Marginal costs and demand are as shown in the figure.

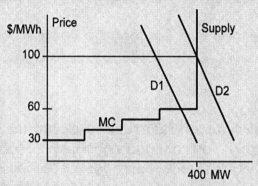

If demand is given by D1,
1. The competitive price is $60/MWh.
2. Any higher price indicates market power.
3. If the market is competitive, no supplier has $MC < \$60$/MWh.

If demand is given by D2 and the suppliers are price takers,
1. The market price (*P*) will be $100/MWh.
2. No generator will have a marginal cost of less than $100/MWh.
3. No market power is exercised at this price.
4. *P* is greater than the cost of the last unit produced ($60/MWh).

In both cases the marginal-cost rule for competition is

$$MC_{LH} \le P \le MC_{RH} *$$

This is sufficient to determine the competitive market price and output.

* $MC_{LH}$ is the savings from producing one unit less. $MC_{RH}$ is the cost of producing one unit more and is considered arbitrarily high, or infinite, if another unit cannot be produced.

*Chapter 1-7*
# Market Structure

*The work I have set before me is this . . . how to get rid of the evils of competition while retaining its advantages.*

Alfred Marshall
(1842-1924)

**P**OOR MARKET STRUCTURE POSES THE GREATEST THREAT TO THE HEALTH OF POWER MARKETS. "Structure" refers to properties of the market closely tied to technology and ownership. The classic structural measure is a concentration index for the ownership of production capacity. The cost structure of an industry, another component of market structure, describes both the costs of generation and the costs of transmission.

Most aspects of market structure are difficult to alter and some, such as the high fixed costs of coal-fired generation, are impossible. But power markets contain some unusual technology-based arrangements that can easily be altered or that require administrative decisions regarding their operation. These arrangements are part of the market structure and require design just as do the architectural components described in the next chapter.

The notion of **market structure** developed as part of the "structure-conduct-performance" paradigm of industrial organization in the early 1950s. The present discussion, however, is based on the structure-architecture-rules classification of market-design problems presented by Chao and Wilson (1999a) and Wilson (1999). The present chapter extends their definition of structure, particularly in the direction of administered reliability policies.

**Chapter Summary 1-7:** Market structure has a decisive impact on market power and investment. The second demand-side flaw, the ability of users to take power from the grid in real time without a contract (see Section 1-1.5), makes structural intervention necessary. Regulators must trade-off price spikes against involuntary load shedding, thereby largely determining the incentives for investment in generation capacity.

Regulators will continue to control transmission investment either directly or through incentives. The strength of the transmission grid affects the market's competitiveness. Demand elasticity can be greatly enhanced by improved metering and real-time billing, which can dramatically reduce market power and improve the stability of generation investment. Long-term power contracts and supply concentration also play a key role in controlling market power.

Because market structure is typically difficult to affect, it is usually ignored by policy makers. The power market is unusual in this respect, partly because it is new, flawed, and utilizes the grid, a regulated shared asset. The present lack of attention to market structure in the United States is producing unnecessarily high price spikes, boom-bust investment cycle and problems with market power during the short-supply phase of the cycle. Part 2 discusses the structural design issues most crucial to the solution of these problems.

**Section 1: Reliability Requirements.** The system operator buys energy and various grades of operating reserves to balance the system and to provide reliability. The number and type of submarkets used for this purpose are a matter of market architecture, but the reserve requirements and price limits imposed administratively are matters of market structure.

The structure of reliability requirements determine not only short-term reliability, but the height and frequency of price spikes and therefore long-run investment in generation and long-run reliability. These consequences of market structure are often overlooked, so the design of the reliability structure is often inappropriate, sometimes with serious consequences.

**Section 2: Transmission.** The transmission grid determines a significant part of the cost structure of the wholesale power industry. Investment in wires and the structure of access charges both have significant impacts on long-distance trade and thus on the market's competitiveness. This is an aspect of market structure which can and must be influenced by policy.

**Section 3: Effective Demand Elasticity.** Demand appears to be inelastic because it is not given real-time price signals. It would be more elastic if customers had a reason to purchase and use the equipment necessary for responding to price changes. Policy can easily influence two key aspects of market structure, meters and billing, that would greatly increase effective demand elasticity.[1] This would reduce the necessary investment in peak generating capacity, but more importantly, it would curb market power.

---

1. When considering the wholesale market, retail billing is part of structure. When considering the retail market it is simply a market outcome.

**Section 4: Long-Term Contracts.**  The structure of the spot market includes the extent of long-term obligations by suppliers.[2] In new power markets, policy can influence this structural component. It can require vesting contracts for newly divested generation and limit the amount of divestiture so obligations to loads at regulated prices are retained. Increasing the extent of long-term obligations increases the competitiveness of the spot market.

**Section 5: Supply Concentration.**  HHI, the classic structural index, measures the concentration of the ownership of productive capacity. While this aspect of market structure significantly affects market power, it can be difficult to change, though it is easily influenced in new markets by divestiture requirements.

## 1-7.1   RELIABILITY REQUIREMENTS

**The Second Demand-Side Flaw.**  Electricity customers can take power in real time without a contract and cause other customers to be blacked out, although in most cases, they will suffer no disruption of their own. (See Section 1-1.5.) Rotating blackouts are implemented without regard to contracts or consumption levels. At such times, the system operator is faced with a difficult choice; it can pay even more for power or it can blackout more customers. The choice is easily made in favor of reliability when the price is within ten times the long-run average, but when it increases to 100 times normal, the correct choice is less obvious. At some price, every system operator chooses to interrupt customers rather than pay the price.

**Operating Reserve Requirements.**  Deliberate interruptions of service are rare events, but their possibility has an enormous impact on the market. To avoid them, power systems buy several kinds of **operating reserves**, generators that are paid to be ready to provide power at a moment's notice. Together these reserves amount to approximately 10% of load at any given time.

Normally the effect of operating reserves on market price is modest. When available capacity exceeds load by 10% or more, a competitive market will hold the price of power down to approximately the variable cost of the most expensive generator producing power. In the normal operating range, this is under $100/MWh. In a year when system load never exceeds the normal range, prices are modest in spite of the administered operating reserve requirement. In fact prices are so low that generators cannot cover their fixed costs. If this situation were to continue year after year, no new generators would be built, and this would be the right outcome.

**Price Spikes and Investment.**  As consumption increases, supply becomes tighter, and the system operator finds it impossible to maintain a 10% operating reserve margin at all times. When reserves run short, the system operator offers to pay more either for power or reserves and this drives up the price of energy. This dynamic is the source of the high prices that induce investment. The heart of this process is the administrative decision about how much to pay depending on how

---

2. When considering the market for long-term contracts, this is an outcome and not part of the market's structure.

low operating reserves are. This links concerns about reliability with the incentive to invest.

The high prices paid by the system operator to meet reliability requirements control all high prices in the power market. Customers can choose between forward purchases and letting the system operator buy power for them in real time, and they choose the cheaper alternative. This holds the price in all forward markets down to the price that the system operator charges for real-time purchases, and it charges only as much as it pays. So the price paid by the system operator sets a limit on the price paid in all markets.[3] During the intervals when its requirements for operating reserves are not met, the system operator's pricing policy controls the price spikes and high forward prices that induce investment.

**Structure vs. Architecture of the Balancing Market.**  The balancing market keeps supply and demand in balance until the system operator is forced to balance the system by shedding load. This market must be administered by the system operator, but it may include a sizable bilateral component. It may be integrated with the markets for operating reserves, or these may be separate. These are questions of market architecture. At a more detailed level, there are innumerable choices concerning market rules.

The market rules and architecture do not determine the height and duration of real-time price spikes, nor how closely voltage and frequency will be maintained, nor the chance that the system will not recover from an unexpected generation failure.[4] These fundamentals are determined by the *structure* of the balancing market. The rules and architecture determine how efficiently trades are organized, who gets their transaction terminated when reliability is threatened, and how closely prices approximate the competitive level.

The structure of the balancing market is in part determined by the interconnection's reliability authority (NERC) and in part by local design. It is also influenced by the regional regulatory authority (FERC) when it caps real-time prices. The accuracy of balancing and short-term reliability are largely determined by the structure of the balancing market. Less obviously, *price spikes, generation investment and long-term reliability are also largely determined by the balancing market.* From this perspective, current balancing market structures appear haphazard and inappropriate. Part 2 examines these problems and presents methods for designing a better structure.

## 1-7.2  TRANSMISSION

Market structure includes the arrangement and capacity of power lines. Insufficient capacity can cause bottlenecks and local market power while additional capacity can expand the size of the market and reduce market power. The effect of transmission on market power makes transmission expansion more valuable in a competitive than in a regulated market.

---

3. The proposition that a price limit in the real-time market effectively caps the forward markets was well tested by the California Power Exchange.

4. This assumes that the rules are at least modestly functional.

While this structural component can only be changed slowly, it is one whose design cannot be ignored. Unless there is a reversal of the 100-year trend, new lines will be needed for reliability and economy as demand for power continues to grow. These will be built either at the direction of market designers working for the system operator or by for-profit transcos whose incentives, determined by regulators, govern their choice of transmission upgrades.

So far, this structural component has received the most attention, and the design issues have proven to be complex and contentious. Australia, New Zealand, Argentina, Alberta, California and Britain have all provided examples of transmission design issues that have been resolved with only partial success. Alberta's case is particularly interesting because it has highlighted the trade-offs between transmission and generator location. Alberta's for-profit Transmission Administrator chose not to upgrade Alberta's major transmission path and instead offered long-term incentives to new generation investment in locations that would alleviate the need for more transmission. This was quite successful, but two years later the TA strongly favored the now unnecessary upgrade in order to facilitate exports from northern Alberta to the United States.

Alberta's transmission issues did not involve the impact of transmission on market power, but in both Australia and California, this has been a prominent concern. Due to their historic rivalry, California's two largest investor-owned utilities are connected by inefficiently small power lines (Path 15). As a consequence, these frequently inhibit trade, causing increased generation costs and increased market power. Besides such major bottlenecks, there are dozens of smaller ones causing infrequent but locally severe market power. Over half of the generators in California are regulated at times, supposedly because of these transmission limitations.[5] While a must-run classification is sometimes only an excuse for a profitable regulatory "must-run" contract, many cases reflect real transmission constraints.

## 1-7.3   EFFECTIVE DEMAND ELASTICITY

Perhaps the most dramatic structural problem of power markets is the almost complete lack of demand response to fluctuations in the wholesale price. It is conceptually dramatic because it sometimes prevents the intersection of the market's supply and demand curves, a flaw so fundamental it is not addressed in any economics text. Its consequences grab headlines when California suffers blackouts and New York prices surge above $6,000/MWh. It is the flaw that makes market power a major issue in so many power markets that are otherwise well structured. But what is puzzling is that the remedy is straightforward and would pay for itself. That such a cheaply fixed fundamental flaw has been consistently overlooked is due in part to confusion over the supposed costliness of **real-time rates**. But it is also part of the pattern of ignoring structural problems in favor of architectural problems and disputes over market rules.

---

5. "Must-run" generators are paid regulated prices, which are sometimes quite high, when they are required to run due to transmission constraints.

When the cost of delivered power is \$1,000/MWh instead of the normal \$35/MWh, most customers will save only \$35 by consuming a megawatt less. They do not respond to the delivered cost of power because they receive no credit for their response. Their meters do not track their usage in real time. The problem is not so much low demand elasticity as lack of pricing. In addition, because customers do not face marginal-cost prices, most have developed little capacity to respond.

This problem can be addressed either indirectly through architecture or directly as a structural design problem. The indirect approach is to add competitive retail markets to the mix of wholesale markets in the hope that they will change the wholesale market structure by installing more real-time meters and implementing real-time rates. The evidence so far is not encouraging. One direct approach was initiated in California in the spring of 2001 in response to dramatic market failures, many of which were greatly exacerbated by this demand-side flaw.[6]

The direct approach requires installation of more real-time meters and implementation of real-time retail rates. The former is trivial. California extended real-time metering of load from 8 to 13 GW in just a few months at a cost of \$25M, the amount of money one utility was losing per day at the height of the crisis. Real-time rates would be simple if customers were not risk averse, but they are, and so it requires care to implement an appropriately hedged pricing scheme. California has adopted a complex administrative approach to hedging that requires documentation of a customer's past usage patterns and a series of continuing corrections whenever plants and equipment are expanded or changed for reasons not related to energy conservation.

To hedge a customer's bill under real-time rates, the customer's usage should be divided into two parts, a "base" usage and deviations from the base. The base can be defined in many ways, and this determines the complexity of the plan and the properties of the hedge. By defining the base as the average of the customer class, or in some other mechanical way, instead of relying on the past usage of each customer, the plan can be greatly simplified. Also, the extent of hedging can be varied. For instance the pricing plan can guarantee that customer bills will be immune to monthly changes in the *average* real-time wholesale prices. Although this prevents customers from reacting immediately to changes in the long-run price level, they still feel the full effect of hourly price changes (scaled to remove the change in the average monthly price).[7]

The essential point is that effective demand elasticity is a crucial structural parameter that is easily affected by policy, and there are many opportunities for clever market design. Typical values of the parameter are so unfavorable to market structure and improvement is so manageable that no power market should be started without implementing a major real-time rate program.

---

6. This demand-side flaw was largely responsible for California's market power problems and blackouts. This fact is often ignored because the flaw, being structural, is considered akin to a law of nature and thus automatically exempted from blame.

7. This result is based on the following real-time rates. Let Q(t) be the "base" profile that would be charged the flat monthly rate. A real-time customer that actually uses R(t) is charged at the flat rate times [the real-time cost of R(t) divided by the real-time cost of Q(t)].

## 1-7.4   LONG-TERM CONTRACTS

Long-term contracts and regulatory obligations to serve load are an often overlooked aspect of market structure. These can greatly increase competitiveness in the spot market. A generator that has sold 90% of its power forward has only $\frac{1}{10}$ the incentive to raise the price in the spot market as an identical generator that has sold nothing forward.

If a utility sells half of its generating plants, it will become short of generation, need to buy power, and have an incentive to hold down the market price. But, without long-term contracts, the nonutility generators (NUGs) that bought these plants will have a strong incentive to raise the price. Because the NUGs have more ability to raise the price than the utility has to hold it down, the net result can be a significant increase in market power. This happened in the California market.[8]

To avoid this problem, it is both customary and advisable when plants are divested to require that the purchasers sign **vesting contracts** to sell most of the output back to the utility for an extended period of time.[9] This hedges the utility (preventing a California-style bankruptcy) and dramatically reduces the market power of the suppliers who bought the divested generation. The price of the long-term contract can be indexed to fuel costs and inflation but should not be indexed to the spot market wholesale price. Such vesting contracts were used extensively in Australia and have served to keep wholesale prices low under almost all market conditions.

Vesting contracts can dramatically improve market structure at the time of divestiture, but in the long run, the permanent market structure and design will determine the equilibrium level of long-term contracts. This is a topic of great importance that is in need of research.

## 1-7.5   SUPPLY CONCENTRATION

Several indexes have been invented to measure supplier concentration; the best known is the Herfindahl-Hirschman Index (HHI). (See Chapters 4-3 and 4-5.) These indexes are of little use for predicting market power because concentration is only one of several important determinants, but this problem with HHI does not reduce the importance of concentration.

The two most effective methods of controlling concentration are divestiture restrictions and limitations on mergers. Utilities are frequently required to limit the amount of capacity sold to any single investor. For example, purchasers might be forbidden to own more than 5% of the total capacity inside a market's territory. In small markets, the purchasers of particularly large plants might need an exemption from this requirement simply because a single plant exceeds the 5% limit.

---

8. See Borenstein, Bushnell, and Wolak (2000), the California Independent System Operator (2000), Joskow and Kahn (2001a and 2001b), and Wolak, Nordhaus and Shapiro (2000).

9. The vesting contract should not specify which plants are providing the power, only that the new owner of the plant will provide the power. This has the desired effect on market power without unnecessarily restricting plant operation. See Wolak (2000) for the effectiveness of vesting contracts.

Note that the stricter the limit, the less valuable the plants will be. Exercising market power is more difficult for those with a smaller market share. Because a lower limit on market share reduces the profitability of generators, it reduces the price they will fetch when sold. Consequently, the utility required to divest will oppose any reduction in such a limit. It will argue that economies of scale or scope require a purchaser to own a large amount of capacity in order to operate a plant efficiently. These can be important effects, particularly when several plants are located on a single site, but disinterested expert opinions should be sought on the extent of such economies. Many suppliers own very little capacity in a market. Most economies from multiple plant ownership can be achieved by owning plants in different markets.[10]

---

10. Alberta, a relatively small market, has attempted to avoid the conflict between small concentration ratios and efficiency by selling twenty-year contracts for a plant's control and output, while leaving the plant's operation and ownership in the hands of the utilities. It remains to be seen how well this has worked.

# Market Architecture

*The whole world may be looked upon as a vast general market made up of diverse special markets where social wealth is bought and sold. Our task then is to discover the laws to which these purchases and sales tend to conform automatically. To this end, we shall suppose that the market is perfectly competitive, just as in pure mechanics we suppose, to start with, that machines are perfectly frictionless.*

Leon Walras
*Elements of Pure Economics*
1874

A MARKET'S ARCHITECTURE IS A MAP OF ITS COMPONENT "SUBMARKETS." This map includes the type of each market and the linkages between them.[1] The submarkets of a power market include the wholesale spot market, wholesale forward markets, and markets for ancillary services. "Market type" classifies markets as, for example, bilateral, private exchange, or pool. Linkages between submarkets may be implicit price relationships caused by arbitrage or explicit rules linking rights purchased in one market to activity in another.

Architecture should be specified before rules are written, but it is often necessary to test the architecture during the design process, and this requires a rough specification of the rules. Architectural design must also consider the market structure in which it is embedded, which may inhibit the proper function of some designs. Market design should not be rigidly compartmentalized, yet it is useful to consider the market's architecture apart from the details of the rules and the limitations of market structure.

**Chapter Summary 1-8:** A market design or analysis project concerns a collection of "submarkets" which are collectively referred to as the "entire market." (Both will often be referred to simply as markets.) Deciding which submarkets should be created for a power market is the first step in architectural design. Section 1-8.1 briefly discusses day-ahead and real-time energy markets and transmission-rights markets as a prelude to Part 3 which examines these choices in more depth.

Private submarkets range from disorganized to highly centralized, and each has its advantages. There is no simple rule for choosing between types of submarkets

---

1. This chapter owes a great debt to Wilson (1999) and Chao and Wilson (1999a), though it is not intended as a summary of their views.

once one has been included in the design list. By bringing the entire market into focus and specifying the market's architecture, the designer can take account of linkages between markets and decide on the types of submarkets. The architectural approach to design helps with understanding interactions and relationships between submarkets and thus with avoiding the proverbial "Chinese menu" approach.

Linkages are the heart of market architecture, but the science of linkages is not well developed. Timing, location, and arbitrage are the keys to most naturally arising (implicit) linkages, while explicit linkages are limited only by the imagination of the designer. Unfortunately many of these have unpleasant side effects. A careful examination of the market's architecture is the best antidote to inappropriately designed linkages.

**Section 1: Listing the Submarkets.**  The architecture of the entire market includes the list of both designed and naturally occurring submarkets. Many controversies surround the questions of which particular submarkets to include. For example, a day-ahead centralized energy market may or may not be included, and some suggest that the only designed markets to include are those for transmission rights and ancillary services.

Besides the designed submarkets, others already exist or will arise naturally. If these play an important role in the functioning of the entire market, they are part of its architecture.

**Section 2: Market Types: Bilateral Through Pools.**  Bilateral markets can be, in order of increasing centralization, search, bulletin-board, or brokered markets, while mediated markets can be dealer markets, exchanges, or pools. Public centralized markets tend to have certain advantages over private markets that are decentralized: lower transaction costs, quicker transactions, greater transparency of price, and easier monitoring. Decentralized private markets are more flexible while centralized private markets are similar to their public counterparts.

Energy exchanges accept only bids that, according to their bid-in values, at least break even. Pools accept some apparently losing bids. Accepted bids that would otherwise lose money are compensated with a "make-whole" side payment. Exchanges do not make side payments. Pools typically use much more complex bids, though exchanges can also use multipart bids.

**Section 3: Market Linkages.**  Implicit linkages are most often produced by arbitrage, the most important example being the arbitrage-induced equality between a forward price for delivery at time $T$ and the expected spot price at that time. The selection and arrangement of submarkets can take advantage of implicit links as when ancillary-service markets are sequenced so that excess supply in one spills into the next. Alternatively, multiproduct markets can be designed with explicit linkages, sometimes increasing efficiency but also adding complexity.

## 1-8.1    LISTING THE SUBMARKETS

The term **market** is used in many ways and has no strict definition. Tirole (1997) explains that it should not be so narrow as to encompass only a specific product produced at a specific location, nor should it be "the entire economy." He concludes that there is no simple recipe. In practice, the market designer must choose a definition to suit the problem at hand by relying on common sense rather than theory.

At least two categories of *market* are needed. First the designer must define the scope of the design problem which will be termed the **entire market**. "Power market" in the present book refers to such a market. Depending on context this could include only the wholesale market or the retail market as well. This choice demonstrates the need for a second market concept. An entire market typically includes components that are themselves markets. These will all be called **submarkets**. The distinction between an entire market and a submarket is relative, not absolute. In a different context it could be useful to view the ancillary services market as an entire market with submarkets instead of as a submarket of the entire power market.

An entire market may comprise many or few submarkets depending on its degree of vertical integration or unbundling. In other words, the intermediate products used to produce the final products of the entire market can be produced internally by the producer that uses them or can be purchased in a submarket. The first step in mapping an entire market's architecture is to decide on the list of submarkets. This is a key step in the design, requiring careful consideration, and it can be highly contentious. Currently, there is no consensus, even within the most informed circles, as to the best collection of submarkets from which to construct a power market.

Just as for a market, there is no simple recipe for the definition of a submarket. If a day-ahead energy market contains two zones, then different suppliers will sell into each zone and there will be two prices. Clearly there are two products: energy delivered to zone 1 and energy delivered to zone 2. Should these be considered different submarkets? That may prove convenient, but when there are 500 locations with different prices it will be necessary to count it as a single multiproduct submarket.

Conversely, PJM's day-ahead (DA) market sells energy and transmission rights. Although the prices of these are strongly linked, they are such different products they should be considered as supplied by two distinct but closely linked markets. Whether the transmission-rights market belongs on the list is distinct from the question of the energy market.

For a given project, some submarkets need to be designed and some do not. The energy futures market at Palo Verde was not designed as part of the California restructuring process though it is an important part of that power market. Generally, private markets will not be part of a design project of the type contemplated, though, when centralized, these markets are privately designed. Submarkets that are designed as part of the public design process contemplated here are by definition public markets and are almost always centralized. Submarkets that play a significant role in an entire market should be included in the list of markets even if not designed as part of the current project because they are a vital part of the market's architecture. Linkages between designed and naturally occurring or pre-existing submarkets are crucial to the health of the entire market.

## A Public Day-Ahead Energy Submarket

The question is not whether a DA energy market belongs on the list. If the system operator does not provide one, a private market will develop. But from the market designer's perspective, the question of whether the system operator will run a DA energy market is crucial. For this reason the public and private DA energy submarkets should be considered distinct. If a public DA market is added to the list, there will still be the question, discussed in Section 1-8.2, of what type of market it should be.

   If a centralized DA energy market is left off the list, a decentralized market must certainly be included because one will develop and play an important role in the entire market. Even with a centralized DA market, a private bilateral market is likely to develop and should be included.

## Energy vs. Transmission Submarkets

Perhaps the most fundamental controversy concerning which submarkets belong in a power market concerns the question of whether the system operator should operate an energy market or a transmission-rights market. One view holds that the system operator is only needed to operate the grid and sell rights to its use, but it should minimize its role in the market and refrain from trading or pricing energy. A centralized transmission-rights market belongs on the list, but a centralized energy market does not. The extreme version of this view would eliminate not only the centralized DA energy market just discussed but also a centralized real-time (RT) energy market.

   The opposing view holds that such an architecture, plausible in a simplified theoretical world, is wholly impractical. At least in real time, the system operator needs to buy and sell energy directly and needs to set different prices for energy provided at different locations. A real-time locational energy market should be on the list. The extreme version of this view holds that public DA and RT transmission-rights markets are not needed.

## Reasons for Including a Submarket

An entire market typically consists of a set of closely related end-product markets and the intermediate-product markets that feed into them. For a power market, the end product markets might be only a single wholesale electricity market in a particular region. Wholesale markets include all forward, futures, and options markets. The difficult task is to decide which public markets should be created.

While there are no clear-cut rules, several possible motivations should be considered:

1. Nondiscriminatory access.
2. Completeness (trading of a product not otherwise traded).
3. Reduced trading costs.
4. A publically known price.
5. Transparent operation.

Nondiscriminatory access may be guaranteed by the governance structure of a public market and is usually of most importance to small consumers and producers.[2] Completeness is most relevant in the case of public goods such as reliability services. It may also apply to natural-monopoly goods and services such as the unit-commitment service. A public centralized market, such as an exchange, will typically have much lower trading costs than a private decentralized market but not necessarily lower than a private exchange.

A publically known price serves two different purposes. First, it is a required assumption of the Efficient-Competition Result and is quite helpful to traders in making efficient trades.[3] Second it can be used as a benchmark for other transactions, both regulated and private, such as settling financial futures. Transparent operation is essential when market power is a potential problem and needs to be monitored.

There are also drawbacks to public markets, all of which seem related to the lack of proper incentives for regulators. A public market may offer products that are not well designed or are too limited, or transactions costs may be higher than necessary. These possibilities argue both for higher quality public institutions and for the substitution of private ones.

## 1-8.2   MARKET TYPES: BILATERAL THROUGH POOLS

There are two basic ways to arrange trades between buyers and sellers. They can trade directly, one buyer and one seller making a "bilateral" trade, or suppliers can sell their product to an intermediary who sells it to end-use customers. Both bilateral and mediated markets come in several types with bilateral markets usually less organized but with some overlap in this regard.

| Arrangement | Type of Market | | | | | |
|---|---|---|---|---|---|---|
| Bilateral: | Search | Bulletin Board | Brokered | | | |
| Mediated: | | | | Dealer | Exchange | Pool |

Less organized ←———————————————————→ More centralized

*Entire* markets often use a mixture of types. For example, the used car market is a mixture of direct search, bulletin board, and dealer markets. The New York

---

2. Motivations 1, 3, 4, and 5 are discussed in Chao and Wilson (1999a, 29).

3. See the statement of the first welfare theorem in Mas-Collel et al. (1995, 308).

Stock Exchange (NYSE) uses an auction, although when the market is thin, it becomes a dealer market. The term "pool" has a special meaning with regard to power markets. For years, utilities in some regions have organized their production in power pools, some of which used a centralized dispatch. In a deregulated market, a pool is an exchange in which the supply bids are complex, and the system operator carries out a complex calculation to select and pay the winners.

Some markets work better as one type and some as another. In the heat of debate, those favoring bilateral markets often imply that exchanges are in some way like central planning, socialism, or even communism, but these analogies contribute little. Often the right answer is for an entire market to utilize both approaches side-by-side. The long-term energy market utilizes a *bilateral* forward market that trades individualized forward contracts and centralized futures *exchanges* that trade standardized futures contracts. The transaction cost of trading in the forward market is greater but provides flexibility while trading in the futures market provides no flexibility in contract form but is inexpensive.

Bilateral markets can be either direct search markets or brokered markets and can be more or less centralized. If the market is brokered, as is the housing market, the brokers do not actually buy or sell in the market but are paid a commission for arranging a trade. Some forward energy markets, thanks to the Internet, are now organized as bulletin-board markets which are just a partially centralized variety of a direct-search market.

With the exception of the bulletin-board approach, bilateral markets need little design. They require an enforcement mechanism for complex contracts, but this is provided by the pre-existing legal framework.

## Bilateral Markets, Dealers, Exchanges, and Auctions

In **bilateral markets** buyers and sellers trade directly, although this is typically facilitated by a broker. Such markets are extremely flexible as the trading parties can specify any contract terms they desire, but this flexibility comes at a price. Negotiating and writing contracts is expensive. Assessing the credit worthiness of one's counter party is also expensive and risky. For these reasons there is a great advantage to moving toward more standardized and centralized trading when this is made possible by the volume of trade.

A *dealer market* is the most rudimentary type of mediated market.[4] Unlike a broker, a dealer trades for his own account, and usually maintains an inventory. He buys the product and holds it before reselling. There is no brokerage fee, but at any point in time the dealer buys for a price that is lower than the price he sells for. This difference is called the spread.

An **exchange** provides security for traders by acting as the counter party to all trades, eliminating traders' concerns over creditworthiness. Exchanges utilize auctions and are sometimes called **auction markets**; the NYSE is an example. As Bodie et al. (1996, 24) point out, "An advantage of auction markets over dealer markets is that one need not search to find the best price for a good." Because an

---

4. The terms "direct search market," "brokered market," "dealer market," and "auction market" are defined in Bodie et al. (1996, 24).

exchange interposes itself between buyers and sellers, the two halves of the market can operate independently, although they are linked by what is called a double auction. (Auctions are a traditional method of implementing a competitive market and are discussed in Chapter 1-9.)

An exchange can have a number of advantages over a bilateral market. It can reduce trading costs, increase competition, and produce a publically observable price. Depending on design and circumstances, it can also facilitate collusion and generally provides less flexibility than a bilateral market. Power marketers often favor bilateral markets because without an exchange there is more room to earn commissions as brokers and to appropriate the spread when they act as dealers.

Because exchanges are inflexible, they can operate much faster than bilateral markets. Stock exchanges routinely execute trades in under five minutes while bilateral markets take hours to weeks. In power markets, speed is crucial. Catastrophes can happen in seconds and system operators often need to exercise minute-by-minute control. Because of their speed, exchanges can operate much nearer to real time than can bilateral markets. This makes them the obvious choice for the real-time market. Weeks in advance, bilateral markets and dealer markets may play a larger role than exchanges. In between, there is much room for disagreement over which is better.

## Exchanges vs. Pools

If there is to be a public day-ahead market, should it be a pool or an exchange? PJM, NYISO, and ISO-NE all adopted pools, but CA ISO adopted a combination of public exchange (the Power Exchange) and private exchanges and dealers (the other "scheduling coordinators"). Although the California market has performed disastrously, this probably has little to do with its architecture and everything to do with its structure. No conclusion regarding exchanges can be drawn from this evidence.

California's Power Exchange was an **exchange** because it did not use the "make-whole" side-payments which characterize a pool. As is typical of exchanges, it used simple bids. These expressed only an energy quantity and price which meant generators could not take account of their startup costs and no-load costs directly within the bid format. Consequently, they had to manipulate or "game" their bids in some way to avoid a loss. Chao and Wilson (1999, 48) discuss such a circumstance as follows:

> *Gaming strategies are inherent in any design that requires traders to manipulate their bids in order to take account of factors that the bid format does not allow them to express directly.*

In order to avoid this problem, more complex bids, perhaps two or three part bids, are needed. Alternatively a pool could be used.

**Pools** are defined by the existence of side payments. Generators bid their marginal cost and certain other costs and limitation into the pool which computes a price and a set of accepted bids. Some accepted bids are found to lose money because the pool price is not enough higher than their marginal cost to cover their other bid-in costs. The pool makes up for this be granting accepted bidders, that would otherwise lose money, a side payment that makes them whole. Because some apparently losing bids are accepted, losing bidders have no way to verify whether their bid was correctly rejected. Typically pools utilize very complex bids which attempt to comprise a complete economic description of the generation process, but this is not necessary. A pool could be designed with two-part bids.

Although the principle of designing bid formats so that they do not require "manipulation" is a useful one, it need not be carried to extremes. Another principle states that, in a competitive market, competitive forces will induce bidders to represent true costs as accurately as possible within the bid framework. Thus in the Power Exchange, bidders included in their bids expected startup and no-load costs as well as their marginal costs. Chapter 3-9 shows that even in an exchange with one-part bids, where considerable manipulation is necessary, a competitive market will do a remarkably efficient job of dispatching generation. The "gaming" exhibited in a competitive market with an inappropriate bid format is largely beneficial and should probably be described by another name.

## 1-8.3  MARKET LINKAGES

Market linkages, aside from arbitrage, have no standard classification or nomenclature. Nonetheless they are tremendously important to the functioning of the entire market. Linkages can be either explicit or implicit. For instance, the requirement to purchase a transmission right, in order to inject and withdraw power, is an *explicit linkage* between the market for transmission rights and the bilateral market for wholesale energy. An *implicit linkage* causes the price in a forward energy market to approximate the expected price in the spot market during the forward's delivery period. There is no rule that enforces this relationship, only the discipline of arbitrage. Implicit links are not designed, but they are an important part of the architecture and must be reckoned with.

Sometimes when explicit linkages are needed, it indicates that two markets should be merged into a multiproduct market. Implicit linkages occur naturally and are usually helpful; explicit linkages are helpful when they reflect real costs. They are frequently harmful when they reflect a preconceived notion of how the market should operate.

Because power markets are geographically distributed, many of their submarkets are multiproduct markets and contain vast arrays of internal linkages. When the transmission system is congested (or if losses are charged for, as they should be) energy at location A is technically a different product from energy at location B. This is not just an academic definition. A completely unregulated bilateral market will price energy differently at the two locations. Consequently, an energy market is a multiproduct market with internal linkages between the products.

Just as there are spacial linkages, there are also temporal linkages. Market architecture establishes the temporal order of markets, and this order causes implicit linkages to develop between the markets. Both temporal and spacial linkages pervade the architecture of power markets.

## The Arbitrage Linkage of Forward to Spot Prices

In a well-arbitraged market the forward price for delivery at time $T$ will equal the expected spot price at time $T$. Because the market designer will find this common linkage extremely useful in understanding and predicting the behavior of *entire* markets, it is explained in some detail here.

There are two types of **arbitrage** (Tirole 1997, 134). The first, and more commonly recognized type, involves the transfer of a commodity from a high-priced location to a low-priced location. The relevant type for this analysis involves the transfer of demand from a high-priced product to a low-priced product. In the present case, customers can transfer their demand between forward purchases and spot purchases.

Say a customer knows it will need a certain quantity of power at a future time $T$. The customer can either buy an energy future now, at known price $F_T$, to be delivered at time $T$, or can wait and buy the power in the spot market at an as yet unknown price $P_T$. Which is preferable? Payment on the future can be largely postponed until delivery, so interest has a negligible effect. Generally a known price is preferred to an unknown price, but the main effect is that $F_T$ will be preferred if $P_T$ is expected to be higher and vice versa. If $E(P_T)$ is the currently expected spot price at time $T$, this can be summarized as

$$F_T < E(P_T) \text{ causes futures to be preferred.}$$

$$F_T > E(P_T) \text{ causes spot purchases to be preferred.}$$

The result is that whenever $F_T$ is lower, demand for futures will increase and $F_T$ will increase, while, if $F_T$ is higher, demand for futures will decrease, with the result that $F_T = E(P_T)$.

---

**Result   1-8.1**

**The Forward Price Is the Expected Future Spot Price**
The price of a future or forward specified for delivery at time $T$ is approximately equal to the expected spot price at time $T$.

---

This relationship is not exact. Both buyers and sellers tend to prefer the certainty of knowing the futures price. If the buyer's preference is stronger, $F_T$ may be greater than $E(P_T)$, and vice versa. These preferences will depend somewhat on the possibilities for diversifying the risks involved, but altogether these effects are too subtle and too unpredictable to be of interest to the power-market designer. For practical design purposes, $F_T$ can be expected to equal $E(P_T)$.

One immediate consequence is that if the real-time (spot) price is capped at $500, then the day-ahead price will not rise above $500. Of course if there are penalties for trading in the spot market, these must be taken into account. This effect was well documented in California, and it provides a mechanism for capping markets that are not accessible to the system operator.

Besides being a valuable tool for understanding the implications of system architecture, this result can provide practical insight. Seeing current high spot market prices and low future prices, i.e., $F_T < P_0$, many concluded that it was nearly always much cheaper to buy forwards than to wait for the inevitable high spot price. In contradiction of Result 1-8.1, they concluded that $F_T < E(P_T)$, and probably much less. This was one of two factors which led to California's huge purchases of forwards in the spring of 2001. It now appears that purchasing all of California's power requirements for 2004 from an extremely thin market during a power shortage, was not a sure bet. Pre-announcing a determination to buy no matter what and using inexperienced traders compounded the problem. Most likely, California did manage to create a temporary exception to the forward arbitrage linkage, $F_T = E(P_T)$, but in the opposite direction from what motivated the purchases: $F_T$ was probably much greater in California during the spring of 2001 than a rational expectation of future spot prices.

## Locational Price Linkages

Consider what happens if transmission rights are auctioned independently, as they often are. Suppose the right from A to B and from B to C and from A to C are all bid for separately, and the system operator sells them at the lowest set of prices that clears the market, given feasibility constraints on the rights. This would give three prices, $P_{AB}$, $P_{BC}$, and $P_{CA}$, but if the market is working efficiently, these prices will be tightly linked and in particular will sum to zero. Arbitrage is also key to this result, but the point is not the result itself but the fact that spacial linkages of energy prices are strong and important.[5] This is just one of a number of such results.

## Cascading Markets

Generators can provide energy or they can provide reserves, an option to buy energy when more is needed. Various qualities of reserves are graded by the quickness and sureness of their response. This classification provides an unambiguous ordering by value with the best quality of reserve always preferred to the second best, and so on. Suppose there are three such products called R1 (best), R2, and R3, and the system operator requires a certain amount of each.

If no linkage is made between the markets for R1, R2, and R3, they will take place simultaneously but separately. In this case any excess capacity in one market cannot flow into the others to help lower the price in those markets, so the markets should be linked, at least by conducting them in sequence. If the R3 market were conducted first, any excess of R2 might not be bid into the R3 market because it would not yet be known that it was in excess. To achieve a cascade, the markets should be cleared starting with the highest quality. Any surplus R1 reserves could bid into the R2 market and so on. This increases the efficiency of the markets relative to the absence of a cascade or one set up in reverse. Even with a forward cascade these markets will not perform optimally. If there is a shortage in the market for R3 and in total, but not for the first two markets, the price in the R3 market could exceed the other two prices. In this case high-quality reserve units would hold out for a chance at the third market, which could result in an inefficient use of reserves.

---

5. This is explained in 3-1.3 and can be proven from that facts that in general $P_{AB} = -P_{BA}$ and $P_{AB} + P_{BC} = P_{AC}$.

Efficiency can be increased by collapsing the three into a single multiproduct market with strong internal linkages. Such an arrangement will be simpler for the suppliers but more complex for the system operator. The reserve market can also be integrated with the energy market. Every new level of integration brings new complexity to the market clearing mechanism and reduces its transparency, but it has the potential to increase efficiency. Making the proper trade-off between efficiency and internal complexity, with its attendant opportunities for design error and gaming, is a controversial and unresolved issue.

*Chapter 1-9*
# Designing and Testing Market Rules

*Genius is one per cent inspiration, ninety-nine per cent perspiration.*

Thomas Edison
c. 1903

*If Edison had a needle to find in a haystack he would proceed . . . to examine straw after straw.*
*A little theory and calculation would have saved him ninety percent of his labor.*

Nikola Tesla
*New York Times*
1931

U NTESTED MARKET DESIGNS CAUSE REAL-WORLD MARKET FAIL-URES. Suppliers are quick to take advantage of design flaws, especially those that pay $9,999/MWh for a product that is worth less than $5/MWh.[1] Currently, many if not most, market designs are implemented without any explicit testing.

Although the most serious market flaws typically arise from structural problems, while architectural problems rank second in importance, problems with rules are the most numerous and their cost can be impressive. The design of rules is more art than science, but economics offers two guiding principles: mimic the outcome of a classically competitive market, and design markets so competitors find it profitable to bid honestly. Simplicity is another virtue well worth pursuing but notoriously difficult to define.

**Chapter Summary 1-9:** In a pay-as-bid auction, a coal plant bidding its variable cost of $12/MWh would be paid $12/MWh, while in a single-price auction it would be paid the system marginal cost which might be $100/MWh. In this case many would object to paying the $100 competitive price to the "inexpensive" coal plant and seek to improve on the competitive model. Pay-as-bid is one suggestion. The result is gaming and, probably, a very modest decrease in price and a modest decrease in efficiency. Ironically, if pay-as-bid succeeded as its advocates hope, it would put an end to investment in baseload and midload plants. In the long run this would dramatically raise the cost of power. The pay-as-bid fallacy illustrates the topics of the first three sections: the danger in attempting to subvert competition, the benefits of "incentive compatible" design, and the relevance of auction theory.

---

1. One of several design flaws that produced this outcome was prohibiting the California ISO from substituting a cheaper, better product for a more expensive, poorer product (Wolak, 1999). Also see Brien (1999).

That testing is the key to successful design is well understood by engineers until they design markets instead of equipment. It is not well understood by policy makers or economists, and the results are predictable. Rigorous testing, though worthwhile, is expensive so a simple "bottom-line" test should always be conducted first. This only requires building the simplest relevant model computing the cheapest possible production costs, and then computing the costs under the proposed design. If these are much different, reject the design. This test cannot prove a design will work, but it can save the cost of a rigorous test or a real-world failure.

**Section 1:  Design for Competitive Prices.**  Competitive prices sometimes include a scarcity rent much greater than needed to cover the concurrent fixed cost payment. Frequently this inspires attempts, such as FERC's advocacy of pay-as-bid auctions, to redesign the market to pay a price below the competitive level. If such a scheme were to succeed, it would cause reduced and distorted investment. Fortunately most such schemes are largely subverted by market forces.

**Section 2:  Design to Prevent Gaming.**  Rules that induce truth telling are called "incentive compatible" and often provide good market designs. Some bidding rules force suppliers to submit bids that do not reflect their true costs; such rules induce gaming. The pay-as-bid auction design is an example. Gaming usually causes inefficiency, the importance of which needs to be evaluated on a case-by-case basis.

**Section 3:  Auctions.**  Exchanges and pools use auctions to determine market-clearing prices. The four main auction types all produce the same revenue when bidders are buying for their own use or selling their own product. There is a slight efficiency advantage for a "second-price" auction which is incentive compatible. Recent work in auction theory shows that these results do not generalize to the multi-unit auctions with elastic and uncertain demand characteristic of power markets. In this setting, results are ambiguous, but pay-as-bid auctions tend to inhibit market power, sometimes at the cost of reducing welfare.

**Section 4:  Testing Market Rules.**  Every market design should undergo at least minimal testing before use. A minimal "bottom-line" test consists of three steps: (1) model the market with and without the design in enough detail to compute the design's impact on production costs, (2) find the minimum possible cost of delivered power, and (3) find the cost of delivered power when the market operates under the proposed rules. If the design raises costs significantly, it fails the test. Such a test cannot prove that the design will work well in the real world, but it often shows it will fail under even ideal conditions, a useful, if disappointing, result.

**Section 5:  Technical Supplement — Example of a "Bottom-Line" Test.**
A proposed charge for transmission access is tested and found to induce generation investment in a pattern that increases the total cost of delivered energy. Since the

proposed design fails to perform under even the simplest network conditions, it could have been rejected without expensive testing.

## 1-9.1   DESIGN FOR COMPETITIVE PRICES

Markets should be designed to produce competitive prices, but especially in power markets, competitive prices sometimes appear disconcertingly high. While most designers remain loyal to the ideal of competitive prices, many decide to redefine them to be lower at times, and some decide that they are just not right.

Recently FERC proposed a pay-as-bid auction design in the hopes of holding prices below their competitive level.[2] Several early proposals for the California market also had this intent. Even the PJM market contains elements of this flaw. These initiatives reflect a basic misunderstanding of the role of *scarcity rent* (defined in Section 1-6.6).

### Scarcity Rent

If a competitive generator sells 10,000 MWh for $500,000 and its variable operating costs are $100,000, then its scarcity rent is $400,000 which is not an unlikely outcome for a low-variable-cost coal plant when the market price is set by the marginal cost of a high cost gas turbine. Scarcity rents are necessary for suppliers to cover their fixed costs, which, as explained in Chapter 1-3, can be thought of as a constant flow of cost equivalent to the cost of renting the power plant. During the period in which the plant produced its 10,000 MWh, its fixed costs might have been $200,000. In this case the scarcity rent was twice what was needed to cover fixed costs during that period of operation. Such discrepancies are common, and they convince many that competitive prices are often too high.

---

*Fallacy*   1-9.1   **Scarcity Rents Are Unfair**
If the price paid to generators always equals system marginal cost, generators with lower variable costs will be paid too much.

---

This conclusion ignores the fact that in power markets, scarcity rents fluctuate dramatically. If they are to equal fixed costs on average, they must be higher sometimes and lower other times. Typically they are lower most of the time but are occasionally much higher. Ignoring this leads to a fallacy which states that paying a low-variable-cost plant the marginal cost of an expensive plant is unfair. This was succinctly expressed by former FERC Chairman Curt Hébert.[3]

---

2. The Blue Ribbon Panel that examined this scheme for California's Power Exchange concluded: "In sum, our response is that the expectation behind the proposal to shift from uniform to as-bid pricing—that it would provide purchasers of electric power substantial relief from the soaring prices of the electric power, such as they have recently experienced—is simply mistaken. . . .   In our view it would do consumers more harm than good." (Kahn et al., 2001, 17)

3. From p. 4 of his concurrence with FERC (2000b).

> *If the market clearing price for the final increment of needed capacity is, say, $100 MWh, why should a supplier who bid a lower figure receive the same value as that afforded to the supplier of [the] higher-priced increment?*

There are two answers to this question. First, as explained below, if low bidders were paid less, they would raise their bids. The Blue Ribbon Panel (Kahn, 2001), which analyzed FERC's pay-as-bid proposal for the California Power Exchange, focused entirely on this first answer. It explained what would have gone wrong had the proposal been implemented and it used auction theory for its analysis.

The second answer and the root of the problem lies in the desire to hold prices down to their short-run limit. Notice that the concern is with a "supplier who bid a lower figure." Suppliers that bid low are lower-variable-cost, baseload suppliers, and they bid low in order to guarantee they will run. The chairman's question was why, with their low variable cost, they should be paid as much as the high-priced supplier. The answer is they have higher fixed costs and need more scarcity rent to cover them.

Many schemes have been proposed to hold prices down to variable cost, and given sufficient regulatory authority they can be effective. Consequently it is important to understand that reducing price in the short run will increase it in the long run. This is not true if prices remain at or above the competitive level, but the competitive price pays higher and lower cost producers exactly the same when they both are needed at the same time. FERC's scheme was intended to hold baseload prices well below the competitive level. The Efficient-Competition Result states that competitive price will, over the long run, induce the right investment in both baseload plants and peakers. If regulators reduce the prices paid to baseload plants, in the short run they can get away with it, but in the long run, investors will see that baseload plants cannot cover their fixed costs and will build no more.

## How the Market Fights Back

Markets have ways of subverting the best-laid plans of regulators. Sometimes this causes problems, but in this case, it would prevent most of the damage that FERC's scheme would cause if it worked as intended. The plan was to capture the rent of baseload plants and thereby reduce prices. The method was to pay low bidders their bid, and as Kahn et al. (2001, 5) explain:

> *The critical assumption is, of course, that after the market rules are changed, generators will bid just as they had before. The one absolute certainty, however, is that they will not.*

Markets are composed of clever, highly motivated players who spend a great deal of time and effort discovering the most profitable way to respond to changes in rules. The most fundamental mistake a market designer can make is to treat a market as if it were a machine that does not change behavior when the rules change.

Knowing that bids will change is easy; finding the new outcome of changed rules and changed behavior is more difficult. One approach is to take a close look at why the New York Stock Exchange does not follow Chairman Hébert's preference "that sellers ... be paid what they bid ... rather than the market-clearing price."[4]

## Why the NYSE Pays the Market-Clearing Price

Each stock on the NYSE accumulates bids and offers over night, and as the market opens, these are traded at a single market-clearing price. Say there are three bids to sell 100 shares each at $50, $60, and $70 per share and three bids to buy at $50, $60, and $70 per share. In this case, the Exchange will set a market-clearing price of $60, and the two offers to sell for this much or less will be traded with the two bids to buy at this much or more. All will receive or pay the market-clearing price of $60.

If the NYSE followed FERC's scheme for running the California market, the opening of the stock market would work like this: The NYSE would accept the two bids from those selling stock (the supply or generator bids) at $50 and $60 and pay these two their bid price. They would then compute the average cost to be $55/share and sell the shares at that price to the buyers who bid $60 and $70/share. Instead of a single market-clearing price, the market would have one price for each supplier and a different but single price for all buyers.[5]

The suppliers would then argue that the NYSE should have done just the opposite. They should have averaged the two buyer's bids of $60 and $70 and paid both suppliers $65 for their stock. They have a point. Why should sellers who bid low *receive* only their bid price while buyers who bid high are not required to *pay* their bid price. Are stock buyers more worthy than stock sellers?

There is a more fundamental problem; traders will not sit idly by. The $50 supplier would realize that a $60 bid would have been more profitable and the next day bid $60. Then the FERC's scheme would produce the same result as the NYSE's market-clearing auction. If bidding low is rewarded with a low payment, while bidding the market clearing price is rewarded with the higher market-clearing price, previously low bidders will become high bidders.

| | | |
|---|---|---|
| **Result** | **1-9.1** | **Changing the Market's Rules Changes Behavior** |

## Auction Rules and Competitive Prices

Predicting the effect of rules and rule changes is a difficult matter which is why Section 1-9.4 prescribes testing for all proposed rules. Many of the important rules in power markets, including those just discussed, are auction rules and are the subject of auction theory, an important and rapidly developing branch of economics. A principle goal of auction theory is the design of auctions that produce competitive prices, and while much progress has been made, much remains unknown.

---

4. "My preference is that sellers in California be paid what they bid, regardless of what that bid is, rather than the market-clearing price." From p. 4 of his concurrence with FERC (2000b).

5. Another possibility is to match buyers and sellers by price so the $50 supplier sells to the $50 buyer, and so on. Applying the concept of total surplus from Section 1-5.2 demonstrates that this yields no total surplus, while using a market-clearing price gives a total surplus of $20.

When considering auction rules, two categories of problems should be distinguished: designing for a market with a competitive structure or designing for a market with a monopolistic structure. A competitive structure implies the bidders are small relative to the size of the market, where "small" must be defined relative to the market's demand elasticity. If bidders are large or demand inelastic, the structure is monopolistic. Within a competitive structure, it is reasonable to ask that the auction rules produce competitive prices. In a monopolistic structure, about the best that can be hoped for are auction rules that reduce the exercise of market power relative to alternative rules.

When considering FERC's proposed pay-as-bid rule change, three questions should be asked:

1. If the rule change worked as intended, would market prices be nearer the competitive level?
2. In a competitive market structure, would the rule change bring market prices nearer to the competitive level?
3. In a monopolistic market structure, would the rule change bring market prices nearer to the competitive level?

The first question is about intentions, the second and third are about how the rule would actually work. The answers to these three questions are: (1) No, baseload prices would be reduced far below the competitive level. (2) No, prices would suffer a mild increase in randomness and baseload prices would be reduces slightly below the short-run competitive level. (3) Quite possibly, but this might come with a reduction in welfare. These results are discussed further in Section 1-9.3.

## 1-9.2   DESIGN TO PREVENT GAMING

Markets often require participants to state a price at which they would either buy or sell. The market then selects the buyers who name high prices and the sellers who name low prices. This occurs in private bilateral markets and in public exchanges. If the buyers who bid high have the highest values for the goods being purchased, and if the sellers who bid low have the lowest costs of production, then their trades will be efficient. All those who can produce for less than the market price will trade with all those with values above the market price.

One way to arrange efficient trading is for all the traders to bid prices that are equal to their true costs and values. If all bidders tell the truth, the outcome is efficient. This is not the only way to get the efficient outcome, but it is the most obvious. An economic theorem states that for a very broad class of markets, if any market mechanism can be designed to give the efficient outcome, then there must be one that works by inducing traders to tell the truth.

Economists call a market mechanism that induces truth-telling an **incentive-compatible** mechanism. It not only makes the market efficient but also tends to be easier to discover and simpler to implement. For these reasons economists look for incentive-compatible designs, but sometimes they are too complex or politically unpopular. In this case a less efficient design will usually need to be adopted, but it is generally useful to look for the efficient design first.

---

| **Result** | **1-9.2** | **Design Market Mechanisms to Induce Truth Telling** |
|---|---|---|

If there is a market design that leads to an efficient outcome, then there is a design that induces traders to tell the truth and produces that outcome. Such a design is called "incentive compatible."

---

The pay-as-bid auction discussed in the previous section is not incentive compatible for it induces misrepresentation of variable cost. This is also called "gaming." As a consequence there will necessarily be some inefficiency unless the market price is always perfectly predictable. The single-price auctions of most electricity markets are nearly incentive compatible. Suppliers almost tell the truth when they are competitive (most of the time) but distort their bids when they have market power.

## 1-9.3 AUCTIONS

In the last forty years, economics has developed an extensive theory of auctions. Perhaps because auctioning electricity is a rather new idea, some have suspected that auctions are an invention of theoretical economics. In fact, the Babylonians used them in 500 BC, and Buddhists employed them in the seventh century. Sotheby's was established in 1744, some time before Adam Smith gave economics its start.[6]

William Vickrey, who won the 1996 "Nobel Prize" in economics largely for his work on auctions, classified one-sided auctions into four types.

1. **English:** Buyers start bidding at a low price. The highest bidder wins and pays the last price bid.
2. **Vickrey (second-price):** Buyers submit sealed bids, and the winner pays the price of the highest losing bid. This is also confusingly called a Dutch auction.
3. **Dutch:** The auctioneer starts very high and calls out progressively lower prices. The first buyer to accept the price wins and pays that price.
4. **Sealed-Bid (first-price):** Buyers submit sealed bids, and the winner pays the price that is bid.

These four types of auction can all be used in reverse to buy a product instead of sell it. For example, in a reversed English auction, sellers can call out progressively lower prices until there are no more bids. The lowest bid wins, and the seller sells at that price. Consequently everything said about auctions to sell, holds for auctions to buy, but in reverse.

There are two motives for buying: for a buyer's own use or for resale. In the first case the value of the purchased good is the **private value** placed on it by the

---

6. See Klemperer (1999; 2000a; 2000b) for surveys of auction theory and literature. The first is the most accessible.

buyer. In the second case, the value is the price for which it can be resold. If the purchase is for resale, then all potential purchasers will find the value to be the same, i.e., the market price upon resale. This is called the **common value**.

## The Revenue Equivalence Theorem

The **revenue equivalence theorem**, a key result of auction theory, states that the four types of auction will all produce the same revenue if used in private-value auctions. To help explain how this is possible, consider a comparison of a Vickrey (second price) auction and a sealed-bid (first price) auction.

| | | |
|---|---|---|
| **Result** | **1-9.3** | **Four Types of Auctions Produce the Same Revenue** |

The four types of auctions, English, Vickrey, Dutch, and sealed-bid, all produce the same revenue if the bidders have private values. If they have common values, then their revenues are in the listed order with English producing the most.

Assume the bids in each were $100, $200, and $300 by A, B, and C respectively. In each auction, C wins, but in the Vickrey auction, C would pay only $200 while in the first-price auction, C would pay $300. This makes it seem that the revenue of the first-price auction is clearly greater by $100.

The explanation of revenue equivalence involves the fundamental principle of rule design, Result 1-9.1. Changing the rules changes the behavior of the bidders. To analyze these behaviors assume that the bidders have private values corresponding to the above bids. First, consider bidder behavior under a Vickrey auction. C will bid $300 because he knows that if A and B have bid less he will win and get the best possible price, the price of the next-highest bidder. If C loses, say to a bid of $305, he will be glad he lost, for winning would mean paying $305 for something he values at only $300. If a bidder bids less than her valuation, she increases the chance of losing when she should win and saves no money. If a bidder bids above her private value her only extra wins from biding higher will be in auctions she would rather lose. So all bidders bid their true values, and this design is incentive compatible. Bidder C will win and pay the second price of $200.

| | | |
|---|---|---|
| **Result** | **1-9.4** | **A Vickrey Auction Is Incentive Compatible** |

A second-price auction causes bidders to bid their true value, or if it is an auction to purchase, they bid their true cost.

Now consider a first-price auction. Bidding strategies depend on what is known about others' bids. Assume that C knows that A and B almost surely have values of $100 and $200. Then C knows B will almost surely not bid higher than $200 and perhaps lower. If C is sure enough he will bid $201 or less. So the revenue of a first price auction will be much less than $300 because C will not bid his true value. This auction is not incentive compatible. The calculation of bidding strategies in such an auction is quite difficult, and it is remarkable that the revenue equivalence theorem can cut through this complexity and produce such a clean result. The winner in the first-price auction will pay $200 on average.

One disadvantage of the first-price auction is the complexity of the optimal strategy. Bidders are forced to guess the probability distribution of other bids and then make a complex computation based on this guess. This introduces errors into the bids. For example, in a power market, bidders in a Vickrey auction will bid their marginal cost and that will reproduce the standard **merit order** from which the system operator will dispatch generation efficiently. With a first-price auction, all generators with marginal costs below the expected market price attempt to bid just under the market-clearing price. This produces a cluster of bids that have little to do with the merit order and which differ more because of estimation errors than because of costs. Dispatching the "cheapest" generators from this assortment of bids will produce an inefficient dispatch.

## Revenue Equivalence and the Pay-as-Bid Design

If the auction sells many units of the product, many megawatt-hours of power, and there are many winners, then a second-price auction pays all winners the same while a first-price, sealed-bid auction pays each winner his bid. This gives rise to the terms **single-price** auction and **pay-as-bid** auction. Economists call a "pay-as-bid" auction a **discriminatory** auction. As discussed in Section 1-9.1, FERC recently argued for replacing California's single-price auction with a discriminatory auction on the grounds that it would result in lower prices. If the auction under consideration were buying a single indivisible unit of power of a known quantity, then the revenue equivalence theorem would prove that switching to pay-as-bid would have no affect on revenue, which means the average price paid would be the same.

Unfortunately, electricity auctions are multi-unit procurement auctions in which demand is elastic and uncertain, and the revenue equivalence theorem does not apply. A paper by Frederico and Rahman (2001, 2) demonstrates that for such auctions, in a competitive market structure, pay-as-bid reduces the price paid but also reduces efficiency. They show that under monopoly conditions, "the exercise of market power is more difficult under pay-as-bid, and that firms with market power may react in inefficient ways to a switch to pay-as-bid" thus reducing welfare. They are also optimistic that under oligopolistic conditions pay-as-bid will significantly reduce market power but have not modeled this situation.

These results are complex and ambiguous, and those who believe they see easy answers have not understood the question. Theory may never yield definitive answers, and testing of auction designs may prove the only reliable method of evaluation. Testing such a design for the first time in a 10 billion dollar per year market may not be the most cost-effective approach.

## 1-9.4 TESTING A MARKET DESIGN

In a turbulent environment such as the restructuring of the U.S. electric industry, new market designs are nearly always conceived and implemented without rigorous testing. In fact a complete absence of testing is common; however, few designs are implemented without claims that they produce "the correct incentives" and "efficient outcomes." Unfortunately, when such designs are tested against even the most well-behaved hypothetical situations, they do not always live up to their claims.

A moderately rigorous test would involve a laboratory simulation of the market design. While this is recommended, especially when gambling with sums that can run into the billions of dollars, a more modest course of testing is presented here. No design should be implemented without this minimal level of testing, which is called the "bottom-line test" because it tests the effect of a market design on the total cost of supply. It does not provide a cookbook procedure, but it provides a structure that is often missing, and it sets a minimum standard. The test procedure can be broken into three steps.

---

**Test**

**The Bottom-Line Test of Market Rules**

1. Model the cost functions of the players in the market.

2. Compute the minimum possible cost of serving the target load level.

3. Compute the cost increase of serving the load under proposed rules.

---

If consumer response to price (demand elasticity) is important for the rules being tested, then the bottom-line test must be modified by examining the decrease in total surplus caused by the design instead of the increase in cost. Fortunately, most market rules can be tested adequately under the assumption of completely inelastic demand. Then consumer benefit is unaffected by market rules and drops out of the test procedure. Occasionally, adding demand elasticity can simplify a model by removing "knife-edge" discontinuities in behavior. In this case the net-benefit test should be used.

Lack of demand elasticity confers market power on suppliers, and most designs are not efficient in the presence of market power. The bottom-line test is only intended to test designs for competitive markets, so to counteract the effects of inelastic demand, suppliers should simply be assumed to behave competitively.

## The Least-Cost Focus of Economic Testing

As discussed in Section 1-5.2, least-cost production is the criterion for both short and long-run supply-side efficiency. Statements about sending the right price signals to generators, providing optimal incentives, and being efficient mean nothing if the design does not keep the delivered cost of production low. Conversely, if the design does minimize generation costs, it necessarily sends the right price signals to generators and provides the right incentives.

The supply-side implications of a competitive market design can be tested just by comparing the total cost of supply under the proposed design with the minimum cost of supply in an ideal world. The point is not that the ideal must be achieved before a design is approved, but the ideal provides a benchmark, and cost is the sole criterion. As simple as this test is, it can still include such complexities as environmental concerns by including their costs. (This is the only method of inclusion that makes otherwise implicit trade-offs explicit.)

## The Advantage of Simple Models

Economic modeling is as much art as science, and mistaken conclusions result as often from poor model design as from faulty calculation. Simple models minimize such errors. Those who prefer their ideas not be tested usually argue that simple models cannot test a design adequately and should not be used. Even when a complex model is available, simple models should be used to check the reasonableness of the more complex models, and a simple model is always better than no model.

Simple models can give clear-cut results when they are negative and can save the considerable effort of a more elaborate test. One disadvantage, however, is that while they can lend credibility to a design, they cannot serve as proof that it will succeed in practice. That such testing can prove a design wrong but cannot prove it right may be discouraging, but it is always better to learn of failure during the design phase.

# 1-9.5  TECHNICAL SUPPLEMENT: EXAMPLE OF A "BOTTOM-LINE TEST"

This example demonstrates how to test a design with the bottom-line test and how to avoid spending a lot of money on complex tests when a design has an elementary flaw. Detailed cost calculations are postponed to the end in order to focus on methodology. This example was an actual proposed design, and the inappropriate test was carried out.

## A Test of Two Transmission Charging Rules

The design to be tested consists of two rules that are intended to charge generators for wires in such a way that new generators are induced to build in the optimal location. The first rule specifies that the power flows caused by a generator are calculated by assuming its power flows proportionally to every load on the system. (This is not a law of physics but only an accounting rule.) If a generator injects 100 MW and there are two loads on the system, of 2000 and 3000 MW, then 20 MW flows to the first load and 30 MW to the second, no matter where the two are located relative to the generator.

> **Rule 1.** Power is assumed to flow proportionally to every load.
> **Rule 2.** Flows are charged their proportional use of each line's capacity times the line's cost.

Having used the first rule to compute the generator's nominal power flow on every wire in the system, the second rule is used to compute the charge for using each wire. The charge is based on the absolute value of the fraction of a line's capacity used each hour.[7] In other words, if a generator causes a 20-MW flow on a 100-MW line with a fixed cost of $50/h, then the generator is charged $10/h no

---

7. This might be termed an absolute impacted megawatt-mile approach and is a design that has been proposed and inappropriately tested.

matter which way it sends power over the line. This completes the description of the design in question; these two rules are sufficient to compute the transmission access charge for every generator. In all other respects the market is assumed to be perfectly competitive.

The stated purpose of the design focuses only on the linkage between this market and the market for generation plants, and it ignores linkages to the energy market. Although these linkages deserve testing, for simplicity, the example will test only the effect of the rules on the location of new power plants.

These rules were "tested" by constructing a model of all the power flows in the state of California and computing them for every hour on every wire for a given year. Graphs were made showing how much power flowed on wires operating at various voltages. It was claimed that since most of the power flowed on high-voltage lines, and those were the ones being charged for, the market design would "send the right signals." As a consequence of sending the right signals, investors would build plants in the right locations, and the market would operate efficiently. But such "testing" misses the point.

Notice that although some costs may have been calculated, no total minimum cost was calculated (step 2), and no total cost under the proposed design (step 3) was contemplated. The test did not rule out the possibility that generation costs would be double the economic minimum under the proposed market design. The central economic question of cost minimization was simply not considered. These oversights are now remedied by applying the bottom-line test.

## Step 1: Modeling Costs and Benefits

Basic economics always assumes market participants will act to maximize their benefit net of costs, so cost and benefit functions must be specified to calculate the predicted behavior. To perform a "bottom-line test," first determine what costs and benefits are relevant. In this case, the relevant costs are (1) the cost of generation including its dependence on location, and (2) the cost of wires. A useful economic model must include both.  ·

Modeling elastic demand would complicate the computation of benefits and is not relevant to the benefits being claimed. The basic bottom-line test which focuses only on supply costs will be sufficient. The first test of a design should be as simple as possible, and the model needs at least one wire and two generation locations to evaluate the design. These are specified in the final sub-section which uses the details of the model to compute costs.

## Step 2: Computing Minimum Cost

The bottom-line test requires computing the minimum cost of producing and delivering the power required to meet a fixed level of demand. In principle this can be done by trying all ways of producing the required output. Generally a little mathematics will simplify the process. In the present example, the minimum cost of delivered power is found to be $362,000/h. For a particular specification of the test model, step 2 always gives a simple answer that is just one number. No excuse should be accepted for a more complex or less precise answer although it may be useful to test several variations of the model (each will have its own one-number result for step 2).

## Step 3: Computing Cost Under the Proposed Rules

The behavior of suppliers must be considered when computing the effect of the rules on the total cost of delivered energy. Suppliers will react to market rules, and this reaction must be predicted. Economics enters the test process at this point by dictating the assumption that suppliers will maximize their profits. This may not be true when a rule is first put in place, but with practice, suppliers should learn how to adjust their behavior to the rule and maximize their profit. The profit maximizing assumption is key to economics, and though not precisely true, is generally the best approximation available.

In the present model, taking into account the profit maximizing behavior of generators, the total cost of delivered power will be $376,000/h. Step 3 should always produce as simple an answer as does step 2.

## Interpretation

The result of this example can be summarized as a $14,000 (4%) increase in cost caused by the proposed rules. This might be tolerable, but a qualitative inspection of the model raises more concerns. The rules cause all of the generation in the system to locate at the load center and consequently cause an eight-fold increase in the cost of transmission. This is unreasonable, and the rules should clearly be rejected because they have failed such a simple and reasonable test.

The benefit of the bottom-line test is dramatic. By applying it first, the expense of a complex study can be saved. Additionally, it provides an intuitive understanding of the design's flaws that would be difficult to extract from a more complex simulation. The following subsection builds the model and uses it to compute the costs for steps 2 and 3, as well as to develop the intuition of the interpretation.

**Step 1: Building the Model and Computing Costs.**   Assume there are two cities in the market, and they are connected by a transmission line as shown in Figure 1-9.1. Assume for simplicity that generation can be located near either city, and the cost is the same: $10/MWh for capacity and $20/MWh for energy. Assume peak load on the system is 16,000 MW with 12,000 MW located at city A and 4000 MW located at city B. Assume the average hourly load is 10,000 MW, assume that line capacity costs $4/MWh, and for reliability purposes assume the minimal line capacity is 500 MW. This is a sufficient model of the relevant costs.

**Step 2: Finding the Minimum Cost.**   Because the above model assumes that generators cost the same at either location, the cheapest arrangement is to locate them where they are needed, that is 12,000 MW of generation at A and 4000 MW of generation at B. In this case, no power line is needed to facilitate trade. The total long-run cost of power in this model is found by summing three components as follows:

**Figure 1-9.1**

Minimum cost locations with flows calculated using proposed Rule 2.

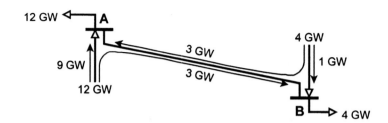

$$
\begin{array}{lll}
16{,}000 \text{ MW} \times \$10/\text{MWh} &= \$160{,}000/\text{h} & \text{for generation capacity} \\
+\quad 10{,}000 \text{ MW} \times \$20/\text{MWh} &= \$200{,}000/\text{h} & \text{for energy} \\
+\quad\ \ 500 \text{ MW} \times \$4/\text{MWh} &= \$2{,}000/\text{h} & \text{for line capacity}
\end{array}
$$

for a total of $362,000/h. This is the minimum total cost of serving the load.

**Step 3: Finding Total Cost under the Proposed Rules**   Under the proposed transmission pricing rules, generators will pay different transmission charges depending on their location. In particular, if they were located in the minimal cost arrangement just described, the generators at A would have $\frac{1}{4}$ of their flow assigned to the line from A to B because by Rule 1, $\frac{1}{4}$ of it goes to the 25% of load at B. The generators located at B would have $\frac{3}{4}$ of their flow assigned to that line. (This second power flow is in the opposite direction and cancels the first, but this does not affect the charges.) Because charges are proportional to flow, the generators at B are charged three times more than the generators at A. Whenever a generator is retired at B, it will be replaced with a generator at A because, counting transmission charges, A is the cheaper place to locate.

The long-run consequence of this dynamic is that all new generators will locate at A and all expansion of existing generators will take place at A. In the long run, all generating plants will be located at A. Consequently a 4 GW transmission line will be built to supply B's peak load. That is eight times bigger than the 500-MW line required for reliability in the least-cost configuration. This increases the total cost of energy by the cost of the extra transmission line. An extra 3500 MW of line costs an extra $14,000/h. This brings total cost under the proposed design to $376,000/h.

*Part 2*

# Reliability, Price Spikes, and Investment

# Reliability and Investment Policy

*When, by building theories upon theories, conclusions are derived which cease to be intelligible,
it appears time to search into the foundations of the structure and to investigate how far the facts
really warrant the conclusions.*

Charles Proteus Steinmetz
The Education of Electrical Engineers
1902

R<small>ELIABILITY, PRICE SPIKES, AND INVESTMENT ARE DETERMINED BY</small>
**REGULATORY POLICIES.** Because these policies impinge on market structure
rather than architecture, they have been overlooked too often as debates focused
on "nodal pricing," "bilateral trading," or on market rules. The result has been a
chaotic pricing policy and disaster in the Western U.S. markets. Part 2 assumes
away two major problems, market power and transmission constraints, to focus
exclusively on the structural core of a contemporary power market. The goal of
Part 2 is to explain the major policy options and their implications. This requires
an understanding of the causal links between policy controls and the key market
outcomes—reliability, price spikes and investment. Both controls and outcomes
are diagrammed in Figure 2-1.1.

Supply and demand characteristics comprise a market's core structure, but in
a power market these are unusually complex. The supply side cannot store its output
so real-time production characteristics are important, and two demand-side flaws
interact detrimentally with this characteristic. Consequently, the market cannot
operate satisfactorily on its own. It requires a regulatory demand for a combination
of real-time energy, operating reserves, and installed capacity, and this demand
must be backed by a regulatory pricing policy. Without this reliability policy, the
power system would under-invest in generation because of the demand-side flaws.
Reliability policy is the part of the structural core that can be affected immediately
by design. The demand-side flaws can also be affected by policy, but these design
changes take longer to implement.

Without the demand-side flaws and reliability policy, Figure 2-1.1 would
represent a normal market; demand and supply conditions would feed into the
market and determine prices. These would determine new investment which would

**Figure 2-1.1**

The structural core of a power market determines reliability, price spikes, and investment.

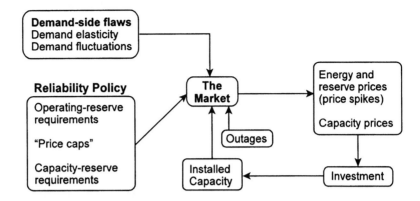

determine the supply conditions and reliability. When the demand-side flaws have been sufficiently ameliorated, short-run reliability policy will no longer be required to play its current role of providing the major incentive for long-run investment in generating capacity. Price spikes will be controlled by demand elasticity and power markets will operate normally. Chapter 2-4 discusses the threshold at which this becomes possible.

**Chapter Summary 2-1:** Energy price spikes and, in some markets, installed-capacity prices induce the investment in generation which determines reliability. The duration of price spikes is roughly determined by accepted engineering rules of thumb, but no engineering and no market determines the height of the spikes. Their height is determined by often-murky regulatory policies that limit what a system operator will pay when the system runs short of operating reserve. Thus, neither determinant of investment, duration or height, is market driven.

Many regulatory policies would produce the right level of investment, but policies that employ extremely high, short-duration price spikes should be avoided because of their side effects. Such spikes cause investment risk, political risks, and increased market power. Correctly designed policies can ameliorate these and other side effects while inducing any desired level of investment.

**Section 1: Why Price Regulation is Essential.** The second demand-side flaw makes it possible for customers to avoid long-term contracts and purchase in real time from the system operator. Because the system operator sells at cost, its customers will never be charged more than the system operator's price limit. Consequently, they are not willing pay more in any market. If system operators had no policy of paying sufficiently high prices for enough hours per year, generators would not cover their fixed costs and would not invest. Thus, the market, on its own, would underinvest and reliability would suffer. System operators do have pricing policies, and these determine the height and duration of price spikes.

**Section 2: The Profit Function.** Regulatory policy determines the market's price when demand, including the regulated demand for operating reserves, exceeds total available capacity. These price spikes determine the short-run profits of generators, and the expectations of these profits induce generation investment. Investment increases **installed capacity** (ICap), which reduces both price and profits. This feedback loop (shown in Figure 2-1.1), which is controlled by reliability policy, determines the equilibrium level of ICap and thus long-run reliability.

The profit function summarizes the information needed to find the equilibrium ICap level. It takes policy into account and plots expected short-run profits (of **peakers**, for example) as a function of ICap. The equilibrium ICap level occurs at the point on the profit curve where short-run profits just cover the fixed costs of peakers. The short-run profit function derived from the combination of energy and capacity prices shows what level of ICap and reliability a given set of policies will produce.

**Section 3: Side Effects of Reliability Policy.** Many different policies produce the same optimal level of ICap and reliability, but they have different side effects according to the steepness of their profit functions. A steeper profit function increases risk and facilitates the exercise of market power. By choosing policies that produce low, long-duration price spikes, a flatter profit function can be achieved.

**Section 4: Inter-System Competition.** Competition between system operators militates against low price spikes. Any system operator with a low price limit on energy and reserves will find its operating reserves purchased out from under it at crucial times. Regional coordination of pricing policies can avoid such competition and its negative side effects.

**Section 5: Demand-Side Effects of Price Limits.** The ideal solution to the investment/reliability problem would be sufficient price elasticity to keep the market-clearing price well below the value of lost load at all times. A high price limit encourages demand elasticity, but this consideration must be weighed against the increased risk and market power associated with high limits. A better solution might be to price demand and supply separately during price spikes.

## 2-1.1   PRICE REGULATION IS ESSENTIAL

The second demand-side flaw, the system operator's inability to control the real-time flow of power to specific customers, necessitates the regulator's role in setting prices (see Section 1-1.5).[1] Without this flaw, the system operator could simply enforce contracts. Instead, it must buy power to balance the system and maintain reliability.

Although the system operator must play an active role in present systems, that role might be restricted to setting a price that equated supply and demand—in other words, a price that cleared the market. If this were possible, the system operator, though active in the price-determination process, would have no control over price and would not be a price regulator. In most hours of the year, the system operator does play such a passive role.

Because of the first demand-side flaw (a lack of real-time metering and billing) demand is so unresponsive to price that a simple market-clearing role is not possible in all hours. If installed capacity is at an optimal level or lower, there will be times during which demand exceeds supply and load must be shed. NERC estimates this will happen for about 1 day out of 10 years. At such times, no price will clear the market.

The system operator could continue to pay the highest nominal marginal cost (left-hand marginal cost) of any generator after the market failed to clear. This would still constitute regulatory price setting, but it would be a minor intervention. With only this minimal intervention, investment would suffer a serious decline. The competitive market-clearing price, when supply exceeds demand, is the physical marginal cost of generation which is usually less than \$200/MWh.[2] For this price to support investment in new peakers, demand would need to exceed supply (forcing blackouts) for more than 200 hours per year, not the 2.4 hours that NERC considers optimal.

---

*Fallacy*   [2-4.1]   **The "Market" Will Provide Adequate Reliability**
Contemporary markets, with their demand-side flaws and negligible demand elasticity, would grossly underinvest in generation without regulatory price setting.

---

The minimal price intervention that would produce a reasonable level of reliability is known as value-of-lost-load (VOLL) pricing. This has been studied

---

1. See also Ruff (1999, 28), and FERC (2001b, 4).

2. Some will object that for the last few percent of supply, marginal cost climbs rapidly to unlimited heights as wear-and-tear and the possibility of significant damage increases in the emergency operating range. This is correct, but it does not solve the present dilemma. These marginal costs cannot provide a benchmark for setting the market-clearing price when demand exceeds supply. They are not verifiable, and even if they were, just before the market fails to clear they easily exceed the value of power to consumers, VOLL. If price were set to the last marginal cost before failure, consumers would pay more than the power is worth and stimulate overinvestment in generation. Chapter 2-4 also discusses the possibility of using the high demand-sided bids, found in some more advanced markets, to set price. This is similarly inappropriate.

and implemented by Australia's National Electricity Market (NECA, 1999b). This approach recognizes that the system operator must purchase power on behalf of load when demand exceeds supply and instructs it to pay $V_{LL}$, the value of additional power to load, whenever some load has been shed (during a partial blackout). This is sensible, and ignoring risk and market power and given the first demand-side flaw, it produces an optimal outcome.[3] It induces exactly the right level of installed capacity, which minimizes the sum of the cost of that capacity and the cost of lost load.

Implementing VOLL pricing requires a regulatory determination of $V_{LL}$ because the market cannot determine it. This value will determine the height of the aggregate price spike, and the duration of the price spike will be determined by the regulator's decision to set this price when, and only when, load has been shed.[4] The Australians estimated $V_{LL}$ to be about $16,000 US but set their price limit at only about $10,000 US. Both the height and average annual duration of the price spikes under $V_{LL}$ pricing are determined by regulatory policy and not by any market mechanism.

In the United States, system operators take a different approach. In compliance with NERC guidelines, they set operating reserve requirements which cover "regulation," spinning reserves and nonspinning reserves which together amount to roughly 10% of load. Instead of waiting until load must be shed to raise price, a shortage of operating reserves is deemed to be sufficient reason to pay whatever is necessary. This results in high prices whenever demand exceeds about 90% to total available supply, an occurrence far more common than load shedding. In this way, U.S. reliability policy determines a much longer duration for price spikes.

Determining the height of energy price spikes is more complex because system operators compete for reserves. If one operator is willing to pay a high price, it can buy up the reserves of its neighbor and thereby force it to pay a high price to acquire its own reserves. This complication is discussed in Section 2-1.4, but the essence of regulatory pricing is better understood by considering a single isolated market.

The engineering approach considers requirements for operating reserve to be sacred and avoids assigning a maximum price that should be paid to comply with them. The suggestion is to "pay what is necessary." Before markets, this caused no problems, but when prices exceed a few thousand

---

### What's a Regulator to Do?

Prices must spike to pay the fixed costs of generators, or there will be no new investment. The highest price should occur when load has been shed, but even then the system operator should pay no more than the power is worth. Academics estimate this to be more than $1000/MWh but less than $100,000. The market gives no answer. Australian regulators pick a value near $10,000 and thereby determine investment.

U.S. regulators, correctly wishing to reduce market volatility, choose a lower price limit and pay that price when reserves are low not just when load is shed. FERC's price limit and NERC's operating-reserve requirements combine to determine the long-run investment incentives—not by design but by chance.

When the demand-side is fully functional, elasticity-based price spikes will correctly determine investment. Spikes will be low and broad and the market stable. Present policy should seek to mimic this long-run, fully-functioning, competitive behavior. Policy should not deliberately mimic the volatility of a market with a demand-side that is still 98% frozen.

---

3. The outcome is optimal provided VOLL is set correctly. Although this is extremely difficult to estimate, Section 2-5.4 shows that market efficiency is not too sensitive to misestimation and that there is no more accurate method available for determining the optimal amount of reliability.

4. The aggregate price spike is defined in Chapter 2-2, and is the upper part of the annual price-duration curve.

dollars per megawatt-hour, system operators understandably begin to have second thoughts.

As a result, all four ISOs have requested FERC's approval for "price caps." These are often, and more accurately, termed purchase-price caps. In fact they are simply limits on what the system operator will pay in order to comply with requirements for operating reserve. They are not price caps or price controls of the type implemented in other markets. They do not tell a private party that it cannot charge another private party more than $P_{cap}$. System operators are not allowed to make a profit, so they charge only as much as the power costs them, which is never more than their price limit. Because of the second demand-side flaw, when the system operator refrains from paying more than a certain price limit, no private party will pay more as it can always take power in real time without a contract.

## The Price-Cap Result

**Result    [2-4.6]**    **The Real-time Price Limit Effectively Caps the Entire Market**
If a system operator never pays more than $P_{cap}$ then it will never sell power for more. Because of the second demand-side flaw, customers of the system operator (typically, load serving entities) can always wait and purchase power in real time for $P_{cap}$ or less. Consequently, they will never pay more in any forward market, and all power prices are, in effect, capped at $P_{cap}$.

Purchase-price limits, openly implemented by the system operator, typically determine the highest market price for the real-time market and for all forward markets, but it is not unusual for the system operator to make "out-of-market purchases" at prices above the official "price cap." These prices are not determined by the market because the demand for out-of-market purchases is typically caused by operating reserve requirements.

While there are sound engineering reasons for operating reserve requirements, these reasons do not extend to price determination. In particular, they do not say operating reserves are worth $2,000/MWh or any similar value whenever the reserve requirement is not met. Any price paid "out of market" is set by some regulatory, nonmarket process no matter how informal that may be.

In conclusion, the limits on what a system operator will pay for power cannot be set by the market. They are set by complex processes that sometimes involve formal regulation and sometimes involve judgments by system operators. In any case, the high prices paid for reserves determine through arbitrage the prices in other markets (see Sections 1-8.3 and 2-4.5). Thus, system-operator policy determines the height of price spikes in all submarkets of a power market. Whether investment is induced purely by "market-driven" price spikes or by a "regulatory" ICap requirement, it is, in reality, primarily determined by a regulatory policy. Neither approach is more market-based than the other.

## The Regulatory-Price-Spike Result

**Result      [2-4.4]      Regulatory Policy Determines the Height and Duration of Price Spikes**
Because of the second demand-side flaw, markets will not set energy prices higher than the system operator's price limit. Consequently this regulatory limit determines the height of price spikes. The average annual duration of spikes is controlled by the application of this limit to the operating reserve requirement.

The essential difference between the Australian and the United States approaches to the determination of price spikes is not that U.S. spikes are lower and of longer duration. The key difference is that Australia has determined the shape of its price spikes deliberately by calculating what is required to produce an installed-capacity level that will produce a target reliability level (some three to five hours per year of load shedding). The United States has made no such effort. In fact, price-spike *duration* is set by short-run engineering considerations, while the *height* of spikes is typically set by a political process that is concerned mostly with market power. Very roughly, NERC controls the price-spike duration, while FERC controls the height. But "NERC regional adequacy and operating reserve criteria do not consider costs" (Felder 2001) and FERC does not consider operating reserve limits when setting price caps. In addition, capacity requirements are often added to the market without any thought for how these mesh with the defacto price-spike policy. In short, Australia's approach is deliberate, while the U.S. approach is a matter of chance.

## 2-1.2   THE PROFIT FUNCTION

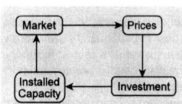

Profit drives the investment that is key to the market equilibrium described by Figure 2-1.1. The core of that equilibrium process is the loop of causal links from the market through prices, investment, installed capacity and back to the market. This circularity determines a long-run market equilibrium.[5] If prices are high, investment will be encouraged, and installed capacity will increase.

With more ICap, prices will fall. The system is in equilibrium if ICap, combined with exogenous factors, causes prices that are just profitable enough to cover the fixed costs of the installed capacity.

Energy and capacity prices both contribute to profit and thus help determine the equilibrium level of ICap. Because separate policies affect energy and capacity prices, these must be coordinated if the right level of ICap is to be induced. The intermediate goal of these policies is to produce the right profit level at the right level of installed capacity, and the combined effect of energy and capacity prices on profit is what matters. This combined effect is computed by finding the profit function produced by each policy and summing the two functions.

---

5. If prices are not capped and demand is too inelastic, there may be no equilibrium.

| Result | [2-8.1] | **Energy and Capacity Prices Together Induce Investment** |
|---|---|---|

Investment responds to expected short-run profits, which are determined by energy prices and (if there is an installed-capacity requirement) by capacity prices. Regulatory policies determining these prices need joint consideration.

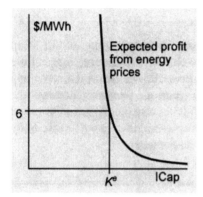

A profit function plots expected profits as a function of the ICap level. A low ICap level produces shortages and high energy prices which provide a high profit level. When ICap is high, shortages are very infrequent and prices are rarely high. This leads to a low expected annual profit. At some level in between, expected prices are just high enough to produce a profit level that would cover the fixed cost of a new peaker, about $6/MWh of installed peaker capacity.[6] This is the equilibrium level of ICap, $K^e$.

The figure at the left shows a profit function for a peaker derived from energy prices. As ICap drops below the equilibrium level, profits increase rapidly, and as it rises above the equilibrium level, shortages of capacity become rare and profits rapidly fall to zero. In most markets they fall to zero much more quickly than shown. The precise shape of this curve depends on reliability policy, on the nature of demand fluctuations, the frequency of generation outages and on demand elasticity. The important points are that the curve depends on regulatory policy and the curve can be calculated.

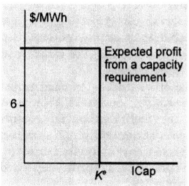

An installed capacity requirement produces a very different profit function. When ICap is below the required level, ICap-profits for all generators are set by the ICap-requirement penalty, and when ICap is above the required level, profits are zero. (A small correction should be made for market power.) Adding profit from the energy market to profit from the capacity market produces a total profit function from which an equilibrium value of ICap can be determined.

The equilibrium value of ICap, $K^e$, is not automatically the optimal value. Policy must be adjusted so that the profit function of a peaker crosses the profit level (approximately $6/MWh) required to cover the fixed costs of a peaker at the optimal ICap level. Any profit function that does this, no matter what its shape, will induce the right level of ICap. Thus there are many "optimal" policies to choose from. Available policy parameters include the system operator's price limit when load is shed, the price limit when **operating reserve requirements** are not met, the required level of operating reserves (which can be increased), the required level of installed capacity, and the penalty for being short of capacity.

---

6. See Chapter 1-3 for an explanation of units. $/MWh can be converted to $/kW-year by multiplying by 8760/1000.

| **Result** | **[2-6.1]** | **Many Different Price Limits Can Induce Optimal Investment** |
|---|---|---|

If the system operator pays only up to $P_{cap}$ but no more any time operating reserves are below the required level, a low value of $P_{cap}$ will suffice to induce the optimal level of installed capacity. The higher the reserve requirement, the lower the optimal price limit will be. Capacity requirements can further reduce the optimal level of $P_{cap}$.

One other policy choice deserves attention. The engineering notion of an absolute reserve requirement, which worked well under regulation, makes little sense in a market. The demand for operating reserves, like the demand for anything else, should be described by a downward-sloping demand function. The fewer the megawatts of reserves, the more they are worth. Introducing such a demand function increases policy choices and has the added advantage of increasing the elasticity of demand in a market where even small increases matter.

## 2-1.3   SIDE EFFECTS OF RELIABILITY POLICY

Many policies will produce a profit function that determines the correct equilibrium level of installed capacity, but this does not mean they are all equally desirable. The first goal is the correct average level of installed capacity, but that is not the only criterion for a well-functioning power market. If ICap is right on average but fluctuates too dramatically, the excess *un*reliability in years of low ICap will more than offset the excess reliability in years of high ICap. In fact any value of ICap other than the optimal value causes a reduction in net benefit, so no matter what the average value, fluctuations produce a sub-optimal result.

But two phenomena other than reliability deserve attention—risk and market power. Even if the ICap were exactly right every year, so that reliability was optimally maintained, profits would fluctuate. The profit function only records expected, or average, profits. Actual profits can vary dramatically. A 2% increase in demand is equivalent to a 2% reduction in ICap in terms of its effect on profit. So if the profit function can be very steep at the equilibrium, any small unexpected increase in demand will cause a large increase in profits.

A VOLL pricing policy is designed to produce about three hours of extremely high prices (equal to $V_{LL}$) every year. But some years, unusual weather, generator outages, or unexpected demand growth will cause 10 or 20 hours of VOLL pricing. If ICap has been set correctly, then profits must average out correctly which implies there must be five years of zero profits for every year with 18 hours of high prices. Ten years of zero short-run profits—losses, from a business perspective—would not be out of the question. Investors will consider such a market very risky and will demand a risk premium for investing in it.[7] Moreover, because building power plants takes years, 2 high-profit years might well occur consecutively. In this case,

---

7. Long-term contracts between new generation and load can reduce this risk premium, but, so far, long-term contracting has meant only 2 or 3 years in most cases and the equilibrium level of contract cover is still unknown.

there may be political repercussions from the observation of high profits coupled with the pain of high prices.

A profit function that is extremely steep also encourages the exercise of market power. Withholding 2% has the same impact on profits as a 2% reduction in ICap. So profit functions that are very steep reward withholding most handsomely. Again, the VOLL pricing policy fits this description.

A flatter (less steep) profit function can be obtained by designing price spikes that are lower and of longer duration. Instead of $15,000/MWh for 4 hours per year, one can design price spikes that reach only $500/MWh for 120 hours per year. This would be done by setting a purchase price limit for the system operator of $500 and requiring that it be paid whenever the system was short of operating reserves (instead of waiting until an actual blackout). If this still does not produce high prices for 120 hours per year, the operating reserve requirement could be increased.

In addition, part of the fixed costs can be recovered from the capacity market if the correct capacity requirement and the penalty are implemented. If half the fixed cost is recovered in the capacity market, the height of price spikes can be cut in half for any given duration of high prices. Some policy combinations produce flatter profit functions and these reduce risk and the exercise of market power. If properly designed, these policies will cause no reduction in the equilibrium level of installed capacity. Consequently, it is the side effects of reliability policy, risk and market power, which should determine the structure of the policy and thus the shape of the profit function.

---

**Result**   **[2-6.4]**   **Reliability Policy Should Consider Risk and Market Power**
The right average installed capacity will provide adequate reliability, but two side effects of reliability policy should also be considered. Infrequent high price spikes increase uncertainty and risk for investors. This raises the cost of capital and, in extreme cases, causes political repercussions. The possibility of extremely high prices also facilitates the exercise of market power.

---

## 2-1.4   INTER-SYSTEM COMPETITION

Consider an isolated competitive market in which $200/MWh is sufficient to call forth all available operating reserves. Suppose $500/MWh is sufficient to purchase all power available from these sources during an emergency. As these prices are too low to stimulate the appropriate level of investment given most operating reserve policies, the system operator has imposed an ICap requirement and enforced it with a moderate penalty. This provides capacity owners with an additional source of profits, which is just sufficient to induce optimal investment.

This policy provides a relatively smooth profit function which reduces risk for investors and discourages the exercise of market power. While engineers may consider $200/MWh an inadequate effort to "buy reserves at any cost," the fact remains that a higher price would procure no more in the short run. Even a supplier with market power would not withhold reserves in the face of a firm price limit, because there would be no possibility of profiting from a price increase. In the short run, paying more would not increase reliability, and, in the long run, it would only serve to increase it above the optimal level.

If this isolated market were connected to another, identical, market, these conclusions would change. Now both markets will find that they can increase short-run reliability by paying more for operating reserves. When reserves are short, say only 5% instead of the required 10%, either system can obtain a full complement of operating reserves by outbidding the other.

To be specific, assume each market has 10,000 MW of load and 10,500 MW of capacity, but system A has purchased 500 MW of operating reserves from system B as well as 500 MW of reserves from its own suppliers. System B now has zero operating reserves while system A has its required 10%. Suppose a generator selling 500 MW of power to market B experienced a **forced outage**. Market B will then be forced to shed 500 MW of load. A similar outage in Market A would cause only a reduction of operating reserves from 1000 to 500 MW.[8]

By bidding above market B's price limit and purchasing its reserves, market A has increased its reliability, something that was impossible when the two markets were isolated and not in competition. Of course, market B will soon realize what has happened and retaliate by bidding above A's price limit. The end result of such competition will be high limits on the price of operating reserves and energy. This will make the profit functions steep and will increase risk and facilitate the exercise of market power. This outcome was acknowledged by FERC in its July 2000 ruling approving identical price caps of $1,000 for the three Eastern ISOs. It correctly ruled against a $10,000 price cap for the NY ISO on the grounds that this would tend to undermine PJM's $1,000 price cap.

---

**Result**    **[2-9.1]**    **Competition Between System Operators Induces High Price Spikes**
If two markets with different price spike policies trade energy and operating reserves, the one with higher price spikes will gain in reliability and save money relative to the other. This will force the low-spike market to use higher price spikes. If regional price limits are not imposed, inter-system competition will lead to reliability policies with undesirable side effects.

---

## 2-1.5    DEMAND-SIDE EFFECTS OF PRICE LIMITS

Low, long-duration price spikes can induce the same optimal level of installed capacity as high narrow price spikes, but the side effects of risk and market power will be reduced. Besides these supply-side effects which argue against high price spikes, there is a demand-side effect that argues for high price spikes. High prices encourage price responsiveness both directly and through learning. Some loads

---

8. It may be suggested that under some emergency agreement system A would allow system B to purchase power from system A, thereby avoiding load shedding by B and reducing A's reserves below its required level. But this implies that A's reserves are in effect also B's reserves. In this case, there is no reason for either to spend extra on reserves as long as it knows the other will purchase them and share them. This contradicts observed behavior of system operators who do pay extra to obtain their own reserves.

will not find it worth curtailing their use of power until prices exceed $1,000/MWh. But, once a load has learned to respond, it may find it sensible to respond to price increases that would not trigger a response before the necessary control equipment is installed and such responses become routine.

Although these effects are important, they do not negate the conclusions based on the supply-side effects, though they suggest a trade-off. The demand-side response obtained by utilizing extremely high price spikes with no warning time may be little more (or even less) than the demand response obtained by lower, more frequent price spikes. These may be more readily anticipated by day-ahead markets, thereby giving loads a day's lead time to plan their response and implement cost-mitigating procedures.

Another possibility is to set demand-side prices higher than supply prices during price spikes. This would require a balancing account to keep the two revenue streams equal in the long-run, but that is a simple matter.

A similar problem arises from the possibly high prices required to wring the last MW of production out of generators. The price of power supplied by generators operating in their emergency operating range could be exempted from the normal energy price limit. Suppose the normal energy price limit were $500/MWh, and the maximum output at this price from generator G were 400 MW over the previous year. Then any output above 400 MW could be subject to a price limit of $2,000/MWh, without significantly steepening the profit curve.

Once the ideas of calculating profit curves and considering the side effects of their shape have been understood, many design possibilities can be considered and evaluated. In the long run, when demand has become quite elastic even in the extremely short timeframes needed to replace spinning reserves, these design considerations will no longer be needed. Before that time, much can be gained by taking full advantage of the multitude of policies compatible with an optimal average level of installed capacity. A pricing policy should be selected from among these based on its ability to minimize the collateral damage from risk and market power.

# Price Spikes Recover Fixed Costs

*Indebtedness to oxygen*
  *The chemist may repay*
*But not the obligation*
  *To electricity.*

Emily Dickinson
(1830–86)
The Farthest Thunder That I Heard

**W**HAT PRICES WILL COVER A GENERATOR'S FIXED COSTS WITHOUT OVERCHARGING CONSUMERS? Short-run competitive prices perform this service and, in addition, induce the right level of investment in every type of generation technology. This does not solve the long-run problems of power markets because demand-side flaws prevent contemporary markets from determining competitive prices.[1] Subsequent chapters will discuss regulatory policies that compensate for these flaws, while this chapter focuses on how competitive power markets will work once the flaws are eliminated.

Although not completely accurate, competitive analysis is useful for dispelling two fallacies concerning fixed costs. The first asserts that prices equal to marginal cost (competitive prices) cannot cover fixed costs. The second asserts that although they can, they will do so only when the market is seriously short of capacity.

**Chapter Summary 2-2:** Short-run competitive prices would recover fixed costs for peakers and baseload plants alike. They would stimulate investment in the mix of technologies that produces the required power at least cost. Demand will occasionally push prices well above the *average* marginal cost of any supplier but not above the right-hand marginal cost. These price spikes can be summarized with a price-duration curve that facilitates computation of the competitive outcome.

**Section 1: The Fixed-Cost Fallacy.** This fallacy asserts that short-run competitive prices (marginal-cost prices) will prevent generators from recovering their fixed costs. These are covered *not* because short-run competitive forces set price equal to marginal cost, but because, if they were not covered, investors would stop building plants while demand continued to grow. This would cause shortages and

---

1. Power markets are termed "competitive" if their supply side is competitive even when demand-side flaws prevent the market from determining competitive prices.

higher prices. The long-run competitive forces ensure price spikes will be high enough to cover fixed costs. This dynamic is very robust and works even with a price cap. Once a lack of investment causes shortages, the lower price spikes will last long enough to cover fixed costs.

**Section 2:  Optimal Price Spikes for Peakers.**   The weak version of the Fixed-Cost Fallacy asserts that marginal-cost prices will be high enough to cover fixed costs only because of a shortage of installed capacity. A model that includes fluctuating demand, two kinds of generators, and marginal-cost prices shows that investors will minimize the total cost of production by building the right amount and right mix of generation. Individual price spikes can be summarized by the price duration curve. The portion of this curve that lies above the variable cost of a peaker is called the *aggregate price spike*, and it measures the annual scarcity rents received by peakers.

**Section 3:  The Lumpiness of Fixed Costs.**  Fixed costs would not be "lumpy" if generators could be scaled down as much as desired, but a 10-MW coal plant is not 50 times cheaper to build than a 500-MW coal plant. In effect, generators come in discrete sizes or "lumps." When a fractional number of plants is optimal, lumpiness interferes with the efficiency of competition. This inefficiency declines in proportion to the square of the number of plants in the market with the result that it is entirely negligible for markets of even modest proportions.

## 2-2.1   THE FIXED-COST FALLACY

A common fallacy asserts that if a generator always prices output at marginal cost it will fail to cover its fixed costs. This notion is based on the true observation that in the normal output range, marginal cost is always less than the long-run average cost of generation including the cost of the plant's capital investment. Section 1-5.3 refutes the fallacy in a general context, and the argument presented there applies without modification to the power industry. But the fallacy causes enough confusion to warrant restating the analysis in the context of the wholesale power market.

The Fixed-Cost Fallacy should not be confused with the true proposition that when generating capacity is excessive, competition will cause generators to fail to cover their fixed costs. This is not a problem with competition or with electricity markets. This is the way Adam Smith's "invisible hand" signals that there should be no new investment.

**The Fixed-Cost Fallacy**

| *Fallacy*  2-2.1 | **Marginal-Cost Prices Will Not Cover Fixed Cost** |
| --- | --- |
| | If the price paid to generators always equals their (physical) marginal cost, they will fail to cover their fixed costs. |

**Figure 2-2.1**
Continuous marginal
costs and a nearly flat
(elastic) demand curve
for one supplier.

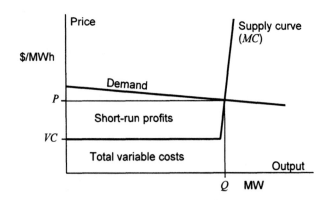

**Modeling a Generator**   Generators will be assumed to have continuous supply
curves that are extremely steep in the final emergency operating range which starts
at their "maximum," or "full," output level (see Figure 2-2.1). This model produces
the same results as one using a truly vertical supply curve but has the advantage
of being more realistic with a well-defined marginal cost at all levels of supply.
(Readers unfamiliar with the treatment of marginal cost when demand intersects
a vertical supply curve should review Chapter 1-6.) The constant value of marginal
cost to the left of full output will be referred to as the generator's variable cost.
Startup costs will be ignored throughout Part 2.

   The demand curve faced by a single supplier in a competitive market is simply
a horizontal line at the market price. At any price above the line, demand for a
supplier's output is nil, while at any below-market price demand is much more than
the supplier's maximum possible output. Such a demand curve is shown in
Figure 2-2.1, but a small slope is added for realism. In most markets the price an
individual supplier can charge is at least a little sensitive to its output.

**How Generators Earn More than Variable Costs.**   Because supply equals
demand, the supply in Figure 2-2.1 is $Q$, and at that point marginal cost has risen
to equal price, $P$. But the first MWh produced, and each succeeding MWh up to
"maximum" capacity, has a cost of less than half that much. As a consequence,
average variable cost is about half of $P$.

   Figure 2-2.1 also shows that revenue, $R$, is exactly $P \times Q$. Because revenue is
greater than total variable cost, $TVC$, the generator has money left over to cover
fixed costs. This is termed *short-run profit*, *scarcity rent*, or more technically,
*inframarginal rent*. (Section 1-6.1 discusses scarcity rent.)[2]

$$\text{Scarcity rent} = \text{short-run profit} = R - TVC$$

   If the market price is high enough, the generator will cover its fixed costs, but
if the market price is too low, it will not. The real-world situation may seem more
complex because prices fluctuate, sometimes dramatically, but this changes little.
If the average price is high enough, the generator will cover its fixed costs, and,
if not, it won't. The only question is what determines the average market price.

---

2. Short-run profit is less than scarcity rent when startup and no-load costs are taken into account, but
those are ignored in Part 2.

**Why the Competitive Price Exactly Covers Fixed Costs.** If marginal-cost prices did not cover fixed costs, investors would choose to build no more generators. As demand grew and generators wore out, the market would tighten causing price to rise. On the other hand, if marginal-cost prices more than covered fixed costs (which includes a normal risk-adjusted rate-of-return on capital) investors would build generation, supply would outstrip demand, and the price would fall. When the market price more than covers fixed costs, price declines; when price fails to cover fixed costs, it increases. Consequently, price converges toward the point at which fixed costs are exactly covered. Once there, it has no inherent tendency to change.

Frequent external disturbances, such as changes in demand or the opening of a new plant, push the market out of equilibrium. But investors make their best estimates of what will be needed, and they err on the high side as often as they err on the low side. Although marginal-cost prices do not exactly cover fixed costs at all times, they cover them on average. This is all that can be expected and all that is needed.[3]

| | | |
|---|---|---|
| **Result** | **2-2.1** | **In the Long-Run, Suppliers Recover Their Fixed Costs** |

In a long-run competitive equilibrium, generators recover their fixed cost and no more, even though price equals physical marginal cost ($P = MC$) at all times and for all generators. Revenues that help cover fixed cost are call short-run profits.

| | | |
|---|---|---|
| **Result** | **2-2.1b** | **Restatement for Supply Curves with Vertical Segments** |

In a long-run competitive equilibrium, generators recover their fixed cost, even though the market price is competitive and satisfies $MC_{LH} \leq P \leq MC_{RH}$ at all times and for all generators (see Section 1-6.3).

This discussion of fixed-cost recovery does not depend on any of the details of cost functions or even on the market being short-run competitive. It depends only on the ability of generators to enter and leave the market. Consequently, the result holds automatically for all of the upcoming models and for all of their generalizations to more complex cost structures. It only fails if there are barriers to entry. Then the generators recover more than their fixed costs; they recover enough to pay for crossing the barrier.

## 2-2.2  OPTIMAL PRICE SPIKES FOR PEAKERS

The previous section showed that competitive prices will cover fixed cost, but that does not prove them optimal. A weaker version of the Fixed-Cost Fallacy admits that long-run forces will ensure that fixed costs are covered, but it claims this would lead to a serious shortfall in generation capacity under competitive pricing. This

---

3. Chapter 2-6 considers the problem of excessive fluctuations in the short-run profit level and the level of investment. Large enough fluctuations will cause significant inefficiencies even if capacity is right on average. Investors can also over- or under-collect total fixed cost for a project if there are technological surprises after the plant is built; but on average, fixed costs will be recovered.

**Figure 2-2.2**

Two technology screening curves showing capacity factors for which each technology is optimal.

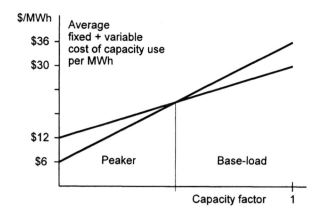

section models a market with two kinds of generators and fluctuating demand. It shows that price spikes caused by short-run market clearing induce the right level of investment for both technologies even though they face the same market-clearing prices. This result extends to any number of technologies.

## The Two-Technology Model

**The Supply Side.**  The supply-side of this market is modeled as having two available technologies, base-load generators and peakers with costs as shown in Table 2-2.1.

**Table 2-2.1**  Costs of Available Technologies

| Technology | Fixed Cost per MWh | Variable Cost per MWh |
|---|---|---|
| Peaker | $6 | $30 |
| Baseload | $12 | $18 |

Note that fixed costs have been expressed in the same units as variable costs.[4] As explained in Chapter 1-3, this is necessary for constructing the screening curves which are presented in Figure 2-2.2.

Screening curves, described in Sections 1-3.3 and 1-4.2, plot the total cost of generation as a function of the generator's capacity factor, its average output divided by its maximum output. (This is not the average cost of the energy produced, but the average cost of the capacity used.) If the generator produces nothing, then total costs equal fixed costs, which means the screening curve starts out at the level of fixed costs. As the capacity factor of a generator increases, the total variable cost of output increases proportionally and is added to fixed costs to produce the screening curve, a linearly increasing total cost curve. The plotted cost is the average hourly cost of a MW of capacity utilized at the indicated capacity factor. These curves can be used to decide which generator is the cheapest source of power to serve load for any given duration because serving a load of duration $D$ results in

---

4. According to Table 2-2.1, if a 5-MW base-load plant is used for 100 hours to produce 200 MWh of energy, the total cost in dollars will be $5 \times 100 \times 12 + 200 \times 18$.

**Figure 2-2.3**

Supply and demand for
the two-technology
model.

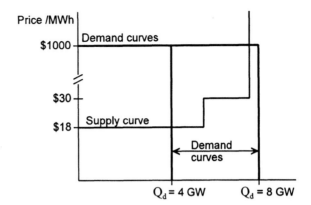

a capacity factor, *cf*, equal to *D*. For example, in Figure 2-2.2, the peaker technology is cheaper for all loads with duration less than 0.5 (4380 hour/year), while the base-load technology is cheaper for all loads with durations greater than 0.5.

**The Demand Side.** Two types of information concerning demand are needed: load duration and price elasticity (see section 1-4.1). Figure 2-2.3 illustrates a demand curve that shifts back and forth daily relative to one possible supply curve. Demand shifts in a linear manner that makes it equally likely to be found at any level from 4000 MW to 8000 MW, assuming it is not curtailed by a high price.

The load-duration curve of Figure 2-2.4 depicts the interaction of supply and demand. The flat spot at the top of this curve occurs if available capacity is less than 8000 MW, which will prove to be the case in equilibrium. Once generating capacity is exhausted, demand is limited by high prices. This is possible because the present chapter assumes that demand is sufficiently elastic to clear the market at all times as is required for a competitive market.

The flatness of the demand curve at $1,000/MWh is unrealistic, as a real demand curve would be downward sloping. It could be interpreted as a price-capped market, but, in this model, it is used only as a simplification that has no impact on the qualitative results. The horizontal portion should be interpreted as demand elasticity that is reflective of the value of power to consumers. They simply do not want any power if it costs more than $1,000/MWh.

## The Regulatory Solution

Faced with this market, the traditional regulatory solution would be to set price to average cost, build enough generation capacity to cover the full 8000 MW of peak demand, and use the screening curves to determine that 6000 MW should be base-load capacity and 2000 MW should be peak capacity. This last conclusion is determined by reading the screening curves to find that the trade-off point is at a duration of 0.5 and then reading the load-duration curve to find that, at this duration, load is 6000 MW.

The regulatory solution finds the optimal level of baseload capacity but sets too high a level for peaker capacity. Regulators have traditionally set price at average cost and this has prevented high prices from dampening demand when

**Figure 2-2.4**

The load-duration curve
flattened by high prices
when load is limited.

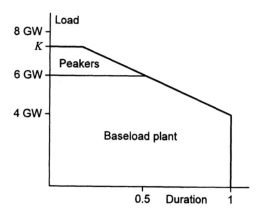

power was scarce.[5] Consequently, peak demand is higher than is socially optimal. If the costs of peak usage were paid for entirely by the users of peak power, consumers would consume less on peak. Because regulators use average-cost pricing, customers pay more at all other times and more in total than they would pay if given the relevant choices.

## The Optimal Solution

The optimal solution is the same as the regulatory solution except that it takes into account the high cost of serving peak load and the willingness to pay for this service. The optimal system will spend some time with load exactly equal to generation capacity.[6]

The duration of the flat load peak will be denoted by $D_{PS}$ because at these times there is a price spike with price greater than the variable cost of a peaker. If another megawatt of peak capacity were added it would produce power for only a fraction, $D_{PS}$, of the year. The average cost of this energy, $AC_E$, including the fixed cost of capacity used to produce it is

$$\text{Average cost of peak energy} = AC_E = (\$6/D_{PS} + \$30)/\text{MWh}.$$

(See Equation 1-3.2.) Note that this is not the average cost per MW of using peak capacity, which is plotted in a screening curve, but the average cost per MWh of energy produced. The shift in approach is necessary because a consumer's value is given in $/MWh of energy consumed. It only makes sense to add a megawatt of peaker capacity if the energy it produces costs no more than its value to consumers. According to the assumed demand curve, the value of power to consumers is $1,000/MWh, so the condition for optimal peaker capacity is

$$6/D_{PS} + 30 = 1000.^{[7]}$$

---

5. Regulation has not set $K$ high enough to provide perfect reliability, but that is a different point and the focus of later chapters.

6. To include reserve operating margins in this analysis, it would be necessary to specify a demand curve for operating reserves. This is discussed in subsequent chapters.

7. If the market price were not constant during times when all operable capacity is in use, the right side of this equation would be the average market price during these times.

**Figure 2-2.5**

The aggregate price spike
of a typical
price-duration curve.

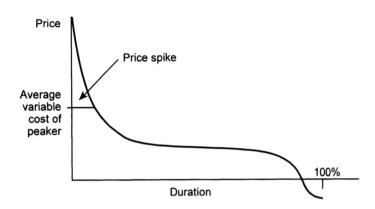

This equation can be solved to find that $D_{PS}$ equals $6/970$, or $0.62\%$ of a year, or 54 hours per year. If a peaker is needed less than 54 hours per year, it is not worth buying. Because the load-duration curve is linear, a duration of 0.62 corresponds to a load that is just 25 MW below the potential peak of 8000 MW. The flat peak is at 7975 MW. Total generating capacity should equal this peak load. Although consumers would be willing to pay up to $1,000/MW hour to avoid the curtailments implied by this shortage of capacity, this would not cover the (mainly fixed) cost of providing the extra capacity. Because baseload capacity has already been determined to be optimal at 6000 MW, the optimal peaker capacity is 1975 MW.

## The (Aggregate) Price Spike

Finding the competitive solution requires using the long-run equilibrium condition that says prices will be at the level that just covers fixed costs. Whether the price is too high or too low cannot be determined from the average price because the width and height of price spikes matter. The individual price spikes are most conveniently summarized by the price-duration curve (Figure 2-2.5), which is constructed in the same way as a load-duration curve (Section 1-4.1) but by using hourly prices instead of hourly loads. As before, duration is measured as a fraction between zero and one which can be interpreted as the probability of finding a price at or above a certain level.

A generator recovers fixed cost only when price is above its variable cost. All such prices during the year are summarized by the part of the price duration curve above this cost. The higher the price, the more fixed cost is recovered, so the area under the price duration curve and above the variable cost line gives the average rate of fixed cost recovery for the year. For example, if the peak of the price-duration curve were $1,050/MWh, and it dropped linearly to a generator's variable cost of $50/MWh at a duration of 5%, its average fixed cost recovery for the year would be $1,000 \times 0.05/2 = \$25$/MWh, a generous sum.

Peakers are often used as a litmus test for whether a market is providing sufficient short-run profit to induce investment, but this is not necessary. Any commercially viable generation type would serve equally well when the market is in equilibrium, but this book will follow the convention as it causes no harm. The part of the price-duration curve above the variable cost of a peaker will be termed the **aggregate price spike** and is often shortened to "the price spike." In 1999, PJM estimated this revenue to be approximately $60,000/MWy or $6.85/MWh, and, in 2000, estimated it at $3.40/MWh (PJM 2001, 13–14).

| | |
|---|---|
| **Definitions** | **The (Aggregate) Price Spike**<br>The (aggregate) price spike is the section of the market's price-duration curve above the average variable cost of the most expensive-to-run, but still investment-worthy, peak-load generator.<br>**The Price-Spike Revenue ($R_{\text{spike}}$)**<br>The area of the price spike is the price-spike revenue. It is the average hourly scarcity rent that would be earned by the most expensive investment-worthy peaker during the year in question. It can be measured in $/MWh. |

## The Competitive Solution

Determining what generation capacities would be induced by a perfectly competitive market requires the use of Result 2-2.1. Fixed costs are exactly covered in a long-run competitive equilibrium. This means short-run profits (scarcity rents) must exactly equal fixed costs. The trick is to find the short-run profits because fixed costs are known from Table 2-2.1.

As shown in Figure 2-2.3, demand can fall into three different regions classified by which plants are marginal. As Figure 2-2.6 shows, if base-load is marginal the market price is $18; if peakers are marginal the price is $30/MWh, and if the marginal unit of load cannot be served, the price rises to the cap at $1,000/MWh. Figure 2-2.6 shows the price-duration curve for this model. The price spike is the shaded area above $30/MWh, and this provides scarcity rent for both generators. It is the only scarcity rent earned by peakers, so their equilibrium condition is: the fixed cost of a peaker equals the price-spike revenue, $R_{\text{spike}}$.

Baseload plants have two sources of scarcity rent, the price spike and the shaded rectangle between $18 and $30/MWh, which has a duration of $D_{\text{peaker}}$, the duration of time when peakers are running. This observation provides the equilibrium condition for base-load plants, which is given in the Result 2-2.2.

| | |
|---|---|
| **Result 2-2.2** | **Long-Run Equilibrium Conditions for Two Technologies**<br>In the long run, peakers and baseload plants must cover their fixed costs from short-run profits (inframarginal or scarcity rents). This implies two equilibrium conditions:<br>$$FC_{\text{peak}} = R_{\text{spike}}$$<br>$$FC_{\text{base}} = FC_{\text{peak}} + (VC_{\text{peak}} - VC_{\text{base}}) \times D^*_{\text{peaker}}$$ |

The peaker's equilibrium condition implies

$$\$6 = \$970 \times D^*_{PS},$$

**Figure 2-2.6**

The price-duration curve
of the two-technology
example.

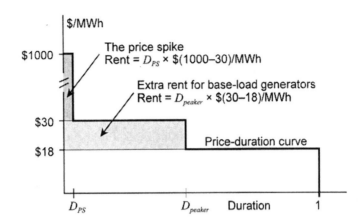

where $D_{PS}$ is the duration during which the price is at the cap. The base-load
equilibrium condition implies

$$\$12 \ = \$6 + \$(30 - 18) \times D^*_{peaker}$$

These two equilibrium conditions can be solved for the optimal durations and
$D^*_{peaker}$, which completely determine the price duration curve and can be used with
the load-duration curve to determine the equilibrium capacities of the two technologies. The solution is $D_{PS} = 0.62\%$, $D_{peaker} = 50\%$, base-load capacity = 6 GW, and
peaker capacity = 1975 MW. These are exactly the values found for the optimal
solution. Consequently, a competitive market induces exactly the optimal levels
of capacity for both technologies.

### The Weak Fixed-Cost Fallacy

*Fallacy*   **2-2.2**      **Marginal-Cost Pricing Causes a Capacity Shortage**
If generators are paid marginal-cost prices, they will cover their fixed costs only
as the result of a serious and socially detrimental shortfall in generation capacity.

**Result      2-2.3**      **Marginal-Cost Prices Induce the Optimal Mix of Technologies**
Short-run competitive prices, which equal marginal costs, provide incentives for
investment in generation technology, which lead to an optimal level and an
optimal mix of generation technologies.

## 2-2.3   THE LUMPINESS OF FIXED COSTS

Poorly behaved costs can interfere with the competitive process and invalidate the
Efficient-Competition Result (Section 1-5.1). Natural monopoly conditions,
discussed in Section 1-1.1, were the first example. This condition is sometimes
confused with the existence of fixed costs because these cause the average production cost of a given generator to decline as average output increases. If an industry's

average cost declines as total output increases, it is a natural monopoly. A generator's average cost always declines until its output nearly reaches its capacity limit. As long as efficiently sized generators have capacities much smaller than the size of the industry, their individual declining costs will not cause the natural-monopoly condition.

**The "lumpiness" problem.** The consideration of natural monopoly leads to the question of how small the efficient-sized generator would need to be relative to the size of the market to avoid a problem of inefficiency. For a true natural monopoly, one supplier must be able to produce the industry's efficient level of output more cheaply than two or more suppliers. But what if three can produce it more cheaply than two but cannot stay in business with competitive prices because they cannot cover fixed costs? This is sometimes called the integer problem because it would disappear if fractional suppliers were possible. Sometimes it is called a lumpiness problem because supply comes in discrete lumps which may not match the level of demand.

Suppose generators could be built only with a capacity of 1000 MW and that demand had an inelastic peak at a level of 2200 MW. Three plants would be needed to satisfy demand, and failing to build three would cause costly shortages. With three plants, excess capacity would cause the competitive price to equal variable cost and all generators would fail to cover their fixed costs. This would lead to a duopoly not because two plants are cheaper, but because competitive prices fail to cover fixed costs for the optimal number of plants—three.

This impact of lumpiness is dramatic, but it occurs in an unrealistically small market. Typical markets have at least 100 generators, and most markets in the United States have thousands, including those that can participate via interconnections. To gain a more realistic view of the problem, consider the example of Section 2-2.2, with the added stipulation that base-load plants always have a capacity of 1000 MW while peakers always have a capacity of 100 MW. That market will have had a peak demand of 7975 MW of which 6000 should be served by six baseload plants and 1975 should (ideally) be served by 19.75 peakers.

**Lumpiness of base-load plants.** First consider the unlikely possibility that lumpiness problems caused one too many base-load plants to be built. The seventh base-load plant would serve a load slice with an average duration of 43.75% at a cost of

$$1000 \text{ MW} \times (\$12/\text{MWh} + 0.4375 \times \$18/\text{MWh})$$

which comes to $19,875/h. Serving this same load with peakers would cost only

$$1000 \text{ MW} \times (\$6/\text{MWh} + 0.4375 \times \$30/\text{MWh})$$

This comes to $19,125/h. The extra cost of $750/h is less than 0.4% of the total cost of serving the market (approximately $200,000/h).[8] Inefficiency from this type of lumpiness declines in proportion to the square of the number of generators in

---

8. The effect is small because the total cost function is a smooth function of the number of base-load generators. Near its minimum value, it is essentially flat and a deviation from the optimal generation set produces only a second-order effect on cost.

the market. For a real market with hundreds of generators, the inefficiency would be many times smaller.

**Lumpiness of peak-load plants.**  A realistic analysis of the peaker situation will need to take account of the randomness in real markets. Building 19 peakers would result in supply being short of demand for a duration 4 times greater than is optimal (2.5% instead of 0.62%), and this would cause peakers to over-recover fixed costs by a factor of four. These high profits would entice the entry of another peaker which would drop short-run profits to zero. This would stop entry as demand grew. With free entry and the uncertainties of a real market, short-run profits would average out to the level of fixed costs.[9] But lumpiness would prevent the right (fractional) number for peakers from being built, and this would cause some inefficiency.

Given the integer constraint, 20 peakers should be built, but for short-run profits to equal fixed cost, there must be only 19 peakers in the market 25% of the time. With only 19 peakers, some load will be unserved, relative to variable cost, for a duration of 2.5%, and the amount unserved will average 50 MW. At the value of $1,000/MWh indicated by the demand function, the average value of unserved load comes to

$$\text{Value of unserved load} = \$(0.025 \times 50 \times 1000)/h$$

which is $1250/h. This loss is offset by the savings of not building the peaker, which is $600/h and the saving of not producing for the load peak, which is only $37.50/h. The net loss is about $600/h, but this occurs only for the 25% of the time when capacity is under-built. The average loss from the lumpiness of peakers is about $150/MWh or less than $\frac{1}{10}$ of 1% of the total cost of power.

---

**Result**   **2-2.4**   **Inefficiency Caused by the Lumpiness of Generators Is Negligible**
In a power market with more that 4 GW of peak load and a 400 MW connection to a larger market, inefficiency caused by the lumpiness of the generators is well below 1% of retail costs. This inefficiency declines in proportion to the size of the market.

---

**A realistic assessment of lumpiness.**  The base-load calculation overestimated the lumpiness problem by assuming that the market would *always* build the wrong number of base-load plants. The peaker calculation overestimated by assuming that load duration is far more sensitive to total generating capacity than is realistic. (Real load-duration curves have a narrow spike, not a shallow linear slope.) Many other factors enter the picture and, in general, the more flexibility that is added in other dimensions, such as demand elasticity or the availability of smaller plants, the smaller the problem becomes.

A more intuitive approach is to ask how much would be gained if fractional plants could be built. If the Eastern Interconnection could add 0.75 more peakers instead of either zero or one, would this allow a finer tuning of installed capacity

---

9. This cannot be seen from a deterministic analysis, but with the uncertainties of real markets and the difficulties of coordination between competitors, it is impossible to maintain an excess-profit equilibrium.

and a more efficient market? No one believes that optimal capacity can be estimated with even $\frac{1}{100}$ of this precision. Transmission constraints may exacerbate the problem a bit, but the flexibility to build fractional plants would make a detectable difference only in the smallest markets.

In conclusion, competitive prices would cover fixed costs and induce the optimal mix of generation technologies with a great deal of precision. The problems of power markets do not arise from fixed costs but from the difficulty in achieving competitive prices in the face of demand-side flaws.

*Chapter* *2-3*
# Reliability and Generation

*Edison's design was a brilliant adaptation of the simple electrical circuit: the electric company sends electricity through a wire to a customer, then immediately gets the electricity back through another wire, then (this is the brilliant part) sends it right back to the customer again.*

Dave Barry

$\mathbf{R}$ELIABILITY IS AT THE HEART OF EVERY DEBATE ABOUT ENERGY PRICE SPIKES. If these are large enough, they induce the investment that provides the generating capacity necessary for a reliable system. In some markets capacity requirements also play an important role. A shortage of installed generating capacity is not the only cause of unreliable operation, but because it is the one most directly related to the operation of the wholesale markets, it is the only one considered in Part 2.[1]

Two aspects of reliability are always contrasted. *Security* is the system's ability to withstand sudden disturbances, while *adequacy* is the property of having enough capacity to remain secure almost all of the time. Part 2 focuses on adequacy and assumes that security requirements will be met if the system has adequate planning reserves. Requirements for operating reserves, which are intended to provide *security*, are of interest here mostly because of their role in raising price, stimulating investment, and thereby contributing to *adequacy*. This role is often overlooked because, under regulation, these requirements were unrelated to adequacy.

**Chapter Summary 2-3:** Operating reserves can be purchased directly by the system operator, but the market must be induced to provide adequate planning reserves. The first step in analyzing the market's effectiveness is to find what determines the optimal level of installed capacity. Under a simple but useful model of reliability, installed capacity is found to be optimal when the duration of load shedding is given by the fixed cost of a peaker divided by the value of lost load.

**Section 1: Operating Reserves and Contingencies.** A contingency is a possible or actual breakdown of some physical component of the power system. Typically, some operating generator becomes unavailable, leaving the system

---

1. Both distribution and transmission outages typically cause more loss of load.

unbalanced with demand greater than supply. System frequency and voltage begin to drop. If sufficient operating reserves are available so lost power can be replaced within 5 to 10 minutes, the system operator will usually not need to shed load.

**Section 2: Adequacy and Security.**   Reliability has two components. Adequacy is a matter of installed generating capacity and does not fluctuate from minute to minute. Security refers to the system's ability to withstand contingencies, and system security *can* change from minute to minute. Immediately after a contingency, and before operating reserves have been replenished, the system is much less secure. Though security depends on generation and transmission, only generation will be considered. A power system has adequate generation if it has enough to keep it secure in all but the most extraordinary circumstances.

**Section 3: The Simple Model of Reliability.**   Out-of-service generation may be thought of as a load on the system and added to actual load. If this "augmented" load is greater than installed capacity ($L_g > K$), then load must be shed. Other times, load is shed for reasons unrelated to $K$. The simple model assumes that, except for these unrelated reasons, loss of load is *exactly* $L_g - K$ or zero when this is negative.

Lost load is assumed to cost customers a constant amount per MWh lost. With this additional assumption, the optimal amount of installed capacity is the amount that makes the duration of load shedding equal to the fixed cost of a peaker divided by $V_{LL}$.

**Section 4: The Fundamental Reliability Question.**   The amount of lost load might be greater than the difference between $L_g$ and $K$. For instance, cascading failures might be triggered by a high value of $L_g - K$, and these cause much more than $L_g - K$ to be shed. The extent of the discrepancy between the Simple Model of Reliability and reality is an engineering question that deserves attention.

## 2-3.1   OPERATING RESERVES AND CONTINGENCIES

Power systems experience frequent disturbances such as short circuits or loss of generators or transmission lines. The possibilities of such events are called **contingencies,** and often the term is misapplied to the events themselves. Such events reduce the amount of generation available to serve load, and for present purposes it will be sufficient to imagine a disturbance as a generator experiencing a **forced outage**.

When this happens, frequency and voltage immediately begin to drop. If they drop too far for too long, load must be shed to rebalance demand with the diminished supply. In the first instant after a forced outage all remaining generators increase their outputs with the extra power coming from the kinetic energy of their rotation, not from their fuel source. As this energy is depleted, they slow down and the system frequency drops as does the voltage. These changes cause loads to draw less power, and the system comes into balance at a lower frequency. This takes only a matter of seconds. The decline in system frequency is immediately detected by frequency censors, and the ones with excess capacity (those that provide regulation and spinning reserve) begin to ramp up.

If there are *not* enough spinning reserve generators to ramp up and restore system frequency in the required 5 to 10 minutes, then the system operator is required to shed load. Thus spinning reserve and load shedding are substitutes for the service of restoring system frequency. (**Shedding load** will always mean the involuntary disconnection of load.) Just how perfectly load-shedding can substitute for reserves is considered in Section 2-3.4.

Operating reserves include **spinning reserves** and several lower qualities of reserves. Typically, 10-minute spinning reserves are the first defense against a contingency. They are supplied by generators operating at less than full capacity. Because they are already synchronized with the AC system (spinning), they respond without delay, but their response is gradual. They can typically ramp up at some constant rate, such as 4 MW per minute. In this case they are given credit for supplying 40 MW of 10-minute spinning reserve because they can increase their output by that amount in 10 minutes. Other types of operating reserves are 10-minute and 30-minute nonspinning reserves. Generators that are off line but can be started quickly can supply nonspinning reserves.

## 2-3.2   ADEQUACY AND SECURITY

**Operating reserves** are required to maintain system **security** by handling short-term disturbances to the system. **Planning reserves** are required to maintain system **adequacy** by meeting annual demand peaks. These two types of reserve are considered the basic inputs to the generation side of system reliability. (The transmission side of reliability is not considered in Part 2). Although security and adequacy are distinct concepts, they are closely linked. A system with adequate capacity can maintain enough security to reduce periods of involuntary load shedding to 1 day in 10 years. A system that maintains security for all but one day in 10 years must have **adequate installed capacity**.[2]

| | |
|---|---|
| **Definitions** | **Security**<br>The ability of the electric system to withstand sudden disturbances such as electric short circuits or unanticipated loss of system elements (NERC,1996).<br><br>**Adequacy**<br>The ability of the electric system to supply the aggregate electrical demand and energy requirements of the customers at all times, taking into account scheduled and reasonably expected unscheduled outages of system elements (NERC,1996). |

The two concepts are not simply different views of the same problem. With an inappropriate policy on operating reserves, the system will have insufficient

---

2. For an economically informed discussion of reliability, see Felder (2001).

security in spite of adequate capacity. Nonetheless, it is relatively cheap to maintain sufficient operating reserves in an adequate system, and it is relatively expensive to provide for system adequacy.[3] Adequacy is the crucial economic problem, while the problem of security is economically secondary yet complex.

To focus on the problems of investment in generation and system adequacy, Part 2 assumes that the operating reserve policy is always effective in providing as much security as possible (up to the normal standard) given the system's installed capacity. There are many such policies, but all must provide a certain minimum level of reserves. The Simple Model of Reliability, described in the next section, assumes that load will never be shed unless it exceeds the amount of operable generation capacity. Section 2-3.4 considers possible exceptions to this assumption.

## 2-3.3   THE SIMPLE MODEL OF RELIABILITY

Part 2 will rely on the Simple Model of Reliability which assumes a well-defined level of installed generation capacity, $K$, and generation outages, $g$, both measured in megawatts. **Load,** $L$, is defined as the economic demand for power, the amount of power that would be consumed if the system were operating normally. Because operating reserves are assumed to be available whenever needed within the limits of installed capacity ($K$), it does no harm to define them to be all installed capacity that is neither out of service nor serving load. Thus,

$$\text{Operating reserves} = OR = K - g - L$$

Because load is defined here to be economic demand (see Section 1-4.3), it can exceed supply. In this case there is lost load, $LL$, and there is served load, and the sum of the two is load, $L$. Lost load is measured by the extent to which $OR$ is negative, and is zero whenever $OR$ is positive; thus,

$$\text{Lost load} = LL = \max(-OR, 0)$$

In reality, load may be shed even when $OR > 0$, and when $OR$ is negative, more than $LL$ may be shed. Such excess service interruptions include those caused by distribution faults and cascading outages. The simple model assumes that excess interruptions are not correlated with the level of operating reserves. If they are not correlated, they are not affected by increasing $K$, and are a constant cost which can and will be ignored. Thus $LL$ does not include all lost load, but within the simple model it includes all that can be affected by a policy that controls installed capacity, $K$. The lack of correlation between excess loss of load and $OR$ is the key assumption of the Simple Model of Reliability and is discussed further in Section 2-3.4. With this assumption, *nominal* lost load, $LL$, is easily computed and is all that matters.

Defining **augmented load,** $L_g$, as $L + g$, simplifies the definition of $OR$ to $K - L_g$. Operating reserves are just installed capacity less augmented load. Taking a generator off the system is electrically equivalent to placing an equal load on line at the

---

3. This assumes an isolated system. Two neighboring systems, competing for each other's operating reserves, will bid prices up and make their procurement expensive.

**Figure 2-3.1**

The Simple Model of
Reliability.

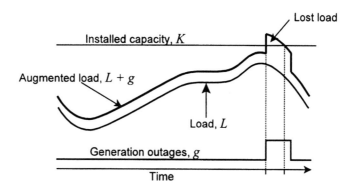

generator's point of connection. Thus, a generator's outage is equivalent to an equal
increase in load at its point of connection.

---

**Assumption 2-3.1**   **Load is Shed Only to the Extent Necessary: $LL = \max(L_g - K, 0)$**
Systematic (nominal) lost load ($LL$) equals the amount by which augmented load,
$L_g$, exceeds installed capacity, $K$. Augmented load includes economic demand,
$L$, and generation out of service, $g$. "Systematic" refers to losses that are correlated with $L_g - K$.

---

Figure 2-3.1 shows how load is lost when augmented load exceeds $K$, typically
because of an unexpected forced outage. There are also **planned outages** for
maintenance. Although outages contribute to augmented load, they cannot be "lost;"
only real load can be shed when augmented load exceeds capacity. To account for
this, $g$ should be considered to contribute to the bottom of $L_g$ rather than to the top.

Augmented load, just as real load, can be represented by a load-duration curve
as depicted in Figure 2-3.2. Generation outages are shown at the bottom to indicate
that they cannot be part of lost load, but they are shown with a dotted line because
they are not represented by a load-duration curve; instead, their correlation with
$L_g$ is depicted.[4] The peak of $L_g$ is above $K$, and this causes load shedding. The area
of the region labeled $LL$ gives the expected annual average load shedding, $\overline{LL}$, in
MW, while $D_{LS}$ is the duration of load shedding and has a typical value of 0.03%
(.0003) corresponding to 1 day in 10 years.

## The Fundamental Reliability Result

Figure 2-3.2 shows that given well-functioning security procedures, generation
adequacy (represented by installed capacity, $K$) is the fundamental determinant
of reliability. The greater is $K$, the smaller is the area of lost load. Increasing $K$
reduces the cost of lost load but increases the cost of serving load. This cost
trade-off determines the optimal value of $K$.

---

4. Because unplanned generation outages cause an increase in $L_g$ when they occur unexpectedly, they
correlate with $L_g$. Because planned outages are scheduled at times of the year when $L$ is low, they have a
second broad peak for low values of $L_g$.

**Figure 2-3.2**

An augmented-load-duration curve.

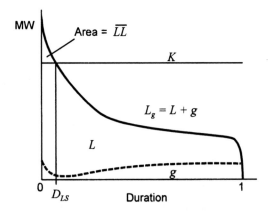

Lost load, $LL$, fluctuates from minute to minute and cannot be used directly to compute the necessary amount of installed capacity which can change only very slowly. Instead, an average value of $LL$ is needed. The cost of both instantaneous lost load, $LL$, and average lost load, $\overline{LL}$, can be found by multiplying these megawatt values by the value of lost load, $V_{LL}$, measured in \$/MWh. The resulting cost is measured in dollars per hour (\$/h). Increasing $K$ by 1 MW would reduce the area $\overline{LL}$ by $D_{LS}$ MW where $D_{LS}$ is the duration for which the augmented load is more than $K$. This would reduce the average cost of lost load by $\$(V_{LL} \times D_{LS})$/h.

From Equation 1-3.1, the average cost of adding (peaker) capacity is given by $FC_{peak} + D_{LS} \times VC_{peak}$. Because $D_{LS}$ is so small (0.0003), the fixed-cost term dominates. Most of the cost of serving peak load is the fixed cost of purchasing the peaker, so the role of $D_{LS}$ in determining the cost of adding 1 MW of peak capacity can be ignored.

As $K$ increases $D_{PS}$ decreases. For low values of $K$, $V_{LL} \times D_{LS}$ will be greater than $FC_{peak}$, and it will cost less to increase $K$ than will be saved by the reduction in lost load. For high values of $K$ the reverse is true, and at the optimal $K$, the cost saved equals the cost of installing another megawatt of peak capacity. The condition for optimal $K$ is

$$V_{LL} \times D_{LS} = FC_{peak}$$

Policy can control neither $V_{LL}$ nor $FC_{peak}$, so it must control $D_{LS}$. The optimal value of $D_{LS}$ is given by

$$D^*_{LS} = FC_{peak} / V_{LL}$$

A reliability policy that induces investment when, and only when, $D_{LS} < FC_{peak} / V_{LL}$ will be optimal, at least according to the Simple Model of Reliability.

Result 2-3.1 is important because it characterizes the optimal equilibrium based on very little information. It is true regardless of the probability distribution of generation outages or the shape of the load-duration curve or the shape of daily load fluctuations.

| Result | 2-3.1 | **Optimal Duration of Load Shedding Is** $D_{LS}^* = FC_{peak}/V_{LL}$ |
|---|---|---|

In a power system that satisfies Assumption 2-3.1, the optimal average annual duration of load shedding, $D_{LS}^*$, is equal to $FC_{peak}/V_{LL}$, where $FC_{peak}$ is the fixed cost of a peaker and $V_{LL}$ is the cost per MWh of lost load (VOLL). Only load shedding that is affected by $K$ is included in $D_{LS}^*$.

## 2-3.4   THE FUNDAMENTAL RELIABILITY QUESTION

Assumption 2-3.1 leads to load-shedding Result 2-3.1, which gives a fundamental characterization of the optimally reliable power system. This characterization states that the optimal duration of nominal load-shedding is given by $FC_{peak}/V_{LL}$, which is independent of all system characteristics except the two named costs. Nominal load shedding, $LL$, was defined to be $L_g - K$ when the difference is positive.

This is a claim that markets may put to the test. For example, under regulation, the energy price does nothing to limit peak demand so the load-duration curve under regulation is often sharply peaked. This means a system with five hours of load shedding might have only 30 hours with spinning reserves under 4%. But a market will dramatically flatten the peak of the load-duration curve by raising price whenever reserves are low. In this case, a system with five hours of load shedding might have 300 hours with spinning reserves below 4%.

It seems possible that the market system might have significantly less reliability even though it had the same value of $LL$. In particular, it might be more susceptible to cascading failures as the result of spending more time with low reserve (e.g., less than 4%) or because of more short-duration load-shedding incidents.

Another possibility is that the control of load-shedding is not precise enough, either with regard to the amount of load shed or how quickly it can be shed. In this case load shedding may be greater than $L_g - K$ simply because it cannot be controlled accurately. The technology for such control, however, would seem to be readily available and inexpensive. If so, any discrepancy between Assumption 2-3.1 and reality may be quickly overcome.

This discussion should not be read as an argument for or against Assumption 2-3.1. The point is simply that the question of its validity deserves more attention and can be evaluated from a purely engineering perspective.

| Question | **The Fundamental Reliability Question** |
|---|---|

Does Assumption 2-3.1 hold, with reasonable accuracy, in actual power systems?

If this assumption is far wrong, then VOLL pricing as well as several of its less high-priced relatives are wrong in theory, quite possibly by a significant amount. As long as load-duration curves remain close to their familiar shapes, this will probably cause little trouble. But once increased demand elasticity moves them into unfamiliar territory, a better understanding of reliability may become essential.

*Chapter 2-4*
# Limiting the Price Spikes

*We don't know what the hell it is, but it's very large and it has a purpose.*

Dr. Heywood Floyd
2001, A Space Odyssey
1968

**W**HEN SUPPLY CANNOT EQUAL DEMAND, THE MARKET CANNOT
DETERMINE A PRICE. When contemporary power markets have enough installed
generation capacity (ICap) to prevent this market failure at all times, they have so
much that generators cannot cover their costs. This contradiction occurs only for
certain combinations of supply, demand curves and load-duration curves, but when
it does occur, the market fails; there is no long-run equilibrium. Most current power
markets may well satisfy the conditions for this failure.

The demand-side flaws are the cause of this market failure, but they need not
be eliminated in order to remove it. Failure can be prevented simply by reducing
the severity of the first demand-side flaw—by increasing demand elasticity. Until
this is accomplished, the system operator must set the market price when the market
fails to clear. At these times, some regulatory rule must be adopted for determining
what price to set. If price is set only for these few hours, it must be set extremely
high, so it may be better to set price more often and lower. Such details are considered in subsequent chapters. The purpose of this chapter is to explain why all current
power markets need and have price limits.

FERC approved price limits of $750, $500 and $250/MWh for California
between 1998 and August 2001. In the summer of 2000 it reduced the NY ISO's
limit from its previously approved level of $10,000 to $1,000/MWh bringing it
in line with PJM's limit. A year later it limited prices indirectly in the West to
roughly $100/MWh. In between it suggested that what Western markets really
needed was no price limit at all. The Australians tell us prices must be capped at
the value of lost load, which they put at between $15,000 and $25,000/MWh AU.
The new electricity trading arrangement in England allows much higher prices and
promptly set a record of over $50,000/MWh. Settling on a reasonable policy will
require understanding the nature of failure in power markets.

**Chapter Summary 2-4:** Most, if not all, contemporary power markets would fail to have a long-run equilibrium were it not for the regulatory setting of real-time prices by the system operator. This price setting is needed only when the market fails to clear and sets no price of its own, or when it sets a price greater than the value of power to customers. The system operator must purchase power on behalf of loads and should set a price no higher than the average value of power to consumers. The second demand-side flaw, one of the two causes of the market failure, makes it impossible for customers to buy more-reliable power in advance than they can in real time. Consequently, the system operator's price limit caps the entire power market.

---

**Why Power Markets Fail
(and What Caps Their Price)**

Three conditions are necessary for failure: (1) inelastic demand, (2) inelastic supply, and (3) volatile demand. Given the proper realization of these conditions, imagine starting with too much generating capacity. As capacity decreases toward an equilibrium level, short-run profit will increase, but before it becomes great enough to cover fixed costs, the supply and demand curve will fail to intersect. This is market failure. Price is pushed toward infinity, but even an infinite price cannot clear the market. Current power markets probably satisfy the conditions required for this failure, which is why they have price caps.

A less extreme form of the failure may occur. The clearing price may rise above $V_{LL}$ (the average cost to customers of load being shed) before short-run profit covers fixed costs. In this case there may be a long-run equilibrium but it would be less efficient than one that caps price at $V_{LL}$ and sheds load.

With the first failure, the system operator *must* set price because the market cannot; with the second it *should* set price because the market sets it at more than power is worth. Because of the second demand-side flaw (Section 1-1.5) no customer will pay more than the highest price the system operator pays for power. So the limit on the system operator's *price limit* caps the price in all energy markets. This occurs through arbitrage and without any regulatory restrictions except the limit on how high a price the system operator should pay.

---

**Section 1: Normal Market Operation with Limited Demand Elasticity.** In spite of the demand-side flaws, contemporary power markets could operate normally and efficiently with a modest amount of demand elasticity, provided the load-duration curve does not have too sharp a peak. In this case no regulatory price intervention would be needed or desirable.

**Section 2: Market Failure with a Steep Load-Duration Curve.** By replacing the rather flat peak of the load-duration curve of Example 1 with one having a more realistic and sharper peak, Example 2 illustrates the possibility of market failure. In this example, the market has no long-run equilibrium unless the system operator sets a price in the real-time market whenever the supply and demand curves do not intersect and load must be shed.

**Section 3: Suppressing the Balancing Market to Avoid Regulating Price.** Some suggest suppressing the system operator's balancing market so the price will not need to be regulated. The only practical approach is to set the price in that market high enough that the market is not used. This avoids the problem of setting the wrong regulated price by effectively setting it so high that all trade is stopped and then claiming there was no regulation because there was no trade.

**Section 4: Setting Price to the Last Clearing Price.** When the system operator must set price, a simple market-based rule is desired. To this end, setting price to the highest demand bid, or more generally to the last market-clearing price, has been suggested. Either rule can produce a price that is too high or too low because neither rule relies on relevant information. Such rules can be less efficient than simply building so much generation that price never rises above the marginal cost of a peaker.

**Section 5: Real-Time Price Setting Caps the Forward Markets.** The highest price that the system operator will pay when it must set a price is termed $P_{cap}$, and this limit should be no higher than the value of lost load. Because the system operator must break even on its trades, it will charge no more than $P_{cap}$ for real-time sales. Because of the second demand-side flaw, there can be no advantage to buying power in advance. As a consequence, the system operator's limit, $P_{cap}$, will limit the price of power in all real-time and forward markets, both public and private.

**Section 6: Technical Supplement: The Condition for Failure.** This section gives the technical condition for the failure of a long-run equilibrium to exist without regulatory price intervention. It also gives the condition for the equilibrium to be as efficient as possible, given the first demand-side flaw, when an equilibrium exists and price is not regulated.

## 2-4.1   NORMAL MARKET OPERATION WITH LIMITED DEMAND ELASTICITY

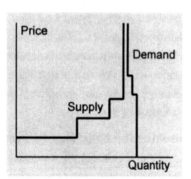

Chapter 2-2 analyzed a power market which behaved normally—one in which the supply and demand curves were bound to intersect. But real power markets have limited elasticity and so the intersection of supply and demand is not guaranteed.[1]

**Example 1.** A slight modification of the model used in Section 2-2.2 will make such a failure to intersect possible and will serve as Example 1. The supply side of the model remains unchanged with two types of suppliers both of which produce at a constant marginal cost up to their capacity limit. At this point their marginal cost rapidly climbs to infinity. Their fixed costs, as in Chapter 2-2.2 are $6 for baseload plants and $12/MWh for peakers.

Maximum demand, demand at a low price, fluctuates uniformly between 4000 and 8000 MW. One hundred megawatts of that load is willing to curtail at $1,000/MWh, but the remainder has no ability to respond to price. Suppose all unpriced load values its consumption of power at exactly $10,000/Mwh (this does not affect the demand curve). The supply and demand curves are shown in Figure 2-4.1.

---

1. Borenstein (2001a) identifies "the fundamental problem with electricity markets [is that] demand is almost completely insensitive to price fluctuations and supply faces binding constraints at peak time." and predicts "short-term prices for electricity are going to be extremely volatile." Green (1998) observes "If customers cannot react in this way [to price], and "random" power cuts are needed, then there is no limit to the price that the generators could set, if they were allowed to set it after the shortage has appeared."

**Figure 2-4.1**

Example 1, a flawed market with an efficient equilibrium.

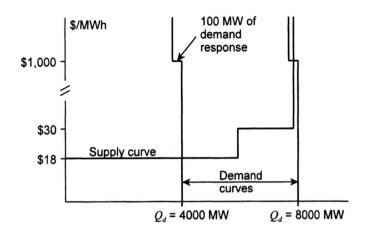

Using Result 2-2.2, the equilibrium conditions are found to be

$$6 = D_{PS} \times (1000 - 30)$$

and $$12 = 6 + D_{peaker} \times (30 - 18)$$

Solving these gives $D_{PS}$ = 0.62% (or 54 hours/year) and $D_{peaker}$ = 50%. The load duration curve (Figure 2-2.4) indicates equilibrium baseload capacity will be 6000 MW and optimal peaker capacity will be 1975 MW for a total of 7975 MW. This is the same solution as found in Section 2-2.2 because the left part of the demand curve, which used to be horizontal but is now vertical, plays no role. In the long-run equilibrium, the right-most vertical part of the supply curve is never found to the left of the 100-MW region of demand elasticity.

In this market, even though the supply and demand curves both become vertical and would fail to intersect with a 1% (75/8000) increase in demand on peak, the market will build enough generation to prevent this. The system operator's role is limited to clearing the market, which is always possible. There is no need to place any limit on the price the system operator should pay. This market suffers from neither demand-side flaw although both are present and could be exposed by different circumstances. Price is not regulated, and the market is fully competitive and efficient.

---

**Result    2-4.1    A Small Amount of Elastic Demand Can Make the Market Efficient**
Even if demand can be reduced less than 2% by a price increase, this could produce a long-run market equilibrium. Such an equilibrium could be more efficient than any price-regulated version of the market in spite of the presence of both demand-side flaws.

---

## 2-4.2   MARKET FAILURE WITH A STEEP LOAD-DURATION CURVE

**Example 2.**   The above example uses a linear load-duration curve, while power markets typically have sharply peaked curves. In that example, load spent 2.5% of the time within 100 MW of the peak. Suppose, more realistically, that load only spent 0.2% of the time within 100 MW of the peak. Algebraically, the peak region of this load duration curve and the curve from the previous example can be expressed as follows:

Example 1: % Duration = (8000 − load)/40.

Example 2: % Duration = (8000 − load)/500.

If the market built 7975 MW of capacity, as in Example 1, peakers would earn scarcity rents (positive short-run profits) only about 0.05% of the time for an average fixed-cost recovery of roughly $0.50/MWh (0.05% of $1,000/MWh). This is far too little, so investment would stop.

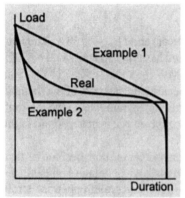

As installed capacity dwindled, short-run profits would increase. If installed capacity declined below 7900 MW, there would be times at which the supply and demand curves would fail to intersect. The maximum price-spike revenue that can be generated by this market without the market failing during peak-load hours is 0.2% × (1000 − 30), or $1.94/MWh.[2] This is too little to cover the fixed costs of peakers, $6/MWh in these examples.

With this level of price-spike revenue, neither peakers nor baseload plants would cover their fixed costs and investment would stop. Plant retirements would decrease supply and load growth would increase demand with the result that the market would fail to clear for an increasing fraction of the year. This is market failure. When the supply and demand curves do not intersect the market cannot determine price, and the system operator cannot set price to the market-clearing price because there is no such price.

This leaves only two possibilities: (1) the system operator can set a regulated price whenever demand exceeds supply at every price; or (2) the system operator might avoid trading at times of peak demand. The first approach admits that the real-time market has failed and provides a regulatory patch. The second approach admits it has failed and shuts it down whenever it fails to clear. The hope for the second approach is that some substitute will be provided in the forward markets. That possibility is explored next.

---

2. Two tenths of one percent is the duration corresponding to the maximum price-response of load, which is 100 MW. The maximum clearing price is $1,000/MWh, and the variable cost of a peaker is $30/MWh.

| | | |
|---|---|---|
| **Result** | **2-4.2** | **Too Little Demand Elasticity Can Cause the Real-Time Market to Fail** |

Even if demand is sufficiently elastic for the market to produce a long-run equilibrium without the help of regulated prices, it may possess no such equilibrium if the load-duration curve has too sharp a peak relative to the elasticity of demand.

## 2-4.3 SUPPRESSING THE BALANCING MARKET TO AVOID REGULATING PRICE

Normally the system operator conducts a real-time market, also known as a balancing market, for the purpose of maintaining system frequency, voltage, stability and reliability in general. The previous section demonstrated that the investment dynamics of some power systems will bring their ICap level into the region of market failure where the system operator is forced to set a real-time price if it continues to operate the balancing market.

Some suggest that this failure can and should be avoided by effectively shutting down the system operator's balancing market, at least in times of supply shortage (private real-time bilateral trading would continue). While this suggestion may seem too irresponsible to be taken seriously, and Part 3 argues that it is not a practical approach to providing reliability, the influence of this suggestion on the design of the new British market, and even within FERC, requires that it be addressed.

The market could be shut down either permanently or only when it failed to have a clearing price. This section answers two questions: (1) Is this practical? (2) Would it be superior to keeping the balancing market open and regulating its price when necessary?

### The Impracticality of Suppressing the Balancing Market

Besides the difficulty of providing the balancing services required for reliability with only a private, bilateral real-time market (see Part 3), the second demand-side flaw makes withdrawal of the system operator from the real-time market quite difficult. The system operator could take either of two approaches when attempting to remove itself from this market: (1) It could shed load; or (2) it could charge a high price for taking power during a shortage.

One approach to withdrawing from the market would be to require balanced contracts and physically curtail any load whose supplier fails to meet a contractual obligation equal to its load. This is physically impossible, however, in contemporary power systems as noted by the second demand-side flaw.[3]

**The First Approach.** Consider a time when demand exceeds supply and the system operator is attempting to suppress its balancing market by shedding load.

---

3. Even if possible, it would be complicated because power injections cannot be physically matched to specific loads. If two loads have a contract for 100 MW with a particular supplier and are using that much power, and the generator is injecting only 150 MW, there is no physical indication of which load is short of power. The complexity of this problem increases when loads contract with multiple suppliers.

Customers will be disconnected involuntarily, and this will leave connected load exactly equal to operating capacity. Only by coincidence would each customer possesses a forward contract for exactly the amount of load left connected. This is true for two reasons: (1) load shedding disregards contracts, and (2) knowing this, customers have no reason to try to match their contracts exactly to their expected load (see Section 2-4.5).

As the matching is imperfect, some customers will be oversupplied and some under-supplied relative to their contracts. The system operator will be taking power from some suppliers whose customers do not need all their power (perhaps because their load was shed) and giving it to other consumers who have contracted for too little. Load shedding does not result in balanced bilateral contracts but leaves most customers trading some power with the system operator. This requires the system operator to set a price on those trades. The first approach does not accomplish its objective.

**The Second Approach.**   The second approach charges a high price whenever demand exceeds supply. Knowing the price will be high, load will pay a high price in forward markets to avoid having to buy power from the system operator's balancing market when that market is short of supply. The combination of high forward prices and high balancing prices will lead to investment in generation and a reduction in the hours during which the balancing market will fail. Charging a very high price during failures does not result in shutting down the system operator's market when that market fails to clear but does reduce the time during which failure occurs. Sometimes, excessively high prices, unrelated to cost, are termed penalties and this approach could be viewed as imposing a severe penalty on trade with the system operator during periods of load-shedding.

This approach is successful at reducing the period of market failure. With a sufficiently high penalty/price that period might be reduced to zero. Then it would be the same as the regulatory price setting approach discussed in the previous section, but in an extreme form.

## Is There an Advantage to Suppressing the Market?

The first approach is unworkable, the second is equivalent to setting such a high regulated price or such a high penalty that the market shuts down. This prevents the market failure of Example 2 by regulating the market out of existence.

But this market only disappears when load-shedding disappears, because load-shedding unbalances the trades and thus forces trade with the system operator. Forcing an absolute avoidance of load shedding, as will be seen in the following chapter, is equivalent to using too high a value for $V_{LL}$ in VOLL pricing. This forces the market to provide too much reliability.

Because power would necessarily be purchased in forward markets, though perhaps only day-ahead forward markets, price spikes would necessarily be lower and of longer duration. This might mitigate the side effects caused by volatility.

Perhaps the main drawback to suppressing the centralized real-time market is the associated loss of efficiency. This is a valuable market because it allows last-minute corrections to be made very efficiently. This is discussed in more detail in Chapter 3-3.

---

**Result    2-4.3**

**Suppressing the System Operator's Balancing Market Is Inefficient**
Suppressing the system operator's balancing market is equivalent to setting the regulated price in that market extremely high—high enough to induce so much generating capacity that load would never be shed. This prevents the market failure by regulating the market out of existence. It is inefficient to induce absolute reliability, and the loss of the centralized real-time market also reduces efficiency.

---

**Conclusion to Example 2.**  Result 2-4.3 shows there is no way to avoid the market failure of Example 2. If the system operator does not intervene, installed capacity will decline below its optimal value and continue to fall until supply fails to intersect demand for some part of the year. Even at this level it would continue to fall if the system operator did not intervene. The only two possible interventions are to set a price or shut down the real-time balancing market. Result 2-4.3 shows the latter to be inefficient. This leaves setting a price as the only method of providing an efficient level of reliability. As long as the first demand-side flaw is sufficiently severe, the market cannot solve the reliability problem on its own.

## The Reliability Fallacy

**𝓕𝒶𝓁𝓁𝒶𝒸𝓎  2-4.1**

**The "Market" Will Provide Adequate Reliability**
Contemporary markets, even with their demand-side flaws and negligible demand elasticity, could still provide reasonable investment incentives and a reasonable level of reliability if regulators and engineers would refrain from setting or limiting prices.

---

Given the market flaws, the system operator must set the price whenever supply fails to intersect demand. In present markets this will mean setting price most of the time it is above the left-hand marginal cost of the most expensive generator. Thus the entire aggregate price spike will be determined by regulator policy. As will be seen in subsequent chapters, there is considerable choice as to how high the price spike will be and how long it will last, but the conclusion at this point is simply that it must be determined by the system operator if an efficient level of reliability is to be achieved.

## The Regulatory-Price-Spike Result

**Result    2-4.4**

**Regulatory Policy Determines the Height and Duration of Price Spikes**
Because of the second demand-side flaw, markets will not set energy prices higher than the system operator's price limit. Consequently this regulatory limit determines the height of price spikes. The average annual duration of spikes is controlled by the application of this limit to the operating reserve requirement.

---

## 2-4.4   SETTING PRICE TO THE LAST CLEARING PRICE

Example 2 demonstrated that some markets cannot sustain sufficient installed capacity (ICap) to prevent market failure. In these markets, as long as market clearing is assured, there is so much ICap that prices and short-run profits are too low to cover the fixed costs of the installed generation. Retirement of older plants and load growth will ensure that, in equilibrium, ICap is low enough that for some fraction of the time demand will exceed supply at any price. During these times the system operator is forced to set a regulated price, which, if properly selected, will be high enough to cover the fixed costs of the optimal quantity of ICap.

While setting an optimal price is desirable, so is lack of regulatory interference with the market. Although the system operator must set some price, perhaps it could set one related to the market in some natural and transparent way. One candidate for this was proposed for ERCOT, the Texas ISO. Its rule is to let the highest demand bid set the price *after* the market fails to clear as well as just *before* it fails to clear. In the present example, this would mean leaving the price at $1,000/MWh after demand exceeded supply.

This price-setting rule can be generalized to require the system operator to keep the price at the last market-clearing price after the market fails to clear, regardless of whether that price was set by a supply bid or a demand bid. In the present example this change would make no difference.

## Example 3: Analysis of a $1,000/MW Price Limit

**Peak Durations**

Three durations related to peak load have been used.

$D_{LS}$   Duration of load shedding:   $L_g > K$
$D_{PS}$   Duration of the price spike:   $L_g > K - OR^R$
$D_{peaker}$  Duration of peaker use:   $L_g > K_{base} - OR^R$

Chapter 2-2 considered a market in which demand only became elastic at $1,000/MWh, so $D_{LS} = 0$.

Under VOLL pricing, $D_{LS} = D_{PS}$. In other cases:

$$D_{LS} < D_{PS} < D_{peaker}.$$

Instructing the system operator to use the last market-clearing price once demand exceeds supply will mean keeping the price at $1,000/MWh in the present example. This will define Example 3 which differs from Example 2 only in that the system operator provides this regulated price whenever the market fails to clear. No regulator intervention was made in Example 2. As determined in Section 2-4.1, the price must stay at this level for a duration of 0.62% in order to cover the fixed costs of generators. In Example 1, this equilibrium was optimal because the curtailment of load was voluntary and based on price. This was a standard competitive equilibrium with the supply curve always intersecting the demand curve.

In Example 3, if ICap is 7691 MW, load will exceed it for a duration of 0.62%. As shown in Figure 2-4.2, this will force load to be reduced by 309 MW on peak through a combination of 100 MW of demand elasticity and 209 MW of load shedding. As a simple check on the efficiency of this equilibrium consider the cost of load shedding. With a value of lost load of $10,000/MWh, load shedding can be computed to cost an *average* of $4,380/h (every hour, not just during peak hours). With a fixed cost of $6.00/MWh, 309 MW of peaking capacity would cost

**Figure 2-4.2**

Equilibrium ICap and load shedding in Example 3.

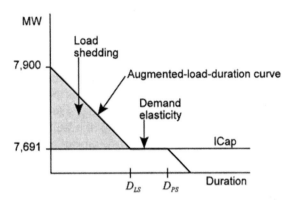

only $1,856/h. Installing this much capacity would completely eliminate load shedding and any need for the 100 MW of load reduction. For a cost of $1,856/h, more than $4,380/h could be saved.[4] This is neither an optimal policy nor a market equilibrium. Example 3 simply illustrates that the regulatory equilibrium induced by setting price to the last market-clearing price before failure is far from optimal.

The optimal duration of load shedding, $D_{LS}^*$ is given by $FC_{peak}/V_{LL}$ (Result 2-3.1), which in this example is $6/$10,000, or 0.06%. It corresponds to an installed capacity of 7870 MW. In the present example, the policy of setting price to the last market-clearing price when the market fails to clear results in reliability costs that are above optimal. They are even higher than under the naive policy of building installed capacity to the point where neither load shedding nor load elasticity ever comes into play. In Example 3, "market-based" regulation is worse than old fashioned regulation with regard to reliability policy.

## The Last Clearing Price Can Be Too High

When the last clearing price is $1,000/MWh, peakers are paid too little during times when the market fails to clear and load must be shed. Consequently, load shedding must last longer than optimal for the price-spike revenue to equal the fixed costs of peakers.

The choice of $1,000/MWh for the demand-side bid was arbitrary, and the highest bid could easily be $100,000/MWh. If a customer's computer uses 100 W, and the expected interruption would last only 0.1 hour, and the customer is willing to pay $1.00 not to be interrupted, this use of power has a value of $100,000/MWh. Entering one such bid into the market would cause the system operator to set the market price to $100,000/MWh during all periods of load shedding. Because this is higher than $V_{LL}$, peakers would be paid too much during these times and there would be too little load shedding. The system would be overbuilt. The same point can be made another way. Customers as a group should not have to pay more than $V_{LL}$, when power is only worth that much to them on average. The value of lost

---

4.  All values in these examples have been rounded. The exact values and detailed calculations can be found at www.stoft.com. Calculations for Example 3 proceed as follows: $D_{PS}^* = 0.0062$. Load with this duration equals installed capacity, $K$, and $K = 8000 - 50,000 \times D_{PS}^* = 7691$. Maximum load shed = $L_S = 8000 - K - 100 = 209$. Duration of load shedding = $D_{LS} = L_S/50,000$. Cost of load shedding = $D_{LS} \times L_S \times V_{LL}/2 = \$4,380$.

load, $V_{LL}$, serves as an upper limit on what the system operator should pay for power.

This and the previous example make clear that the last market-clearing price, or the highest demand-side bid, is simply irrelevant to correct pricing. The correct price to set when the market fails to clear and load is shed is the cost to consumers of shedding load, $V_{LL}$. This is the best proxy for the demand curve that is missing because of the first demand-side flaw: the "lack of metering and real-time billing" (see Section 1-1.5). (While VOLL pricing is optimal under the Simple Model of Reliability it has two detrimental side effects when considered in a broader context.)

| | | |
|---|---|---|
| **Result** | **2-4.5** | **Do Not Cap Prices at the Highest Demand Bid** |

When the market fails to clear and it becomes necessary for the system operator to set the real-time market price, the last market clearing price should not be used as it is irrelevant and can be either too high or too low. By extension, the highest demand bid should also not be used.

## 2-4.5   HOW REAL-TIME PRICE SETTING CAPS THE FORWARD MARKETS

Example 2 demonstrates that, due to the second demand-side flaw, contemporary power markets may fail to have a long-run equilibrium. Such markets will periodically fail to clear and then will determine no market price. The system operator will be forced to set the real-time market price, and its price setting should have a limit no great than $V_{LL}$. Whatever the limit, it will be referred to as the price cap, or $P_{cap}$, for reasons that will become apparent.

A purchase-price limit on the system operator is not a price cap or price control of the type used in other markets. It is not a restriction on private parties but is simply a rule governing the system operator's actions. The system operator is instructed to pay no more than $P_{cap}$ for power and to charge no more than $P_{cap}$ for power. No unusual restrictions are imposed on private parties by this rule. They can offer to sell power to the system operator or to any other private party at any price they choose. Similarly customers can pay whatever they want for power. In particular, if $P_{cap}$ is $10,000/MWh, any load may contract with any generator at any time to buy power for $20,000/MWh.

Extraordinary rhetoric has surrounded such a price limit. These limits have been treated as a dire threat to investment and a contradiction of free markets. In fact, this limit comes into play only when the system operator is forced to purchase power on behalf of load. In this case, it is natural to limit the amount paid at least to the value of the power to load. Nothing is a more normal part of market operation.

What is extraordinary about the system operator's limit on purchase-price is that through voluntary actions on the part of other purchasers it effectively caps the price in all forward markets. This is not due to regulation but rather to the second demand-side flaw. The market would be better off without this flaw, but its existence does not suggest the need for irresponsible purchasing behavior on the part of the system operator. With or without the flaw, the system operator should pay no more than the power is worth.

## The Second Demand-Side Flaw Caps the Market

Because it is required to break even, the system operator's price limit is the same for selling as for purchasing. Thus if it pays $P_{cap}$ for power it will sell it for no more. Knowing this, all power customers realize they can buy power in the real-time market for at most $P_{cap}$. Given this backstop price, they will pay no more for power in advance. This simple logic caps the price in every forward market even though there is no rule against customers paying more or suppliers charging more. Customers know it would be a waste of money and choose not to pay more than $P_{cap}$.

How does the second demand-side flaw enter this process? Consider a customer with a high value for reliability who might be willing to pay $100,000/MWh during a period of load shedding to be sure of obtaining the necessary power. The customer realizes it would never need to pay more than $P_{cap}$ for power but decides to sign a contract for $20,000/MWh with a supplier who guarantees always to keep a spare 50-MW generator as spinning reserve just for this customer. Eventually the day of reckoning arrives and the system suffers a major outage just as load is rising toward its afternoon peak. The alert customer phones his supplier to check on the readiness of his private spinning reserve and is assured that all is well. The supplier then phones the system operator to say that, if necessary, it is prepared to supply an extra 50 MW of power in case the system operator has any intention of shedding its prized customer.

Unfortunately, there may not be a system operator anywhere that could respond effectively to such a call. Typically the supplier would be told that the system operator does not control which load is shed. When the supplier calls his customer's local utility, which *is* in control of load shedding, he will be told that if the customer is next to a police station he may be in luck but otherwise it is simply a matter of chance and the die has already been cast. When load is shed, contracts between private parties are never taken into account as this is not currently possible. When a local area is shut off, there is no way for the utility to leave one customer in the middle of the area connected. Such a pattern of disconnection would require flipping by hand the main circuit breaker of every customer in that area except the one customer with the special contract.

The inability of contemporary power systems to enforce bilateral contracts in real time is the second demand-side flaw. It prevents the individual purchase of reliable power if that power must be transmitted through the power system. This inability means no customer can pay more and thereby obtain more reliability than it obtains by simply taking power from the system in real time.

This does not mean all customers will wait for the real-time market to buy their power. They can buy it with less risk, and quite possibly a little cheaper, by purchasing it in a bilateral forward market. But if the price in those markets were to exceed $P_{cap}$, customers would shun them and wait to buy from the system operator in real time.

This theory of price-capping forward markets was well tested in California. The price limit in the CA ISO's real-time market varied between $250 and $750 during the life of California's day-ahead market, the Power Exchange. Because this mechanism was not understood, the Power Exchange had its own price cap of $2,500. Although there were small discrepancies explainable by various penalties on real-time trading, the Power Exchange's day-ahead price was always capped quite accurately by the ISO's far lower real-time price limit.

## The Price-Cap Result

| | | |
|---|---|---|
| **Result** | **2-4.6** | **The Real-Time Price Limit Effectively Caps the Entire Market** |

The second demand-side flaw causes the system operator's limit on the price paid for power purchased on behalf of load to effectively cap the real-time power market as well as all forward markets, both public and private. The mechanism of the capping is voluntary arbitrage behavior on the part of power customers.

# 2-4.6   TECHNICAL SUPPLEMENT: THE CONDITION FOR FAILURE

Example 2 demonstrates that a contemporary market can lack a long-run equilibrium were it not for regulatory price-setting when its supply and demand curves fail to intersect. Example 1 shows this failure is not a necessary result of the two demand-side flaws: (1) unnaturally low demand elasticity, and (2) an inability to enforce bilateral contracts in real time.

This raises a question: Under what conditions will the market fail to have a long-run equilibrium without regulatory price intervention.[5] A second question of interest is when will regulatory price intervention be beneficial in spite of the existence of a long-run equilibrium. Both questions can be answered within the framework of the Simple Model of Reliability, which does not take account of the side effects of high price spikes.

Let $K$ be installed capacity and $L_g$ augmented load. Let $L_{g\text{-max}}$ be the maximum possible value of $L_g$. The dependance of $L_{g\text{-max}}$ on $K$ due to demand elasticity is expressed by $L_{g\text{-max}}(K)$. Let $R_{\text{spike}}(K)$ be the price-spike revenue.

| | | |
|---|---|---|
| **Result** | **2-4.7** | **Conditions for the Failure of a Power Market** |

A market without regulatory price intervention will have a long-run equilibrium if and only if there is a $K$ such that $R_{\text{spike}}(K) > FC_{\text{peak}}$ and $L_{g\text{-max}}(K) < K$.

A market with a long-run equilibrium and without regulatory price intervention will have a market clearing price, $P_{\text{max}}$, at all times. If this price is lower then $V_{LL}$, then the best that can be done (ignoring problems of risk and market power) is to let supply and demand determine the price. If $P_{\text{max}}$ is higher then $V_{LL}$ there is no reason to instruct the system operator to pay more than power is worth, so a

5. Besides regulation, a number of other legal considerations set a cap on prices, with bankruptcy being one example. Consequently, real markets may have an equilibrium that relies on such price caps. For example, the equilibrium might involve periodic bankruptcies. Such equilibria are ignored here.

regulatory price limit should be placed at $V_{LL}$. Even with $P_{max} < V_{LL}$, it may be possible to increase efficiency by improving the operation of the demand side of the market. Result 2-4.8 is only meant to indicate that efficiency cannot be increased by setting a price limit because the appropriate price limit would not be binding.

| Result | 2-4.8 | **Conditions for an Efficient Power Market** |
|---|---|---|

**Conditions for an Efficient Power Market**

If a power market has a long-run equilibrium and no regulatory price intervention, its prices will be efficient if and only if $P_{max} < V_{LL}$.

*Chapter* 2-5
# Value-of-Lost-Load Pricing

*Fifteen years ago I used charred paper and card in the construction of an electric lamp on the incandescent principle. I used it too in the shape of a horse-shoe precisely as you say Mr. Edison is now using it.*

Joseph Swan
in a letter to *Nature*, January 1, 1880

*There you have it. No sooner does a fellow succeed in making a good thing than some other fellows pop and tell you they did it years ago.*

Thomas Edison
in reply

SHEDDING LOAD IS AN EXPENSIVE WAY TO CURB DEMAND. It makes no distinction between those who need the power most and those who need it least. Because most customers' usage is not metered in real time, and because most do not know the price, contemporary markets have little ability to ration demand with price. Instead, when it is necessary, the system operator must ration demand by shedding load. In this case, the value of another megawatt of power equals the cost imposed by involuntary load curtailment. This value is called the *value of lost load*, VOLL.

Basic economic theory says it is efficient to pay suppliers the value of supplying another unit of output. Because VOLL is very high, perhaps above $10,000/MWh, this implies a very high price whenever load must be shed. Implementing this policy causes extreme price spikes, but these will be brief and lead to optimal investment in generating capacity and optimal reliability.[1] Although basic theory ignores risk and market power, it provides valuable insights and a basis for discussing more subtle theories of setting energy prices.

**Chapter Summary 2-5:** Because of the two demand-side market flaws, power markets are not yet able to use market forces to determine an appropriate reliability level. Some authority must estimate VOLL or some other determinant of optimal reliability. All such approaches are based directly or indirectly on VOLL, so the consequences of error in the estimated value of VOLL cannot yet be avoided. These consequences are not dramatic and can be reduced by overestimating VOLL. Price risk and market power are negative side effects of VOLL pricing.

---

1. For a discussion of generation adequacy and VOLL pricing in the U.S. context, see Hirst and Hadley (1999).

**Section 1: Valuing Lost Load.** The most straightforward approach to the problem of price caps makes the regulator act as a surrogate for load by setting price when power is being shed. The appropriate price equals the average value that shed load places on power. This is termed the value of lost load, and although it can be well defined, it is very difficult to estimate.

**Section 2: VOLL Pricing Is Optimal in The Simple Model of Reliability.** Under the reliability assumptions of Chapter 2-3, setting price equal to the average value of lost load, $V_{LL}$, will induce competitive suppliers to invest in an optimal level of generating capacity. This does not account for the negative side effects of VOLL pricing.

**Section 3: Practical Considerations.** VOLL pricing is often avoided because of the difficulty of estimating $V_{LL}$, but until the demand-side flaws of present markets are sufficiently corrected, there is no way around this problem. Some reliability level must be chosen, and this is the most accurate method available. Price risk and the exacerbation of market power are both significant problems with VOLL pricing.

**Section 4: Technical Supplement.** If $V_{LL}$ is estimated at $15,000/MWh but differs from the true value, the resulting power system will be inefficient. If the true value is 10 times higher, the resulting excess cost of unreliability would probably be significant but less than 10% of the total cost of power.

## 2-5.1  VALUING LOST LOAD

In the most critical circumstance, when supply has reached its maximum and load is being shed, the system operator must choose how much to offer for additional supply. The standard regulatory choice is to offer to pay the cost of additional generation. The market approach is to offer the value that customers place on not being cut off. This value might be $10,000/MWh while the cost of the last unit of power produced might be only $500/MWh. If the market is perfectly competitive, the cheaper approach is to offer $10,000/MWh and pay this much whenever load is actually shed. This is the price determined by the intersection of supply and demand. Setting the price of energy in the spot market to this price whenever load has been shed is **VOLL pricing**. This result depends on competition to prevent market power, on risks of extreme prices being costless, and on the assumptions of the Simple Model of Reliability described in Section 2-3.3. Even with these restrictions, the result provides a basis for a more realistic analysis and explains why paying more than $V_{LL}$, as suggested by "no-price-cap" advocates, is counterproductive.

**VOLL Pricing is Regulatory.** To apply VOLL pricing, the value of lost load must be determined. Because most customers do not respond directly to real-time prices, there is almost no market information on the value of lost load, and available estimates are highly inaccurate. Consequently, VOLL pricing sets a regulated, not a market, price.

The cost of load shedding is great because load is arbitrarily disconnected by the system operator rather than demand being voluntarily reduced by customers in response to the market price. Blackouts shed high-value loads and low-value loads in proportion, so when load needs to be shed, a load that values its power at $100,000/MWh is just as likely to be shed as one that values it at $200/MWh. VOLL pricing, together with its inevitable blackouts, is highly inefficient relative to a market that could rely on real demand elasticity. The claims for this policy must be judged relative to the current demand-side flaws—it is optimal only as long as they cannot be eliminated.

**The Problems of Estimation.**   The largest impediment to determining VOLL is the dramatic variation in values between customers and from one time to another.[2] The effect of the duration of a disconnection on cost is also typically quite nonlinear. Often the first few seconds are the most costly but in other cases costs grow at an increasing rate with the duration of the outage. Because most of these costs involve no market transactions, they are particularly difficult to evaluate.

---

*Fallacy*   **2-5.1**      **VOLL Cannot Be Usefully Defined**
Because the value of lost load depends on the customer, the time of the loss, and the nonlinear dependence of loss on the duration of the loss, no useful definition of VOLL is possible.

---

In spite of measurement problems, the concept of **VOLL** is well defined. Consider a given system with a given level of installed capacity and a given transmission system. The system will be subject to random load shedding due to contingencies, and each event will have a different average cost per MWh of lost load. Fortunately, the details of this process are irrelevant. Generating capacity is not built to prevent any particular incident in which load is shed but to avoid the long-run average cost of load shedding. This expected long-run average is what is meant by the value of lost load.

Conceptually, if the system could be operated for many years with its present pattern of load and its present installed capacity, then the total loss of value to consumers (loss of consumer surplus) divided by the accumulated MWh of lost load would equal average VOLL. But this average value is slightly different from the marginal definition of VOLL that is needed for pricing policy. **Marginal VOLL** measures the decrease in lost value divided by the decrease in lost load when installed capacity is increased by a small amount.

---

2. Australia's National Electricity Code Administrator reports that "Research by Monash University indicates that the value of lost load to end-use customers is in a very wide range of $1,000 to $90,000/MWh." (See NECA 1999b, 10.)

---

**Definition**        **Value of Lost Load ($V_{LL}$)**
In a given system, let $H$ be the average MWh of load that is shed and $V_H$ be the
average consumer surplus of power consumption. Let $-dH$ be the decrease in $H$,
and $dV_H$ the increase in $V_H$, caused by a small increase in installed capacity. Then,
$V_{LL} = -dV_H/dH$. Technically, this is marginal VOLL, but it is the appropriate
VOLL for present purposes and will be referred to simply as VOLL.

---

$V_{LL}$ is difficult to estimate, and defining the concept carefully does not lessen
this difficulty; however, it does allow the development of a theory of price regulation in the face of market failure. The next section shows that $V_{LL}$ can be the right
price for inducing the investment needed to provide minimum-cost reliability.

## 2-5.2   VOLL PRICING IS OPTIMAL IN THE SIMPLE MODEL OF RELIABILITY

Section 2-3.3 developed the condition for optimal load shedding under the Simple
Model of Reliability, but it did not investigate the market equilibrium. This section
shows that the equilibrium condition is the same as the optimality condition, so
a competitive market will build the right amount of installed capacity. It will
minimize the combined cost of energy and blackouts. Before considering the market
equilibrium, the connection between VOLL and the market's demand curve is
examined.

### Connecting VOLL to the Demand Curve

Markets give optimal outcomes when they are competitive and have demand curves
that truly reflect consumer preferences. Because of the first demand-side flaw
(Section 1-1.5) power-market consumers do not express their true demand and so
the system operator needs to buy power on their behalf. This precludes the optimal
outcome promised by Adam Smith and modern economics for competitive markets,
but the use of VOLL pricing proves to be the best strategy given the limitations
of the market's structure. This result can be understood by investigating the relationship between VOLL and the true consumer demand curve.

Most customers cannot respond to daily price fluctuations, so the short-run
demand curve is unobservable. If consumers were charged real-time prices and
could respond to them without transaction costs, they would use much less power
at sufficiently high prices. As a crude approximation of this unobservable demand
curve, assume that demand for power is zero at $30,000/MWh and increases linearly
to 20,000 MW at the retail price of power (see Figure 2-5.1). The area under this
curve measures the total value of power to consumers, the consumer surplus, in
the sense that consumers would pay that value for power but no more.[3] When the
variable cost of power is subtracted, the result is the total surplus of producing and

---

3. To obtain this revenue from customers, it would be necessary for a "perfectly discriminating"
monopolist to set price at different levels for different MWh of power. This revenue is the most consumers
would pay voluntarily.

**Figure 2-5.1**

The market demand
function and the value
of lost load.

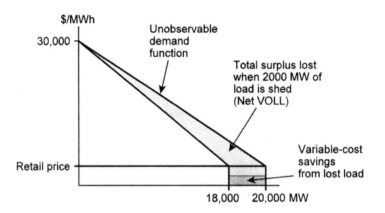

consuming that power (see Section 1-5.2). When load is shed there is a reduction
in total surplus.

When load is shed, customers are disconnected without regard for the value
they place on power. Consequently, the best assumption is that the demand curve
of those remaining on the system is a scaled-back version of the complete market
demand curve. As an example, Figure 2-5.1 shows the demand function scaled back
10%. Since demand of every value is scaled back 10%, the total reduction in net
social value is $30,000,000/h. Dividing this by the 2000 MW of load shedding gives
the (net social) value of lost load, which is $15,000/MWh[4]. Because consumer
surplus is so much greater than the variable cost of power, and because consumer
surplus is such an uncertain value, the distinction between consumer surplus and
total surplus can be ignored.

The reduction in consumer surplus caused by 1 MWh of shed load is $V_{LL}$. When
load shedding is optimal, a reduction of installed capacity would cost consumers
as much in lost value as would be saved by the reduction in capacity. According
to Equation 1-3.2, the average cost of supplying peak energy during the period of
load shedding is $AC_E = FC_{peak}/D_{LS} + VC_{peak}$. Given the approximate nature of the
present calculation and the smallness of $VC_{peak}$ relative to $FC_{peak}/D_{LS}$, the variable-
cost term can be ignored.  Thus, the condition for optimal load shedding is

Lost consumer surplus = savings from reduced capacity

$$V_{LL} = FC_{peak}/D_{LS}$$

Solving for $D_{LS}$ gives the condition for optimal load shedding

$$D_{LS}^* = FC_{peak}/V_{LL} \qquad (2\text{-}5.1)$$

which is exactly Result 2-3.1.

---

4. This is not a long-run calculation but represents the consumer surplus relative to a sudden
disconnection.

## Is the Market Equilibrium Optimal?

Having characterized the optimal duration of load shedding, the market equilibrium under VOLL pricing must now be examined to see how it compares. Result 2-2.2 gives the long-run equilibrium condition for investment in peakers as

$$FC_{peak} = R_{spike}.$$

The right side is the price spike revenue, which in this case is $V_{LL} \times D_{LS}$. Solving for the equilibrium duration of load shedding gives $D_{LS}^e = FC_{peak}/V_{LL}$, so equilibrium and optimal load shedding are the same.

---

| Result | 2-5.1 | **Within the Simple Model of Reliability, VOLL Pricing Is Optimal** |
|---|---|---|

Under the Simple Model of Reliability defined in Section 2-3.3, which ignores both risk and market power, VOLL pricing would induce optimal investment in generating capacity and thus optimal reliability. This assumes an optimal short-run security policy.

---

## 2-5.3   PRACTICAL CONSIDERATIONS

The first concern with VOLL pricing is the difficulty of estimating $V_{LL}$. Because such estimates are probably accurate only to within a factor of 10, they are usually assumed to be useless. But the consequences of misestimation are relatively small and currently unavoidable. If the true value of $V_{LL}$ is 10 times greater than the estimate used to set price, the consequent net cost of excess unreliability might be 10% of the wholesale cost of power (4% of retail costs). If the error were in the other direction, the cost of excess reliability would be at least three times smaller. These values are based on a conservative estimate for the PJM market and an estimated $V_{LL}$ of $15,000. Details are given in Section 2-5.4, the Technical Supplement. A more precise estimate might well indicate that the costs were considerably smaller.

---

| Result | 2-5.2 | **Inaccuracy of Estimation Does Not Rule Out the Use of VOLL** |
|---|---|---|

Some reliability level must be chosen, and as a flawed demand-side prevents the market from determining reliability, no more accurate method of selecting a level is available.

---

Although misestimation of $V_{LL}$ may waste as much as a few percent of total power costs, the use of $V_{LL}$ should not be rejected until a better alternative is available. Instead of estimating $V_{LL}$, $D_{LS}^*$ might be estimated directly and this could be used to determine reliability policy. But $D_{LS}^*$ cannot be estimated without using $V_{LL}$, so nothing is gained. Similarly the optimal value of installed capacity may be estimated. But what makes such a value optimal? Only the fact that it minimizes system costs plus the cost of lost load, so again, no estimate of $K^*$ is possible without an estimate of $V_{LL}$. The only way out of this dilemma is to fix the demand side of the market and let the market determine optimal reliability. Until then, a few percent of inefficiency must be accepted.

Those concerned about the poor quality of the estimate of $V_{LL}$ have sought other rationales and other ways of setting the value of the price cap. The two most popular considerations are supply-side costs and risk.

**Supply-Side Calculations.** Engineers have reliability standards which are not derived from explicit calculations of consumer surplus and so they have no basis in economic theory. Nonetheless, they are derived from common sense and years of experience and should not be discarded until a better approach is available. Engineers have found it costs very little to reduce the duration of shed load to about three to five hours per year and, at such a level, customers do not complain much. The fixed cost of a peaker, which is roughly \$50,000/year, when divided by five hours per year, gives a price of \$10,000/MWh. This is the so-called VOLL price needed to induce investment up to the point where load shedding is reduced to five hours per year.[5]

Of course, this is not an estimate of the value of power to consumers; it is an estimate of what price is required to reduce load shedding to five hours per year. If this is a desirable level of reliability, it is because actual $V_{LL}$ is near \$10,000/MWh. The calculation of VOLL from a choice of load-shedding duration and the price of peaker technology is only reasonable if a lack of customer complaints is a reasonable indicator of optimal load shedding. In other words, the normal engineering procedure that starts with some arbitrary "acceptable" duration of load shedding, say one day in ten years, is simply a crude estimate of the VOLL trade-off.

**Risk.** Risk is another matter. It provides no means of computing an appropriate level for VOLL and no information about reliability. It is, however, an important concern. This is recognized explicitly in the "Final Report" of Australia's Reliability Panel (NECA, 1999b, 6).

> *The core principle that must be met in establishing the level of VOLL has been identified as balancing the ability of the market to clear voluntarily, i.e. for supply to match demand, under all but the most extreme conditions, against risk.*

This "core principle" has nothing to do with establishing the level of VOLL, but risk is correctly seen as an important problem.[6] For this reason, the Reliability Panel proposed that if revenues from VOLL pricing exceed \$300,000/MW in a 7-day period, the price cap should be temporarily dropped from \$20,000 to \$300/MWh on peak and to \$50/MWh off peak (NECA 1999c). In this view, risk is a necessary evil, and it must be limited.

Unfortunately the section of the Final Report that is devoted to risk begins by expressing the opposite view (p. 12).

---

5. This approach to calculating VOLL is suggested in NECA (1999a).

6. If VOLL changed due to a change in customer technology, e.g., battery backups for all computers, this would not affect the trade-off between "voluntary clearing" and price risk. This trade-off is simply unrelated to VOLL and to any meaningful economics.

> *The NECA capacity mechanism report discussed the essential
> role risk and efficient risk management play in an energy only
> market, in particular as a driver for reliability. The reliance on
> commercial price signals to provide incentives to participants
> to influence market behavior centres on risk.*

Although there is no necessary contradiction between these two views (risk could
be good for one reason and bad for another), the latter view is incorrect. Risk to
generators makes investing more expensive, and risk to customers has similar costs.
Also, the inevitable demand-side risk-management contracts mute the accurate
incentives of the extreme VOLL prices. This is not to say that customers are wrong
to avoid risk; they are not. Risk is a cost, not a benefit. The many contractual
responses to the risks of VOLL pricing indicate that market participants go to a
great deal of trouble to avoid the risks it imposes because risk is costly.

---

*Fallacy*   2-5.2      **Risk from VOLL Pricing is Beneficial**
Because the risk of extreme price swings gives the market an opportunity to sell
new products to ensure customers against the imposed risk, imposing these risks
on the market is socially beneficial.

---

The concept of risk is frequently misunderstood, but there is a simple test for
what is caused by risk and what is not. Market participants can be of two types:
*risk averse* and *risk neutral*. A participant who is risk neutral does not respond to
risk (the standard deviation of price) but instead responds only to average prices.
No behavior exhibited by a risk-neutral party can be a response to risk.

The beneficial incentive properties of VOLL pricing are not due to risk but
simply to the costs it imposes and the profit it delivers when price is high. A risk-
neutral consumer will cut consumption when the price is high, and a risk-neutral
investor will build peakers and keep them ready at a moment's notice to capture
VOLL prices. This proves that these beneficial responses are in no way due to risk.
The model in the previous section assumed risk-neutral suppliers and consumers,
and the result was optimal. Had risk aversion been included, the result would have
been suboptimal, due to the costs of risk.[7]

**Market Clearing.**   The various NECA reliability reports frequently refer to "the
ability of the market to clear voluntarily" and, as quoted previously, elevate this
to the "core principle" of VOLL pricing. Market clearing is a key economic principle, but it is not an end in itself. In this case the approach to market clearing can
be carried too far. Because demand cannot yet respond effectively to real-time
prices, it makes sense to have the system operator bid some average value on their
behalf. By bidding $1,000,000/MWh, the operator could ensure that the market
cleared for all but five minutes per year (as an average over many years). This would
not be an improvement. Ideally, there should be some load shedding because it
is cheaper than buying reserve generation. When load is shed, the market has not

---

7. There is one indirect way in which risk may be socially useful. Risk will induce the signing of more
long-term contracts, and long-term contracts reduce market power in the spot market. This conclusion
ignores any impact on long-run market power, so further study is needed.

cleared. Until the market flaws described in Section 2-3.1 are removed, the goal is to have the optimal, not the minimal, number of nonclearing hours.

**Long-Run Incentives.**  Adequate investment in generating capacity is the central goal of VOLL pricing. It is a long-term process that has no need for incentives that change drastically from minute to minute or even from month to month. VOLL price spikes are not needed to encourage adequate generation investment. In fact, VOLL payments are extremely risky and this makes investing more expensive. The riskiness is not due to the hourly price fluctuations but rather to the annual fluctuations in price-spike revenue. If the five hours of price-spikes were made up of 50 independent 6-minute events, revenues would be quite dependable. But they are made up of very few events. Consequently, VOLL price-spike revenue fluctuates dramatically from year to year and is difficult to estimate.

**Short-Run Incentives.**  VOLL pricing is not needed to induce generators to provide spinning reserves. Spin must be provided in advance of a contingency and so before the price increases. When price does jump to $V_{LL}$, generators provide energy, not spin. In other words the possibility of $P = V_{LL}$ raises the **expected** profitability of providing spin. Of course the same effect can be achieved by paying them this expected value with certainty instead of on average. Once generators are spinning, a price of $15,000/MWh will certainly induce them to provide energy but so will a somewhat lower price—probably $300/MWh would work just as well.

VOLL pricing is most advantageous as a short-term incentive for demand reduction, which can require a genuinely high price. Unfortunately, many risk-management schemes will severely curtail this incentive. Many customers will opt not to see the real-time price, and for now most cannot see it. In these cases $V_{LL}$ is useless as a short-run incentive. But well-designed contracts can reduce customer risk dramatically while leaving the VOLL incentive in place, and eventually much more load will be metered. Load may become quite price-elastic, and most of the beneficial reduction in required generating capacity may come from load's response to prices 20 times lower than $V_{LL}$. Still, when load has to be shed, VOLL provides the correct incentive for customers to reduce demand.

**Market Power.**  Perhaps the greatest drawback to VOLL pricing is not risk but market power.[8] Consider 2 possible price caps, 1 at $500 and 1 at $20,000. Consider a supplier with all of its 2000 MW of generation dispatched and an average variable cost of $50/MWh. Assume that the market price is $100, which is the marginal cost of the marginal plant. The supplier in question is earning a short-run profit of $50/MWh on its 2000 MW for a total of $100,000/h.

Suppose load is 18,200 MW and available supply is 20,000 MW. If the supplier withholds 1900 MW it can push the price to either $500 or $20,000, depending on the cap. At $500 it would earn $450/MWh on the 100 MW it still has in production, which comes to $45,000/h—only half of what it earns at full output. At $20,000 it would earn $19,950 on 100 MW, or $1,995,000/h, which is almost 20 times more than it earns at full output. With a $500 price cap it loses profit by

8. Joskow (2001a) makes a similar suggestion. ". . . the combination of relatively tight supplies and extremely inelastic demand means that prices can rise to extraordinary levels and are much more susceptible to market power problems than when supplies are abundant."

**Figure 2-5.2**

An unusually steep load-duration curve.

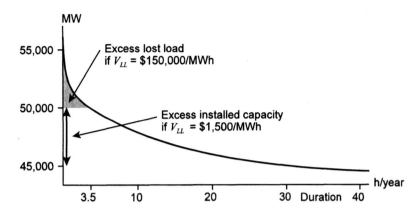

attempting to exercise market power. With a $20,000 price cap it makes enormous profits by exercising market power. VOLL pricing provides strong incentives for the exercise of market power.

## 2-5.4   TECHNICAL SUPPLEMENT

Misestimating $V_{LL}$ may cause a relatively small decrease in the overall efficiency of the power market. To demonstrate this, assume that $V_{LL}$ has been estimated to be $15,000/MWh and consider two possibilities: (1) actual $V_{LL}$ could be $1,500/MWh, or (2) it could be $150,000/MWh. How much inefficiency is associated with each possibility?

Using $6/MWh for the fixed cost of a peaker as in previous examples and the three values of $V_{LL}$, three optimal load-shedding durations may be calculated using Equation 2-5.1.

**Table 2-5.1** Results of Errors in Estimation of VOLL

| VOLL | Duration, $D_{LS}^*$ | | $K$ in MW | Comment |
|---|---|---|---|---|
| $150,000/MWh | $D_{LS}^* =$ | 0.35 h/year | 55,000 | Possible optimal value |
| $15,000/MWh | $D_{LS}^e =$ | 3.50 h/year | 50,000 | Assumed value |
| $1,500/MWh | $D_{LS}^* =$ | 35.04 h/year | 45,000 | Possible optimal value |

These optimal durations must be translated into installed capacities using a load-duration curve. The steeper that curve, the greater the error in the installed capacity level and thus the greater the resulting inefficiency. PJM's load-duration curve will be used for this example after exaggerating its steepness enough to prevent any chance of underestimation. The installed capacity levels corresponding to the calculated durations are shown in Table 2-5.1.

If the true value of $V_{LL}$ is $150,000, then 5000 MW too little capacity would be installed with the result that too much load would be shed. From Figure 2-5.2 it can be seen that excess load shedding would be less than one third of 3.5 h × 5000 MW, or about 7000 MWh. Using the true $V_{LL}$ of $150,000 and dividing by 8760 h/year gives a reliability cost of $120,000/h. If the true value of $V_{LL}$ is $1,500, then

5000 MW too much capacity would be installed with an excess fixed cost of $6/MWh × 5000 MW which equals $30,000/h.

Returning to the first possibility, the cost of the extra lost load is partially compensated for by the reduced cost of installed capacity, again $30,000/h. So the net excess reliability cost in this case is $90,000/h. The total cost of serving load in PJM is about $30/MWh × 30,000 MW, or $900,000/h. If $V_{LL}$ was actually $150,000, while VOLL pricing was based on an estimated $V_{LL}$ of $15,000, the resulting excess reliability cost would be 10% of the total cost of power. Because the steepness of this load-duration curve has been exaggerated, the actual cost of such a mistake in PJM might be considerably less.

In conclusion, it would appear to be safer to over-build than to under-build relative to an estimated level of $V_{LL}$. Either a dramatic underestimation of $V_{LL}$ or a dramatic overestimation of the optimal duration of load shedding (3.5 h/year when 0.35 h/year is the correct value) would result in a significant, though not dramatic, cost of unreliability.

*Chapter 2-6*
# Operating-Reserve Pricing

*By September [1879] a little building at Fourth and Market was completed and two tiny Brush arc-light dynamos were installed. Together they could supply 21 lights. Customers were lured by the unabashed offer of service from sundown to midnight (Sundays and holidays excluded) for $10 per lamp per week. Yet in light-hungry San Francisco, customers came clamoring. By the first of the next year, four more generators with capacity of more than 100 lights had been added. Electricity had come to the West.*[1]

Pacific Gas and Electric<br>Our History

**O**PERATING RESERVES ARE NEEDED TO SOLVE VERY SHORT-RUN RELIABILITY PROBLEMS, BUT THEIR PRICING CONTROLS THE LONG RUN AS WELL. Engineering suggests appropriate levels for operating reserves, but it cannot, on its own, determine what price to pay for them. Surprisingly, their price should depend on the value of lost load (VOLL) and on long-run, more than short-run, reliability considerations. By setting prices to a relatively modest level when the system is short of operating reserves, rather than to the extremely high value of VOLL when the system is short of capacity, operating-reserve (OpRes) pricing can substitute for VOLL pricing. This opens up a wide range of policy options which can be used to solve some of the most pressing problems of today's power markets.

**Chapter Summary 2-6:** A market with random shifts in the annual load-duration curve is examined to compare the side effects of high and low price spikes. High price spikes are found to cause investment risk and to encourage the exercise of market power. Low spikes are just as effective as VOLL pricing at encouraging optimal investment in generation capacity. High price spikes are more useful on the demand side than on the supply side, so different price limits should be used for the two sides of the market.

**Section 1: Less Risk, Less Market Power.** In a market with two load-duration curves, one for "hot" years and one for normal years, short-run profits are found to fluctuate between zero and 400% of normal under VOLL pricing but only

---

1. This plant preceded Edison's "invention of the electric light" and opened three years before Edison's Pearl Street power station—the "first" central station. Charles Brush lit Broadway, New York from a central station in 1880. (See http://www.pge.com/009_about/past/&e=921.)

between 80% and 160% of normal under OpRes pricing. One supplier selling 5% of total ICap into the spot market would exercise enough market power under VOLL pricing to raise the average price of power by more then $50/MWh but could raise it only by $5/MWh under OpRes pricing.

**Section 2: How Can OpRes Pricing Be Better than Optimal?**   VOLL pricing is optimal according to simple economic theory because it offers to pay what power is worth. This is an advantage of VOLL pricing if there are sources of supply or demand that would respond to the $15,000 VOLL price but not to a $500 OpRes price. This advantage must be traded off against the negative side effects of high price spikes. The best solution would be to allow high demand-side prices while keeping supply-side prices relatively low.

## 2-6.1   LESS RISK, LESS MARKET POWER

Markets that do not use VOLL pricing usually have an operating reserve (OpRes) requirement backed by high prices; this will be called **OpRes pricing**. These prices are often higher than needed to attract operating reserves from the local market, but they serve to compete with other control areas for reserves and to induce investment in generating capacity. PJM has never had a generator in its market that would not provide operating reserves for a price of $200/MWh. Yet if PJM runs short of reserves, it offers to pay up to $1,000/MWh to any generator that has bid that high.[2]

   This section compares VOLL pricing with OpRes pricing in three dimensions: (1) equilibrium installed capacity, (2) risk, and (3) market power. To model risk, randomness is introduced into the load-duration curve. Two such curves are specified, one for "hot" years and one for normal years. Hot years could be years that are literally hot or that have unusually high load or generator outages for other reasons. Market power is modeled by assuming that one generator sells an amount of energy equal to 5% of the market's total capacity into the spot market.

### A Model with Risk and Market Power

The present example is designed to have an optimal installed capacity level of $K = 50,000$ MW and is assumed to have a true value of lost load known to be $V_{LL} = \$15,000/\text{MWh}$. The fixed cost of peakers is again assumed to be $FC_{peak} = \$6/\text{MWh}$. According to Result 2-3.2, the optimal duration for load shedding is $D_{LS}^* = 3.5\text{h/year}$. A pair of load-duration curves, one for hot and one for normal years, are designed to give this duration for the optimal capacity level. Near peak

---

2. This statement considers PJM as an isolated system, which it is not, and so ignores the opportunity cost of not selling power out of PJM. Competition between systems is analyzed in Chapter 2-9.

load, these two load-duration curves (with duration measured in h/year) are given by the following equations:[3]

In hot years (25%): $D = 3.5 \times (52 - L_g/1000)^2$, for $L_g < 52$.

In normal years (75%): $D = 3.5 \times (50 - L_g/1000)^2$, for $L_g < 50$.

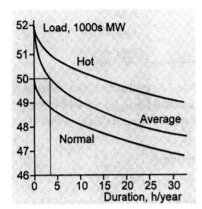

Values of $L_g$ not covered by these equations have a duration of zero. Note that augmented load, $L_g$, which includes generation outages, has been used to determine duration, so these are actually augmented-load duration curves (see Section 2-3.3). In hot years, the peak (augmented) load is 52,000 MW, so the duration of $L_g$ at this load level is zero, but the duration at a load of 50,000 MW is $3.5 \times 2^2$, or 14 h/year. Because only 25% of the years are hot, and because normal years have zero duration at a load of 50,000 MW, the average duration at that load is 25% of 14, or 3.5 h/year. This is the optimal duration for load shedding, which confirms the fact that $K = 50,000$ MW is the optimal level of installed capacity.

One supplier in the market is assumed to have 2500 MW of capacity to sell into the real-time market. With this capacity it could earn the high spot prices generated by VOLL or OpRes pricing, or it could exercise market power in the spot market. It could withhold, for example, 2400 MW in order to cause $L_g$ to exceed 50,000 MW, cause load to be shed, and cause the market price to be set at $15,000/MWh thereby profitably earning $1,500,000/h on its remaining 100 MW of operating generation.

## Equilibrium Installed Capacity

As demonstrated in Chapter 2-4, VOLL pricing induces the optimal level of installed capacity, so comparing the two approaches requires only an evaluation of the installed capacity induced by OpRes pricing. This will of course depend on both the required level of operating reserves, $OR^R$, and on the price paid when the system is short of operating reserves, $P_{cap}$.[4]

Contrary to many predictions that a low price cap would prevent adequate investment, the two policy variables provide more than enough control to induce optimal investment or more. For this example, $OR^R$ will be arbitrarily set to 5000MW and only $P_{cap}$ will be used to design an optimal pricing policy. From Result 2-2.2, the long-run equilibrium condition for installed capacity is $FC_{peak} = R_{spike}$. Price-spike revenue equals $P_{cap}$ times the duration at which price is at this level. Price is at the cap whenever $L_g > K - OR^R$, i.e., when $L_g > 45,000$ MW. The following table presents durations for various useful load levels.

---

3.  Duration measured in h/year is dimensionless (time/time gives a pure number) as required but must be divided by 8760 to produce the numeric value required in formulas using duration. (For example, 1h/year = 1h/8760h = 0.000114.) See Chapter 1-3 for a more complete explanation.

4.  $P_{cap}$ will often be referred to as a price cap, and although it functions as one, it is not a cap in the traditional sense of a legal restriction on the actions of private parties. It is instead a purchase price limit on the system operator. See Chapter 2-4.

**Table 2-6.1** Durations of Load Shedding

| Augmented Load | Duration In h/year | Duration In % |
|---|---|---|
| 50,000 MW | 3.5 | 0.04% |
| 47,520 MW | 33.7 | 0.38% |
| 45,000 MW | 108.5 | 1.24% * |
| 43,050 MW | 196.9 | 2.25% |

Table 2-6.1 shows that with $OR^R$ = 5000 MW and $K$ optimal, price will be set to $P_{cap}$ for 108.5 h/year, so the price spike revenue will be $R_{spike}$ = 0.0124 × $P_{cap}$. To obtain the required price-spike revenue for a long-run equilibrium, $6/MWh, the price cap must be set to

$$P_{cap} = (\text{equilibrium } R_{spike})/(\text{optimal } D \text{ at } P_{cap})$$
$$= (\$6/MWh)/0.0124 = \$484/MWh.$$

more than 30 times less than $V_{LL}$. With this price cap, and an OpRes requirement of 5000 MW, the market's long-run equilibrium level of installed capacity will be optimal. With any higher level of $P_{cap}$, too much generation will be built, more than would be built with a price cap of $15,000/MWh under VOLL pricing.

---

**Result    2-6.1**    **Many Different Price Limits Can Induce Optimal Investment**
If the system operator pays only up to $P_{cap}$ but no more any time operating reserves are below the required level, a low value of $P_{cap}$ will suffice to induce the optimal level of installed capacity. The higher the reserve requirement, the lower the optimal price limit will be. Capacity requirements can further reduce the optimal level of $P_{cap}$.

---

## Risk

Under both pricing systems, price-spike revenues are high in hot years and low in normal years—under VOLL pricing these fluctuations are dramatic. In hot years the market price is set to $V_{LL}$ for 14 hours, which produces an average annual price-spike revenue of $24/MWh, four times the amount needed to cover the fixed costs of peakers. In normal years, augmented load never exceeds capacity and there are no price spikes. Peakers recover none of their fixed costs, and baseload plants recover only a fraction of theirs.

Under OpRes pricing, the duration of high prices is 3.5 × (52 − 45)² in hot years and 3.5 × (50 − 45)², which is 172 and 88 hours per year, respectively. These durations produce price-spike revenues of $9.48/MWh in hot years and $4.84/MWh in normal years, much less variation than occurs under VOLL pricing. While the standard deviation of price-spike revenues is about five times greater than under OpRes pricing, a better appreciation of the difference in profit dynamics can be gained by examining a sequence of years. To this end, Figure 2-6.1 displays a 40-year sequence of price-spike revenues generated by a Monte-Carlo simulation. To

**Figure 2-6.1**

One realization of a
Monte Carlo simulation
of profits under VOLL
and OpRes pricing.

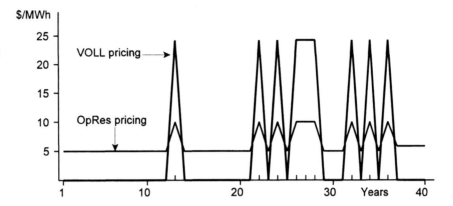

avoid selection bias, the first draw from the simulation has been presented.[5] Notice
that the sequence begins with 12 years in which peakers cover none of their fixed
costs under VOLL pricing. This is rather unlucky. If the market had opened instead
in year 22, generators would have recovered four times their annual fixed costs
in five of the first seven years—a bonanza.

Although the market generating these profits is in perfect long-run equilibrium
every year, investors would surely begin to doubt this after losing all of their fixed
costs twelve years running. Economics may suggest that investors are rational and
will realize this is just a string of bad luck, but the suggestion only makes sense
if they have experienced this sort of phenomenon enough times to learn from
experience—that might take a few hundred years.

Even under OpRes pricing, investors will be discouraged, but the reaction should
set in less quickly as they will recover 80% of their fixed costs in normal years.
Such a market might still require a significant risk premium, and it is worthwhile
to look for a less risky design. Another factor, long-term contract cover, should
be considered, as it mitigates risk under both pricing schemes.

---

**Result      2-6.2          A Lower, Longer-Duration Aggregate Price Spike Is Less Risky**
A short-duration aggregate price spike causes much more year-to-year variance
in short-run profits than does a long-duration aggregate price spike.

---

**Long-Term Contracts.**  Generators typically sell part of their power in long-term
contracts that last anywhere from one to twenty years. If all power were sold under
such contracts and they were rationally and accurately priced, generators would
be fully hedged against the vicissitudes of the spot market. Understanding this point
requires some background in the role and operation of long-term markets.

Both suppliers and consumers have a choice between buying in the long-term
forward markets or buying in the spot market, and both have a preference for buying
in the forward markets so they can lock in a price and avoid the volatility of the

---

5. The initial string of normal years is not an artifact of this being the first draw, as Excel's random
number generator was used extensively before this sample was drawn.

spot market. They may also see some advantages to the spot market, such as flexibility in trading, but the behavior of power markets indicates the motive to hedge is stronger and most power will be sold under at least moderately-long-term agreements. Because both parties can choose the spot market, the long-term price must be quite close to the spot price. Otherwise one party will find its preference for certainty overcome by its preference for a favorable price and will move its trade to the spot market.

Because of this arbitrage condition, long-term contracts will reflect average spot market prices. This is how VOLL and OpRes pricing, which apply only to the spot market, can act with full force on the investment decisions of generators. If the spot price is high, then even though an investor plans to sell no power in the spot market, it knows this high price will be reflected fully in the long-term market in which it plans to sell its power. Of course one year's spot prices do not determine the price of long-term contracts sold at the end of that year. Such contracts are priced according to the expected long-run average of future spot prices. Nonetheless, if VOLL prices are expected to be high on average, long-term contract prices will be equally high.

Because long-term contracts are based on expected long-run averages, they smooth out the year-to-year volatility of the spot market. But if spot prices are extremely low for four years at a time, which would be a common occurrence in the present model, this will inevitably have a significant impact on long-term expectations and prices. Thus, even if the market is in perfect long-run equilibrium (optimal installed capacity) every year, as in the present Monte Carlo simulation, fluctuations in long-run prices must be expected, and they will be more extreme under VOLL pricing than under OpRes pricing. Because of this and the fact that many generators may sell significant amounts of power in the spot market or short-term markets, VOLL pricing will impose more risk on the market than OpRes pricing.

The random fluctuations in long-term prices, caused by runs of good and bad luck in the spot market, will be reflected in long-run fluctuations in investment and in installed capacity. Consequently the market will not always be in long-run equilibrium, and this will cause even greater fluctuations in scarcity rents collected from the spot markets.

## Market Power

In this example, the supplier with market power is assumed to be selling power from 2500 MW of capacity into the spot market. Thus, under VOLL pricing, if load is 47,501 MW, the supplier can withdraw its capacity and force load shedding, thus sending the price to $15,000/MWh. Unfortunately, this leaves no output on which it can earn a profit.

With a price increase of this magnitude, it would need only a little output to break even. Say its variable cost is $30/MWh and the market price is $130/MWh before its exercise of market power. Raising the price to $V_{LL}$ increases its profit per MW sold by about 150 times $[(15,000 - 30)/100]$. Thus it will increase its profits if it can raise price to $V_{LL}$ and retain only 20 MW of output. When load is greater than 47,520 MW, this supplier will exercise market power and raise the price to $V_{LL}$. This load level will occur for 33.7 h/year in normal years which is almost 10 times the optimal duration. Thus under VOLL pricing, the exercise of market power by this supplier will raise the price-spike revenue by almost a factor of 10. The result will be a $52/MWh increase in the average cost of all power sold in this market.

Under OpRes pricing, the same supplier would have more difficulty exercising market power because it could increase profits by only a factor of $(P_{cap} - VC)/(P - VC)$, which comes to 3.54, much less than the 150 times increase for VOLL pricing. Still, the supplier could withhold 77% (1950 MW) of its capacity and profit from the resulting price increase. Thus, whenever load increases above 45,000 – 1950 MW, or 43,050 MW, which occurs for 197 h/year, the price will be set to $P_{cap}$. Since under OpRes pricing, the competitive duration of the price spike is 108 h/year, the increase in price-spike revenue is less than a factor of two—a little less than $5/MWh.

---

| **Result** | **2-6.3** | **High Price Caps Invite the Exercise of Market Power** |
|---|---|---|

High price caps allow a supplier to withhold a greater percentage of its capacity and still profit from the resulting increase in market price. Additionally, the ratio of the duration of maximum prices with and without the exercise of market power is greater for low-duration price spikes. Because high price caps must be designed to produce low-duration price spikes, high price spikes exhibit a greater percentage increase in duration as a result of an equal level of withholding.

---

There are two distinct mechanisms that work together to produce increased withholding of capacity under high price caps. First, a high cap allows a much greater increase in profit from the exercise of market power. This allows the supplier to withhold a greater percentage of its capacity (99.3% vs. 77%) while still experiencing a net increase in profit.

Second, high price caps are necessarily associated with short-duration price spikes; otherwise they would produce enormous excess profits. The percentage increase in price-spike duration is greater when applying a fixed amount of capacity withholding to a short-duration spike than to a long-duration spike. This is a general property of load duration curves. For every additional fixed MW decrease in capacity, the load duration increases by a smaller percentage than for the previous decrease. Thus, high, short-duration price spikes exhibit a greater percentage increase in duration (10 times vs. twice) for a given amount of withholding than do lower, longer-duration price spikes.

This demonstrates what appears to be a general tendency for lower but longer-duration price spikes to be less conducive to the exercise of market power. In the short run, excess market power could result in significant transfers of wealth, but in the long run it, will simply attract new investment and increase the level of installed capacity. The resulting reliability level will be difficult to predict because the extent of market power and the effect of withholding are difficult to predict. This phenomenon is considerably more complex than the reduction of risk under OpRes pricing and deserves further investigation.

## 2-6.2   HOW CAN OPRES PRICING BE BETTER THAN OPTIMAL?

Chapters 2-3 through 2-5 argued that VOLL pricing was the economically sensible approach. It sets the market price to the value that demand would collectively place on power were it able to express its collective (average) valuation. The preceding example suggests that the market price could be set 30 times lower with no ill effects. What desirable property of high, value-based prices has been overlooked in the current analysis?

Economics suggests the use of VOLL pricing because it will send the right signals to all possible suppliers and loads. If $V_{LL}$ is actually $15,000/MWh and some generator has a variable cost of $14,000/MWh, offering to pay $15,000/MWh will provide a net social benefit of $1,000/MWh for every MWh purchased if it is used to reduce load shedding. With $P_{cap}$ set to $500, this opportunity would be missed. The importance of such missed opportunities depends on how much generation has a marginal cost greater than $P_{cap}$.

From the perspective of taking advantage of all possible trading opportunities, VOLL pricing is optimal and OpRes pricing is suboptimal. But if the trading opportunities missed by OpRes pricing are few and if the negative side effects of VOLL pricing are great, OpRes pricing will be superior in the real world even though it is suboptimal in the simplest of economic models.

The demand side of the market may contain more important missed opportunities than the supply side. While generation technology is fairly uniform and known to be capable of producing very little output at marginal costs above $500/MWh (with normal fuel costs), the demand side is far more complex. The very definition of VOLL, as average value, indicates there are many uses of power with higher value. So the spectrum of demand-side values can neither be limited to the low levels associated with the marginal costs of production nor predominately higher than VOLL. Almost by definition, there must be a great amount of demand with values between the low levels of OpRes prices and the high levels of VOLL prices, and all of these are opportunities potentially missed by a low price cap. But the qualification, "potentially," is crucial. If these opportunities were all real, the market would be functioning properly and would not need a price limit. In fact, it would always clear at a price far below VOLL.

VOLL pricing will have a significant advantage on the demand side only if demand is so unresponsive to price that VOLL pricing is needed but responsive enough that a price of $15,000 will illicit significantly more demand reduction than a price of $500/MWh. This is a possibility, and the opportunities on the demand side should not be missed if they can be captured without even more damaging side effects.

**Side Effects.**   The side effects of risk and market power are potentially serious, so the trade-off between these and the advantage of stimulating more demand response should probably favor low price spikes, especially until a new power market has clearly stabilized. Although the California market suffered from numerous flaws unrelated to its core structure, it seems quite likely that the type of profit instability associated with high price spikes played an important role. During its first two years of operation, profits were low and investment was discouraged. In the third year, augmented load in the Western region was high, both because of rapid growth in real demand and a reduction in available hydroelectric output The result was very tight conditions in the California market, mostly because imports dropped by close to 4000 MW. This caused price spikes and these encouraged the exercise of market power which exacerbated both the high prices and the shortages.

---

| Result | 2-6.4 | **Reliability Policy Should Consider Risk and Market Power** |
|--------|-------|--------|

**Reliability Policy Should Consider Risk and Market Power**
The right average installed capacity will provide adequate reliability, but two side effects of reliability policy should also be considered. Infrequent high price spikes increase uncertainty and risk for investors which raises the cost of capital, and in extreme cases, causes political repercussions. The possibility of extremely high prices also facilitates the exercise of market power.

---

## A Separate Demand-Side Cap

The obvious solution to the dilemma of needing high price spikes on the demand side and low price spikes on the supply side is to have separate caps for the two sides of the market. At most times, the market would clear and the two prices would be equal. During a shortage, however, load could be charged more than generation is paid. This would result in a profit for the system operator, which would be saved in a balancing account and returned to load as a rebate spread over as many hours as possible. In effect it would probably just reduce other distorting charges placed on load for such purposes as paying for the fixed costs of transmission and the services of the system operator.

Even though demand-side prices should be higher than supply-side prices, they should still be capped at VOLL or less. High prices on the demand side would encourage load-serving entities to develop real-time rates in order to reduce their costs. Once such rates are in place, high prices on the demand side will serve to encourage the rapid development of price responsiveness, which will eventually make the demand-side cap irrelevant.

*Chapter 2-7*
# Market Dynamics and the Profit Function

*An electrick body can by friction emit an exhalation so subtile, and yet so potent, as by its emission to cause no sensible diminution of the weight of the electrick body, and to be expanded through a sphere, whose diameter is above two feet, and yet to be able to carry up lead, copper, or leaf-gold, at the distance of about a foot from the electrick body.*

Sir Isaac Newton
Samuel Johnson's Dictionary of the English Language, 1755

E conomics focuses on equilibria but has little to say about the dynamics of a market. Once economics shows that a system has a negative feedback loop so that there is a point of balance, it considers its job done. Engineers move beyond this stage of analysis to consider whether a system will sustain oscillations and, if not, whether it is over- or under-damped. Economics understands that investment dynamics can produce "cycles" but has faith that rationality will generally prevent this. It also ignores the noise sources (randomly fluctuating inputs) that keep economic systems excited.

Usually these oversights do not offer much cause for concern. In power markets, however, a 4 or 5% fluctuation in either load or capacity, coupled with the wrong pricing policy, can cause the average annual spot price to triple. Such dynamics cannot be ignored. If they are not corrected at the time of market design, they will be reported later by the press.

Many profit functions determine the same optimal equilibrium value of installed capacity, but this means only that they agree at one point. At other points, they may differ dramatically, and these differences imply different market dynamics. Although the profit function falls far short of providing a theory of those dynamics, it does provide some basic insights which make possible a discussion of the topic.

**Chapter Summary 2-7:** Profit functions can be calculated from the load duration curve and two policy variables: the price cap and the required level of operating reserves. Once calculated they reveal the equilibrium level of installed capacity and give some indication of the market's riskiness and conduciveness to the exercise of market power. Thus the first step in assessing a pricing policy should be the calculation of the associated profit function.

**Section 1: Calculating Profit Functions.** The profit functions give expected short-run profits as a function of installed capacity, $K$. It is most convenient to calculate them for the peaker technology, in which case short-run profit is the same as price-spike revenue. This is determined by the price-cap level and the duration of the price spike. This duration is determined from the load-duration curve and the required level of operating reserves, which triggers the price cap. The profit function is found to be much steeper under VOLL pricing than under OpRes pricing.

**Section 2: Interpreting the Profit Function.** The profit function decreases as $K$ increases, eventually crossing the level of peaker fixed costs. To the left, because short-run profits exceed fixed costs, economic profits are positive and there is an incentive to invest. To the right, profits fail to cover fixed costs, so investment stops, while load growth continues. As a result, long-run equilibrium installed capacity is determined by the intersection of the profit function and the level of fixed costs.

If the function is steep, then an increase in load, or withholding of capacity, both of which mimic a decline in $K$, will cause a large increase in profit. The first implies that steep profit functions are risky, and the second implies that they increase market power.

## 2-7.1   CALCULATING PROFIT FUNCTIONS

Chapter 2-6 modeled a power market and two pricing limits for the system operator: (1) the VOLL pricing limit, and (2) the OpRes pricing limit. Under VOLL pricing the system operator sets the spot-market price to $V_{LL}$ ($15,000/MWh in this example) whenever augmented load, $L_g$, exceeds installed capacity, $K$. Under OpRes pricing, the price is set to $484/MWh (in this example) whenever operating reserves, $OR$, fall below the operating reserve requirement, $OR^R$, of 5000 MW.

The market is characterized by two load-duration curves, one for "hot" years and the other for normal years, with the hot ones occurring 25% of the time and the normal ones 75%. Together these produce a single long-run average load duration curve which will be sufficient for the calculation of the profit function because this function suppresses the year-to-year randomness examined in the previous chapter.

A **profit function** plots long-run average profits for a particular type of generation technology, usually for the peaker technology.[1] In this case, short-run profits are equal to the price-spike revenue. As explained in Chapter 2-6, these fluctuate from year to year due to fluctuations in the augmented-load duration curve. Over the long run, price-spike revenues average out, and that average exactly equals the price-spike revenues that would result from the long-run average load-duration curve. So, although the average load-duration curve is not correct for any particular

---

1. The profit function gives short-run (business) profits, not long-run (economic) profits.

year, it is exactly what is needed for computing long-run average price-spike revenue, $\overline{R}_{spike}$.

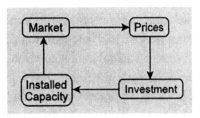

The purpose of constructing the profit function is to understand the feedback loop discussed in Chapter 2-1 which controls investment and installed capacity. The price-spike revenue is the appropriate summary statistic for market prices because it connects prices to investment by a very simple rule. $\overline{R}_{spike}$ encourages investment when and only when it exceeds the fixed costs of a peaker.[2] The other key variable in this feedback loop is installed capacity, $K$. The profit function connects (short-run) profit to installed capacity.

### The profit function: $SR_\pi(K)$

In the model under consideration, demand is not elastic; price spikes are simply a matter of the price being set to the price cap whenever augmented load exceeds a certain threshold. The number of times this occurs, together with the price cap, determines the price-spike revenue. The load-duration curve is essential in making this determination, and it will be represented algebraically as follows:

### The load-duration function: $D(L_g)$

The threshold that $L_g$ must exceed to cause the system operator to set price to the price cap is $K - OR^R$, installed capacity minus the operating reserve requirement. In the case of VOLL pricing, $OR^R = 0$, so price is set to $V_{LL}$ whenever $L_g$ exceeds installed capacity and load must be shed. Whenever $L_g > K - OR^R$, then $K - L_g < OR^R$. But $K - L_g = OR$ by the assumptions of the Simple Model of Reliability (Chapter 2-3), so the last inequality implies $OR < OR^R$, or operating reserves are less than required. This is the condition for setting price to $P_{cap}$ under OpRes pricing. So, under either pricing system,

Price spikes occur when: $L_g > K - OR^R$.

The duration of their occurrence is given by $D(L_g)$, where $L_g = K - OR^R$. This gives the rule for finding price-spike duration:

Price-spike duration = $D(K - OR^R)$

Price-spike revenue is simply the duration of the price spike times the difference between the price cap and the variable cost of the peakers. Thus

$$SR_\pi(K) = D(K - OR^R) \times (P_{cap} - VC_{peak}) \qquad (2\text{-}7.1)$$

This is how to compute the profit function in the present model. Calculation would be more complicated if there were a supply or demand bid above $VC_{peak}$, but the concepts would be the same. Notice that $SR_\pi(.)$ is considered a function of $K$, although according to the right side of Equation 2-7.1 it depends on four different values. Two of the other three are policy parameters, $OR^R$ and $P_{cap}$, while the last, $VC_{peak}$, is a technology parameter. Regulators (including system operators and

---

2. Each type of installed capacity has its own equilibrium condition, and it is possible to need more baseload but less peaker capacity. Focusing on peaker investment is a simplification, but the market provides correct signals for the relative investment in off-peak technologies. Thus, if reliability policy induces the right level of investment in peakers, the market will take care of other types of investment.

engineers) use the policy parameters, consciously or unconsciously, to control the profit function and thereby control equilibrium installed capacity, $K^e$.

The long-run average load-duration curve given in Chapter 2-6 is

$$D(L_g) = .0004 \times [.25 \times (\max(0, 52 - L_g))^2 + .75 \times (\max(0, 50 - L_g))^2] \quad (2\text{-}7.2)$$

The constant, 3.5, used previously has been multiplied by $\frac{1}{8760}$ so that duration is no longer given in hour/year but is simply a number, 0.0004, and thus can be used directly in Equation 2-7.1. Using this function, the profit function for both VOLL pricing and OpRes pricing was calculated, and the results are displayed in Table 2-7.1 and graphed in Figure 2-7.1.

**Figure 2-7.1** Profit functions

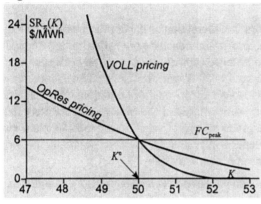

**Table 2-7.1** Profit functions

| Installed Capacity (1000 MW) $K$ | SR$_n(K)$ in \$/MWh | |
|---|---|---|
| | VOLL Pricing $P_{cap}$ = \$15,000/MWh $OR^R$ = 0 MW | OpRes Pricing $P_{cap}$ = \$484/MWh $OR^R$ = 5000 MW |
| 47 | 78.00 | 14.13 |
| 48 | 42.00 | 11.03 |
| 49 | 18.00 | 8.32 |
| 50 | 6.00 | 6.00 |
| 51 | 1.50 | 4.06 |
| 52 | 0.00 | 2.52 |
| 53 | 0.00 | 1.35 |

**Result    2-7.1**    **The Higher the Price Spikes, the Steeper the Profit Function**
A pricing policy that relies on a higher, shorter-duration price spike to stimulate investment is characterized by a steeper profit function.

## 2-7.2  Interpreting the Profit Function

VOLL pricing produces a much steeper profit curve than does OpRes pricing. This steepness is closely related to the negative side effects discussed in the previous chapter. But the main effect of the profits described by these curves is to produce the optimal long-run equilibrium, and the two profit functions show that the two pricing policies would serve this purpose equally well.[3]

**Equilibrium.**   Both profit functions intersect the level of the fixed cost of a peaker, \$6/MWh, when they reach an installed capacity of 50,000 MW. To the left of this they are greater than \$6/MWh, and to the right they are less. For lower levels of installed capacity, the extra profit induces investment, and for higher levels,

---

3. This is only strictly true when $K$ is deterministic. The next chapter shows that when it fluctuates randomly, the value of $OR$ should be a bit higher in order to induce the right average value of $K$.

investment is unprofitable and therefore will not be undertaken. Of course investors look to the future and base their decision on estimates of expected future levels of capacity. However, the main point remains true: Both pricing policies induce the same equilibrium level of installed capacity, and that level can be read from a graph of the profit function.

**Hidden Risk.**   When the profit function reports a short-run profit of $6/MWh, this is *not* the value that will be realized in any given year but, is instead, a long-run average value. As explained in the previous chapter, actual profit in any given year will be either $0/MWh or $24/MWh under VOLL pricing and will be either $4.84/MWh or $9.48/MWh under OpRes pricing. These annual variations are not shown explicitly, yet the profit function implicitly indicates their existence and relative magnitude.

**Implied Risk.**   Risk comes from the interaction of the unpredictability of aug-mented load with the predictable profit function. The price distribution depends on how load fluctuates relative to installed capacity. If $L_g$ is 1% higher in a particular year, it has the same effect on price as if $K$ were 1% lower. Thus, an increase in $L_g$ is equivalent to an equal move to the left on the profit function. If the function is steep, the resulting increase in profit is large; if the function is relatively flat, the increase is small. Thus for a given pattern of year-to-year load fluctuations, a profit function that is 4 times steeper will produce 4 times the variability in short-run profits. Steep profit functions are inherently risky.

**Implied Market Power.**   Although market-power analysis is more complex, the same analysis used with risk can be used to shed light on the problem of market power. Withholding 500 MW is equivalent to decreasing $K$ by that amount, and the average effect can be read from the profit function. If the function is steep, withholding is very effective at increasing profit.

---

**Result     2-7.2     Steeper Profit Functions Increase Risk and Market Power**
For the same level of year-to-year fluctuation in the load-duration curve, a steeper profit function will cause more variation in fixed-cost recovery and will increase the exercise of market power.

---

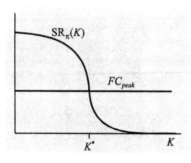

**How Steep Should the Profit Function Be?**   The steeper the profit function, the  greater the negative side effects, but the flatter the profit function, the less incentive there will be for the market to return to its equilibrium value. Because the optimal equilibrium value is not known with great precision, there is nothing gained by using a profit function that provides extremely strong restorative forces. If the optimal $K$ is known to be between 48 and 52 GW with a certainty of only 50%, there is no use in raising profit levels by $50/MWh when $K$ is only half a GW low. On the other hand, some slope to the profit function is absolutely necessary.

In most industries, profits 5% or 10% above the normal rate of return on investment are considered extremely good. A short-run profit function that pays double the fixed costs of generating capital will yield a profit level near 10% above normal. An extra $6/MWh in a market with an average price of $30/MWh would appear to be a reasonable incentive from the perspective of both the customer and the investor. But an excess short-profit level of $78/MWh, as is produced by VOLL pricing when $K$ is 6% (3 GW) below optimal, seems to be more than would be needed or could be tolerated. If this happened in a "hot" year, profits would be twice this high.

One compromise would be to have a profit function that is steep in the vicinity of the optimal level of installed capacity, $K^*$, but does not continue steeply once a sufficient level of incentive is reached. This provides an incentive that becomes strong quickly when the system deviates from optimal $K^*$ but which does not become excessive as $K$ gets lower or the load-duration curve suffers a random upward fluctuation. The next chapter shows how to design profit functions that are steep only at the level of optimal installed capacity.

*Chapter* **2-8**
# Requirements for Installed Capacity

*The shaft is 20 feet long and 6 ¼ inches in diameter. The wheel, which is 56 feet in diameter [is] provided with 144 blades twisted like those of screw propellers. The sail surface is about 1,800 square feet. The speed of the dynamo at full load is 500 revolutions per minute, and its normal capacity at full load is 12,000 watts. The working circuit is arranged to automatically close at 75 volts and open at 70 volts. The amount of attention required to keep it in working condition is practically nothing. It has been in constant operation more than two years.*[1]

Scientific American
December 20, 1890

**P**RICE SPIKES ENCOURAGE INVESTMENT INDIRECTLY; A CAPACITY REQUIREMENT GETS RIGHT TO THE POINT. The capacity approach is defined by two regulatory parameters but, as has been demonstrated, so is the price-spike approach. Both can induce any desired level of reliability while preserving the correct mix of technology, so the choice between them should be based on their side effects.

**Chapter Summary 2-8:** A capacity requirement produces an easily controlled, low-risk profit function. It can be combined with a price-spike approach to produce a profit function that is still relatively low risk while providing high prices at a few crucial times to tap existing high-priced resources. Combining price-spike and capacity-market profit functions does not increase equilibrium profits, but it does increase the equilibrium value of installed capacity unless the two policies are properly adjusted. Adjustment requires taking account of random fluctuations in the level of installed capacity.

**Section 1: The Capacity-Requirement Approach.** All load-serving entities are required to own, or to have under contract, a certain required capacity. The sum of these is the market's installed capacity (ICap) requirement and is typically about 18% greater than annual peak load. A load-serving entity is penalized if it fails to meet its requirement.

---

1. This wind generator was constructed by Charles Brush for his personal use. The article begins by noting that "After Mr. Brush successfully accomplished practical electric illumination by means of arc lights, incandescent lighting was quickly brought forward and rapidly perfected."

**Section 2: Short-run Profits With a Capacity Requirement.** Capacity requirements easily induce sufficient generation by setting a penalty level that is higher than the cost of new capacity. Random fluctuations in ICap help determine the equilibrium level of ICap. This must be taken into account when designing profit functions which show only the deterministic equilibrium.

**Section 3: Combining a Capacity Requirement with a Price Spike.** VOLL pricing produces an optimal average ICap level, and a capacity requirement can be designed to do the same. If these two policies are both implemented, the resulting level of ICap will be too high, but this can be corrected by adjusting policy parameters.

**Section 4: Comparing the Two Approaches.** Price spikes are sensitive to load fluctuations while capacity markets are not. By eliminating this source of risk, the capacity approach makes it easier to control reliability and suppress market power in the energy market. Price-spike systems have the advantage of sending efficient price signals to the demand side of the market and to expensive existing generators. A hybrid system may be best.

## 2-8.1   THE CAPACITY-REQUIREMENT APPROACH

Capacity requirements provide a fundamentally different approach to generation adequacy. Under the price-spike approach, energy revenues cover the fixed costs of generators. Under the capacity-requirement approach, capacity revenues completely cover the fixed costs of peakers and play a crucial role in determining the level of all types of installed generation capacity.

| Result | 2-8.1 | **Energy and Capacity Prices Together Induce Investment** |
|---|---|---|

Investment responds to expected short-run profits, which are determined by energy prices and (if there is an installed-capacity requirement) by capacity prices. Regulatory policies determining these prices need joint consideration.

Capacity requirements are a more direct approach to reliability than assigning high price limits when the system is short of operating reserves. This directness has its advantages. The capacity requirement bears a relatively clear and stable relationship to reliability. A side effect of the capacity approach is also helpful; it eliminates the need for high energy price spikes to induce investment. As explained in Chapter 2-6, these spikes cause investment risk and encourage market power, but the capacity approach has drawbacks of its own.

In a capacity market, demand is set by regulators who usually forget to make it elastic; inevitably, this leads to market power. Also, defining and verifying installed capacity is difficult. A third problem, discussed in the next chapter, occurs whenever a capacity market trades with a price-spike market.

## How a Capacity Requirement Is Implemented

Capacity markets often serve a dual purpose. While originally derived from regulatory requirements designed to assure adequate installed capacity, in a deregulated environment they are asked to perform double duty and provide short-term reliability. This requires rules to prevent "export" of capacity to neighboring systems when these have high prices. This short-term market role and its attendant regulatory complications are ignored in this chapter to focus clearly on the role of capacity markets in securing adequate ICap. This focus is maintained by assuming the market under discussion is isolated and does not trade with other markets.

The first step toward implementing a capacity system is to determine the optimal level of ICap. As explained in Section 2-5.3, this can only be accomplished by trading off the value of lost load against the cost of more ICap. This approach is never taken. Instead the crucial input is the acceptable number of hours of load shedding, an engineering constant shrouded in secrecy but said to be "one day in ten years."[2] With this in hand and knowing the reliability of generators and the variance of load, the amount of spare capacity required to attain this level of reliability can be calculated. But often it is not calculated and a traditional value is selected. Generally the requirement for ICap is in the neighborhood of 118% of expected peak load, though some systems, such as Alberta's, seem to survive on considerably less. This requirement is then divided among load-serving entities, which are retailers and regulated utilities, in proportion to their individual expected peak loads. This produces individual requirements for installed capacity.

These individual requirements must be met by either purchasing generation or contracting for its use. A contract must specify that the generator will be made available to the system if requested. Any load-serving entity that fails to own or contract for the required capacity is penalized. A second penalty applies to any generator that fails to perform when called on. This chapter assumes that the performance penalty is sufficiently effective so that it is rarely needed, and the penalty for insufficient capacity will be referred to simply as "the penalty." In PJM, this penalty has been set at about the level of a peaker's fixed cost, or about $7.38/MWh ($177/MWday when adjusted for forced outages).

## 2-8.2   SHORT-RUN PROFITS WITH A CAPACITY REQUIREMENT

### The Profit Function of a Pure Capacity Approach

If there is no market power in an isolated capacity market, the price of capacity will be either zero or equal to the penalty. If total installed capacity, $K$, is less than

---

2. How a parameter describing consumer economics was derived from astronomical constants remains a mystery. Why should the cost-minimizing value of load shedding equal the time it takes the earth to rotate once times the number of digits on two hands divided by the time it takes the earth to orbit the sun?

required capacity, $K^R$, the price will equal the penalty, $Y$. If $K > K^R$, the price should be zero.[3]

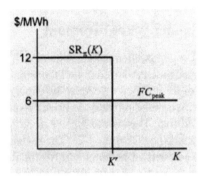

A capacity requirement is intended to induce investment when $K < K^R$, and will be successful if $Y$ is greater than the fixed costs of owning a new peaker, $FC_{peak}$. As with a price-spike policy which determines a profit function by setting $P_{cap}$ and $OR^R$, so a capacity-requirement policy determines a profit function by setting $Y$ and $K^R$. The profit function is the step function shown at the left.

Unlike the profit function for price spikes, there is no level of ICap for which $SR_\pi(K) = FC_{peak}$; profits are either zero or twice this value. In spite of this, the deterministic equilibrium ICap is the required capacity level ($K^e = K^R$), but this equilibrium is maintained by small fluctuations in $K$ that result in capacity earning the penalty price part of the time and nothing the rest of the time. This analysis depends on an understanding of $K$ as a random variable and a new definition of equilibrium installed capacity.

## Equilibrium with Capacity a Random Variable

**Policies that Would Induce the Same Deterministic ICap Produce Different Levels of Reliability**

Random variations in installed capacity, $K$, around the deterministic equilibrium, $K^e$, increase profit when the profit function is steeper on its upside. Bringing profit down to its equilibrium value requires raising average $K$ by an amount determined by the shape of the profit function.

If two profit functions have the same $K^e$, the one with the steeper upside will induce the higher average $K$ and thus more reliability at a higher cost.

Private investors do not coordinate decisions to construct plants, and plant construction times are unpredictable. Consequently, $K$ will fluctuate randomly. A particular profit function will lead to some particular distribution of $K$s, which by definition is the equilibrium distribution. This distribution will have some average value, $\bar{K}^e$, called the equilibrium value of average $K$, or simply average $K$. The equilibrium distribution of $K$ will have an average profit of $FC_{peak}$ just as $SR_\pi(K^e) = FC_{peak}$. But $SR_\pi(\bar{K}^e)$ will not equal $FC_{peak}$, because profit functions are usually not straight lines. This means $SR_\pi(K^e) \neq SR_\pi(\bar{K}^e)$, so $\bar{K}^e$ does not equal $K^e$. The average installed capacity does not equal $K^e$ when the deterministic equilibrium capacity is $K^e$.

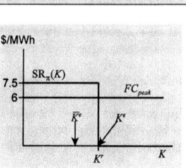

Because $\bar{K}^e$ differs from deterministic $K^e$, two profit functions with the same deterministic equilibrium may have different $\bar{K}^e$s. If a profit function is steeper to the left of $K^e$, then the equilibrium distribution of $K$s must move further right to bring average profit down to $FC_{peak}$.

These concepts can be applied to the profit functions of a capacity market. As long as such a profit function is greater than $FC_{peak}$ to the left of $K^R$ and less than $FC_{peak}$ to the right, its deterministic equilibrium will be $K^e = K^R$. But its equilibrium-average $K$ will differ according to the height of the penalty. If the penalty is just slightly higher than $FC_{peak}$, then $K$ will need to spend much more time to the left of $K^R$ in order to bring

---

3. If (the use of) capacity could be "exported" to a neighboring market, then the price of capacity would reflect the opportunity cost of selling into that market.

average profit up to $FC_{peak}$. Increasing the penalty moves $\bar{K}^e$ to the right and increases reliability.

## Mimicking VOLL Pricing with a Capacity Requirement

Suppose VOLL pricing produces a certain average installed capacity level, $\bar{K}^e$. If $\bar{K}^e$ is also the required level ($K^R = \bar{K}^e$) under a capacity approach and the penalty is set to \$12/MWh, the same average capacity of $\bar{K}^e$ will be induced by the capacity market.[4] In this case, price spike revenue will not be required and the price cap can be set only \$1/MWh above the peaker's variable cost. This would mean a price cap of \$31/MWh for the two-technology model used since Chapter 2-2.[5] The revenue from the capacity market would cover the fixed costs of peakers and would also flow to baseload generators thereby increasing their short-run profits by \$6/MWh, just as when peakers recover their costs from energy price spikes. Consequently, baseload plants will also have an equilibrium installed capacity that is optimal. In short, the capacity-requirement approach can fully substitute for the price-spike approach with regard to inducing the optimal investment in all types of technology.[6]

| | | |
|---|---|---|
| **Result** | **2-8.2** | **A Capacity Requirement Can Eliminate the Need for Price Spikes** |

(This holds only with regard to inducing new generation, and not with regard to the utilization of expensive demand- or supply-side resources.) In an isolated market, or in a region in which all markets have adopted a capacity-requirement approach instead of a price-spike approach, energy-price spikes are not needed to induce investment in generation. The capacity approach will induce the same optimal level of all generation technologies as would an optimal price-spike approach.

## 2-8.3　COMBINING A CAPACITY REQUIREMENT WITH A PRICE SPIKE

If an optimal price spike approach is combined with an optimal capacity approach, the result is a market that builds too much ICap. At the optimal ICap value, generators would cover their fixed-costs twice, once in the energy market and once in the capacity market. This would encourage more investment, and ICap would rise above its optimal level.

Profit functions for VOLL pricing and OpRes pricing, presented in Chapter 2-7, produced the same deterministic equilibrium, $K^e$. Consequently, both were said to be optimal with respect to installed capacity. But the previous section argued that the two profit functions will not produce the same equilibrium ICap in a more realistic model. This means that at least one of them does not induce optimal installed

---

4. This assumes that the distribution of $K$ is symmetrical.

5. In real markets, there are old inefficient generators with higher marginal costs. For instance PJM currently has one with a marginal cost of about \$130, so this would require a price cap of \$131/MWh.

6. Hobbs has proven this result in a stochastic model (Hobbs, Iñón, and Stoft 2001).

capacity. Before judging the results of a combined approach, a better understanding of the VOLL pricing result is required.

**Interpreting Average ICap for VOLL Pricing.**[7]  Because the profit function of VOLL pricing is very steep, $\bar{K}^e$ may be significantly greater than $K^e$, but would $\bar{K}^e = K^e$ be desirable? If these were equal, there would be 1 GW too little ICap as often as 1 GW too much. But the reliability loss from having too little is greater than the gain from having too much because a finite *decrease* in $K$ increases blackouts by more than the same finite *increase* would decrease them. Accounting for capacity costs, this implies a greater net loss from downside errors than from upside errors in $K$. It is better to have $\bar{K}^e$ somewhat above $K^e$.

$\bar{K}^e$ *is* greater than $K^e$, but $\bar{K}^e$ *should be* greater than $K^e$. Do these two effects coincide? Is $\bar{K}^e$ greater by the appropriate amount? In fact, it is. The long-run equilibrium condition still assures that $\bar{K}^e$ will keep price-spike revenue equal to $FC_{peak}$ on average, and this means the duration of load shedding will be the optimal value given by Result 2-3.1. VOLL pricing produces optimal $\bar{K}^e$ even though random fluctuations in $K$ make this different from the deterministic equilibrium value $K^e$.

---

**Result   2-8.3**   **VOLL Pricing Induces Optimal ICap Even When ICap Is Random**
If random fluctuations in ICap are accepted as inevitable, then VOLL pricing induces the correct distribution of ICap. In this case, optimal average ICap will be greater than the optimal ICap in a deterministic world with deterministic $K$.

---

**Combining Approaches.**  Because VOLL pricing is optimal, combining it with an optimal capacity approach would double the profits of generators if they maintained the same (optimal) level of $\bar{K}^e$. This would lead to more investment and a higher level of ICap than justified.

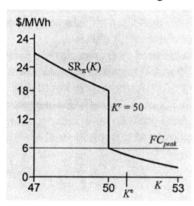

Because of the side effects of VOLL pricing, OpRes pricing may be a better choice for a combination. The figure to the left shows the profit function that would result from using both the OpRes pricing described in the previous chapter and the capacity requirement of the previous section. Note that the profit function still intersects that fixed-cost line at the optimal deterministic value of ICap, 50 GW (50,000 MW).

It might seem that adding the two profit functions must surely result in too high an equilibrium ICap distribution, but that may not be the case. Profit functions that are flatter than a VOLL profit function actually induce too little ICap if a deterministic analysis indicates that they are optimal. Consequently, although combining two of them increases $\bar{K}^e$ above the value induced by either, it may not be too high. It may be that the difference between $\bar{K}^e$ and $K^e$, even for VOLL pricing, is smaller than the errors in estimating $K^e$. In this case, the preceding analysis is not of much importance, except to show that combining a price-spike and capacity market approach should not cause too much overinvestment. Currently little is known about

---

7. This discussion assumes for simplicity that the random distribution of $K$ is symmetrical. If it were not, the conclusion would still hold, although some of the details would change.

the amplitude of the random fluctuations in $K$, but there is growing evidence from the north east of the United States that these may be large.

| | | |
|---|---|---|
| **Result** | **2-8.4** | **Profit Functions Are Additive, but Resulting Profits Are Not** |
| | | Combining price spikes with an ICap market does not increase equilibrium profits; it only induces more installed capacity. |

## 2-8.4   COMPARING THE TWO APPROACHES

### Market Power

When $K$ is just slightly greater than $K^R$, the capacity market is particularly susceptible to market power. Suppose the requirement is $K^R = 50$ GW and one supplier owns 5 GW of uncommitted capacity and has no capacity requirement of its own. Also assume that no other supplier has any market power and that $K$ for the entire market is 50.1 GW. This is essentially the same situation as analyzed for the price-spike approaches in Section 2-6.2.

Competitive bidders would bid a price of zero into a capacity market because there is no cost to suppling capacity. The supplier with 5 GW of capacity would bid just less than the penalty value, $12/MWh, because at this price it could sell 4.9 GW to those with capacity requirements. This strategy raises the price on all capacity in the spot market close to the penalty price. So profits are $6/MWh greater than normal. This may be compared with an increase in profits of $52/MWh under VOLL pricing and about $5/MWh under the example of OpRes pricing.

When $K < K^e$, market power becomes worse under both price-spike approaches but totally disappears under the capacity approach. Once $K < K^e$, the capacity market price is $Y$ under competition, so market power cannot push it higher.

In a capacity market, market power can be neutralized by decreasing the capacity requirement. In this case, setting $K^R$ to 45 GW would bring the price back down to zero. If $K$ then fell below the optimal level of 50 GW, market power would again be exercised and the capacity price would go back up to $12/MWh. The supplier with market power and 5 GW of capacity causes the profit function in a market with $K^R = 45$ to be exactly the desired profit function (the one that results from $K^R = 50$, and no market power).

### Risk

Annual changes in the load-duration function can cause dramatic year-to-year changes in price-spike revenues, but they cause no change in revenues from the capacity market. After observing the 40 years of annual VOLL data displayed in Figure 2-6.1, it is impossible to decide whether expected short-run profits in the last year are approximately $5.40/MWh, the 40-year average, or $9.66/MWh, the average of the last 20 years. Is it better to stay out of the market or is it a very profitable time to invest? This source of uncertainty is entirely absent from markets using the capacity approach.

Investing in markets that are based on a pure capacity-requirement approach is still risky because profits depend on how much other capacity enters the market. But unless the market is grossly overbuilt, demand growth should restore the equilibrium fairly soon. Investors do not need to wait until a favorable combination of low $K$ and exceptionally high load produces a super profitable year, making up for all the years of low profits.

## Demand Response

The weakness of a pure capacity-requirement system is that it produces no price spikes. These are needed to take advantage of demand elasticity and are also needed to stimulate the development of that demand elasticity. Price spikes are also needed to make use of old inefficient generators and the **emergency output** of other generators. A hybrid system can accomplish this while minimizing risk and market power.

*Chapter 2-9*
# Inter-System Competition for Reliability

*In the same open market, at any moment,*
*there cannot be two prices for the same kind of article.*

W. Stanley Jevons
*The Theory of Political Economy*
1879

*There are two fools in every market; one asks too little, one asks too much.*

Russian proverb

**M**ARKETS WITH LOW PRICE CAPS HAVE LITTLE PROTECTION FROM
COMPETING MARKETS WITH HIGH PRICE CAPS. At crucial times, the high-
cap market will buy up the reserves of the low-cap market. This can cause a group
of competing markets to evolve toward a risky high-priced regulatory approach.
Competing system operators are not "led by an invisible hand" to the optimal policy.

This chapter uses previously developed models and tools to investigate what
happens when pairs of markets, operating under different pricing rules, compete
for energy and capacity.[1]

**Chapter Summary 2-9:** Competition between markets with different price caps
will favor the market with the higher cap. The lower-price-cap market will spend
as much on inducing investment but will find its reserves bought out from under
it at crucial times by the high-price-spike market. The result will be competition
between system operators for the higher price cap unless a regional regulator
prevents this. Capacity-requirement markets can solve a similar problem by requiring capacity rights to be sold on an annual basis.

**Section 1: Price-Cap Competition.** A VOLL market and OpRes market,
identical except for their pricing policies, are modeled as able to trade energy and
operating reserves. The result is that when load is expected to be high, more
generators sell power and reserves to the VOLL market. This equalizes the expected
short-run profits in the two markets; as a result the OpRes market suffers reduced
reliability. This will force it to adopt a higher price cap to protect itself from inter-
system competition.

---

1. Cournot anticipated Jevons, Menger, and Walras in discovering "the law of one price" and other
neoclassical results. Jevons (1835-1882), with his publication of *The Coal Question* in 1865, was the first
to draw attention to the imminent exhaustion of energy supplies. In 1870 he invented the "logical piano",
the first machine to solve problems faster than a human.

**Section 2: Competition between Price Spikes and Capacity Requirements.** Pure capacity-requirement markets will suffer the same fate as OpRes markets unless they make their capacity requirement annual. This will protect them, even from VOLL markets.

---

## 2-9.1   PRICE-CAP COMPETITION

Chapter 2-6 developed models of VOLL pricing and OpRes pricing that made use of common supply and demand conditions. Chapter 2-7 extended the analysis by constructing profit functions for the two markets. This section considers what would happen if two of these markets, one VOLL market and one OpRes market, were trading partners but were isolated from all other markets.

Each market has its own system operator which must serve a load described by the load-duration curve of Equation 2-7.2. This is a long-run average curve, and it summarizes many fluctuations in load, both predictable and unpredictable. This section is concerned with predictable fluctuations, such as seasonal ones. For easy reference call the high load season *summer* and call the non-summer *winter*. For simplicity, assume that these seasons shift the load duration curve uniformly. In this case they are equivalent to an equal but opposite shift in installed capacity, $K$.

Because (augmented) load minus installed capacity ($L_g - K$) determines short-run profits, the new profit function during a season is found by shifting the average profit function to the right in the summer and the left in the winter. Assuming for simplicity that $K$ is deterministic, the optimal ICap level in each market is 50 GW as explained in Chapter 2-6. As a benchmark, assume that the two markets each start out with this level of capacity, and consider the case, shown in Figure 2-9.1, in which load is 1 GW above normal in the summer and 1 GW below normal in the winter.[2] Installed capacity does not change by season and is always 50 GW in each market, but profits are high in the summer and low in the winter as depicted.

Trade between the two markets will produce different equilibria in the summer and winter. Assuming the interconnection allows sufficient trade, these equilibria will have two defining characteristics:

1. Profits in the VOLL and OpRes markets are equal.
2. The average capacity sold to the two markets is 50 GW.

If the OpRes market were more profitable, some generation would migrate from the VOLL market to the OpRes market and equalize the profit levels. Then the migration would stop. Equilibrium occurs when profits in the two markets are equal and one market is short as much generation as the other is long. The allocation of generation between the two markets is shown by the dots in Figure 2-9.1. In the summer the average short-run profit of a peaker will be about $9.60 and the VOLL market will have about 1250 MW more generation under contract than the OpRes market. In the winter the OpRes market gets more and the price is about $3.30.

---

2. Obviously, seasonal variations in load are much greater, but they are partially offset by seasonal variations in planned outages. The real effects described here may be much larger than depicted in these examples which are intended only for qualitative analysis.

**Figure 2-9.1**
VOLL and OpRes profit
functions for high- and
low-load conditions.

(Note that these are not
the standard, long-run
average profit functions,
but they do average load
fluctuations over two
different seasons.)

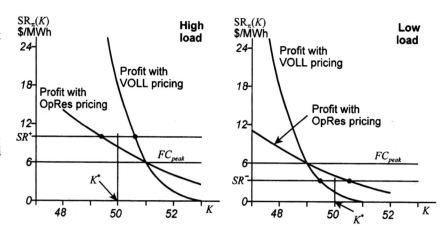

This is a good deal for the VOLL market. As an isolated market it would have
had to pay enough for reserves to provide profits of $18/MWh and would have
had fewer reserves in crucial hours. Reliability would have been less. More detailed
calculations confirm that even when the winter is taken into account, the VOLL
market has increased reliability at a reduced cost, while the OpRes market has
reduced reliability at an increased price.

Trade between markets essentially lets the high-spike market steal the reserves
of the low-spike market. They pay the same price for reserves, but the high-spike
market gets them when they are needed and the low-spike market gets them when
they are not. Moreover, because they are not shared evenly, they are less effective
at preventing blackouts than if they were, so the overall reliability of the market
is reduced.

The result will be that low-spike markets are forced to adopt higher price caps.
This competition between system operators leads to the adoption of regulatory
policies that favor risk and market power. When inter-system competition is ignored
and different markets are allowed to adopt different policies, the result is high-price
spikes even if all would prefer low price spikes. Each one finds it advantageous
to outbid its neighbor and will do so when desperate, at first in secret "out-of-
market" purchases and the later by raising price caps. Only through explicit high-
level regulation can a low-risk policy be implemented.

---

**Result     2-9.1**          **Competition Between System Operators Induces High Price Spikes**
If two markets with different price spike policies trade energy and operating
reserves, the one with higher price spikes will gain in reliability and save money
relative to the other. This will force the low-spike market to use higher price
spikes. If regional price limits are not imposed, inter-system competition will lead
to reliability policies with undesirable side effects.

---

Further analysis reveals that summer must have a duration of 25% and winter
of 75%, the resulting profit level for peakers in only about $5/MWh, less than the
$6/MWh required for equilibrium. This means that competition between the markets
will cause a decrease in investment, and there will be less than the optimal level

of installed capacity. Total reliability will decline both because total installed capacity declines and because it is not distributed optimally between the two markets.[3]

| | | |
|---|---|---|
| **Result** | **2-9.2** | **Trade between Markets with Different Policies Can Reduce Reliability** |

**Trade between Markets with Different Policies Can Reduce Reliability**
When two markets that trade with each other implement different price caps that would provide optimal reliability if implemented in an isolated market or by all markets, the outcome will usually be suboptimal. Trade will reduce average profits in both markets and leave too little incentive for investment.

## 2-9.2   COMPETITION BETWEEN PRICE SPIKES AND CAPACITY REQUIREMENTS

Markets with a capacity requirement will typically contain a submarket (or several submarkets) for trading the capacity that load-serving entities must contract for to satisfy their capacity requirements. The term **capacity-requirement market** will not be used to refer to these submarkets but rather to the entire power market that relies on a capacity requirement to induce a reliable level of investment in installed capacity. The term "capacity-requirement market" is analogous to the terms VOLL market and OpRes market. Each of these markets also has its own pricing limits for the system operator. VOLL markets have extremely high pricing limits, OpRes markets have lower limits, and capacity-requirement markets have the lowest limits because they do not need a price spike to cover the fixed costs of peakers.

Capacity markets can suffer a fate similar to the fate of low-price-cap markets. If capacity requirements are enforced on a daily basis, then on a day when the price-spike market expects a price of $600/MWh, the $12/MWh penalty of the capacity market loses its sting. A **load-serving entity** will offer a generator $11.99/MWh for its capacity and the generator will decide instead to accept $600/MWh for its energy from the market next door.

The situation is more complicated in real markets because capacity requirements are inevitably accompanied by energy price spikes, and the markets have complex rules allowing the export of energy from capacity under contract as well as its recall when needed. Nonetheless, on days when the capacity market is likely to be short of energy so recall of capacity is likely, a higher expected energy price in export markets can easily overcome a modest capacity penalty. This allows the same type of reserve stealing described in the previous section.

The root of this problem is the predictable fluctuations in load that produce the high-load periods and low-load periods described above. So long as these exist, the high-price-spike market will win the reserves at times when it pays most to

---

3. With two markets trading, each would require a lower level of installed capacity because they could share resources. This is one of the main reasons tie lines have been built historically. But this model does not account for emergency sharing. Each market must serve its own load once generators have signed contracts with one or the other. Emergency sharing is not allowed. Within this model, bringing the two systems into trading contact would produce a lower equilibrium level of ICap and less reliability. A more complex model that allowed emergency sharing would have a reduced optimal level of ICap, but it too would demonstrate that the combined system would produce too little reliability.

**Figure 2-9.2**

VOLL and OpRes profit functions for high and low load conditions.

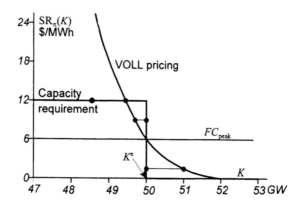

win. A capacity market can solve this problem by annualizing its penalty so that generators must sell their capacity for a year at a time. Then the generator's choice is between a year of revenues from the capacity market and a year of revenues from the external market. In this case there is no possibility of participating in the price-spike market when load is high and in the capacity market when load is low. As in the previous section, this assumes generators must sell into one market or the other, a convenient simplification.

| Result | 2-9.3 | **Capacity-Requirement Markets Need Annual Requirements** |
|---|---|---|

Capacity markets are vulnerable to competition from price-spike markets unless they utilize an annual requirement.

**A Model with Annual Requirements.**  Assume again that there are two markets with a strong tie line connecting them and allowing a considerable amount of trade. One is a VOLL market, and the other a pure capacity-requirement market. The capacity-requirement market, because it has no price-spike incentive, has a penalty of $12/MWh, twice the fixed costs of a peaker. It also has an annual requirement for installed capacity.

**Analysis.**  Figure 2-9.2 shows the profit functions for the two markets. If both markets had 50 GW of capacity, they would share it equally. If the capacity market got any less, its price would be $12/MWh and the VOLL market's expected price spike would be slightly less than $6/MWh. This would attract capacity back to the capacity market.

The pairs of horizontally connected dots show how capacity would be divided between the two markets if the total capacity in the two markets differed from the equilibrium level of 100 GW. If there were 101 GW, the VOLL market would get the extra GW of capacity, and the price in the capacity market would sink to $1.50/MWh—the opportunity cost of not selling into the VOLL market. If there were slightly less than 100 GW, the capacity-market price would rise to $9/MWh and it would still retain its full 50 GW required capacity. The VOLL market would absorb the shortage.

**Figure 2-9.3**

VOLL and capacity-market profit functions for high and low load conditions.

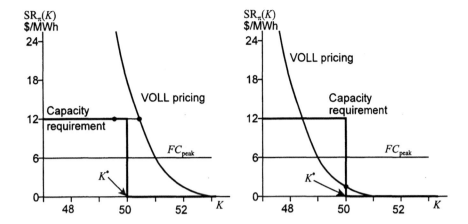

Once capacity was more than about 0.5 GW short, the capacity-market price would rise to its penalty level of $12/MWh and beyond this point it would absorb any further decrease in combined capacity. This should not prove too problematic because whenever capacity is short in the combined system, price rises above the fixed costs of a peaker quite quickly and this should induce investment.

Unlike the two markets in the previous section, neither market has a strong advantage over the other. They compete on an annual basis so the VOLL market cannot buy up reserves only at the times they are most needed.

**Shorter-Term Capacity Requirements.** If the capacity requirement can be met for less than a year at a time, say for three months, then it will have to compete with the VOLL market during high-load and low-load seasons. This is shown in Figure 2-9.3. Just as in Section 2-9.1, the VOLL market wins out during the high-load period. It gets more of the reserves and has a lower price spike than it would otherwise have. It suffers no penalty during the low-load period when both markets fare exactly as they would have if they had not been in contact.

**Complications.** Capacity markets can protect themselves in other ways from external price spikes, so they may not need to use an annual requirement. PJM's $1,000/MWh price cap plus its ability to pay higher prices in "out-of-market" transactions offers significant protection, although it does bring with it the standard side effects of price spikes.

The present examples all consider two markets that are the same size. If the price spike market is bigger, then it will buy up more of the other market's capacity. Small low-spike markets and small capacity-requirement markets are more at risk than large ones.

*Chapter* *2-10*
# Unsolved Problems

*The day when we shall know exactly what "electricity" is, will chronicle an event probably greater, more important than any other recorded in the history of the human race. The time will come when the comfort, the very existence, perhaps, of man will depend upon that wonderful agent.*

Nikola Tesla
c. 1893

LOW PRICES, EVEN THOSE ONLY **200%** ABOVE NORMAL, CAN INDUCE INVESTMENT, BUT THEY CANNOT INDUCE CURTAILMENT OF HIGH-VALUE LOAD. Neither can they induce supply from the odd high-marginal-cost generator nor from a generator that is producing in its emergency operating range. The problems of low price caps have nothing to do with adequate investment in generating capacity. They concern only the short-run responses of high-value load and high-marginal-cost supply. These problems are small compared with the problem of long-term market stability and generation adequacy, but they are worth solving. This chapter defines them and suggests steps toward solutions.

**Chapter Summary 2-10:** Genuinely high-marginal-cost power should be purchased in a market separate from the regular supply market and with its own high price cap. Load reductions, most of which have high marginal values, should be purchased at a price that is allowed to exceed the price cap in the supply market. The (relatively) low price cap of the supply market should be set regionally and enforced strictly to prevent "out-of-market" purchases.

**Section 1: High Marginal Costs and Low Price Caps.** Several small sources of high-marginal-cost power make high prices worthwhile. To avoid disruptive side effects, these should be restricted to an emergency power market that caters only to high-cost sources. The problem is to select only these sources and limit market power under a high price cap. An adequate solution appears to be within reach.

**Section 2: Pricing Supply and Demand Separately.** Allowing demand-side prices to rise above the supply-side cap makes use of existing demand responsiveness and stimulates the development of greater elasticity. This requires a method of setting price higher than any supply bid and of refunding the extra revenue collected.

**Section 3: Price-Elastic Demand for Operating Reserves.** An additional megawatt of operating reserves is worth more when reserves are near zero than when they are near the required level. The demand for reserves should be expressed by an elastic demand function. Estimating an optimal function may be quite difficult, but improving on current practice should not be.

**Section 4: System Operators' Psychology.** Some argue that system operators find it impossible to respect a price cap, especially a low one, when operating reserves are in short supply. If this is true, low price caps would always be undermined by "out-of-market" purchases at prices above the cap. The problem is more political than economic or psychological. A hard, interconnection-wide cap is required to solve the problem, and this can be implemented only by the highest-level regulatory authorities.

# 2-10.1  HIGH MARGINAL COSTS AND LOW PRICE CAPS[1]

Sometimes particularly low price caps are desirable because of problems with either market power or scarcity rents. This complicates the price-cap design, but even quite high price caps need exceptions for peculiar generators.

There are three classes of exceptions: (1) expensive generators that have not been built (e.g., backup generators); (2) expensive existing generators (e.g., old generators); and (3) the expensive emergency output levels of cheap generators. These exceptions are small, perhaps 5% capacity, and represent extremely little energy. Emergency output from baseload and midload generation is the most important of the exceptions and perhaps the most difficult to accommodate. Approximately 3% of PJM's total installed capacity is classified by generation owners as being in the **emergency operating range**. Some of this should become available at a price above the generator's typical variable cost but below a typical OpRes price limit of $250 to $1000/MWh, but some emergency output capability may still be unreported, so the full extent of this problem is still unknown.

Price caps below the variable cost of new peaker technology should be avoided. If such a low cap is needed, the market should be re-regulated. Other price-cap discrepancies could be handled by a separate market for emergency power.

**An Emergency Power Market.** Only capacity accepted into the "emergency power market" would be paid the potentially high prices in that market which would be capped at a higher level than the regular spot market. The design problem is to find a way to accept all high-marginal-cost capacity without accepting much low-marginal-cost capacity and to contain market power in the emergency power market. (In a well-functioning market, such price discrimination would be an exercise of monopsony power, but as explained in the preceding chapters, these

---

1. This is not the type of price cap used in other markets but is instead a purchase-price limit on the system operator as explained in Chapter 2-4.

price limits are explicitly designed to cover fixed costs with optimal installed capacity.)

**Rules for Acceptance.**  Generators should self-select into the emergency market subject to certain limitations, which might include any of the following: the percent of the emergency market's capacity controlled by one entity could be limited; the frequency of entering and leaving could be limited so they cannot just enter whenever hot weather is predicted; or the percentage of emergency capacity accepted from a standard generator could be limited.

The last limitation is perhaps the most important. Emergency capacity might be defined as capacity beyond the maximum output observed while price was under the regular market's price cap during the last year. This might need to be corrected for ambient temperature or other factors.

**Bid Limitations.**  This market should be capped at no more than $V_{LL}$, with the level set to make the proper trade-off between market power and inducement of supply. Requiring longer duration bids should reduce the exercise of market power. If market power is only a small problem, bids may be superfluous, and the system operator can just set prices and let supply respond, in the way load may respond to a high price.

**Extent and Effect of Market Power.**  Any proposed design should be analyzed for market power as the emergency market will be quite susceptible to it. In doing so, it should be remembered that, once load has been shed, having the energy at too high a price (but still below $V_{LL}$) is better than not having it at all.

**Prognosis.**  A useable design should not be difficult to discover. For a low but plausible price cap on the non-emergency market, consider a $250 price cap in PJM. The amount of generation with marginal cost above this level is quite small, perhaps well under 5%. It should be possible to include most of this in the emergency market, and at worst it should all become available by the time the emergency price reaches some high cap, say $10,000. While significant market power might be exercised in this market, as long as the system operator offers to pay no more than the power is worth, consumers will be better off with than without the market.

Consider a generator that can produce 500 MW with a marginal cost of $25/MWh, an additional 5 MW with marginal costs up to $250/MWh, and another 5 MW with marginal costs up to $2,000/MWh. Observing past behavior reveals that 505 MW are available at a price of $250/MWh or below, and so only capacity beyond this level is allowed into the emergency market. Assume that market power is exercised by another supplier and the emergency market is called into play at a price of $3,000/MWh. The generator in question will profit handsomely from the exercise of market power (not its own), but consumers will benefit from the extra 5 MW of emergency power if this is used during a period of load shedding when that power is worth, say, $15,000/MWh. Without the emergency market, this power would have been withheld, exacerbating the market power already exercised.

In summary, the emergency market allows the system operator to make use of high price spikes without suffering the significant side effects caused by high price spikes applied to the entire spot market. Though the potential for increased efficiency is clear, the magnitude of this gain is not. Because emergency power is only needed for a few hours per year and constitutes only a few percent of total power, gains are limited. Although this power can be very valuable, it will also be very costly. A quantitative evaluation of potential efficiency gains is needed.

## 2-10.2   PRICING SUPPLY AND DEMAND SEPARATELY

Low price spikes can call forth any level of investment, but they cannot make use of high-cost generation or high-value load. An emergency power market could take care of the supply side, but the demand side might also operate more efficiently with higher prices. One possibility is to set demand-side prices higher than supply prices during price spikes. This would require a balancing account to keep the two revenue streams equal in the long run.

Most load, like most generation, will be hedged by long-term contracts, but some will be subject to spot-market prices. This load should consist of customers who can buy power more cheaply by accepting the spot price and adjusting their usage to price, thereby taking advantage of cheaper off-peak power. Faced with a volatile spot market, load-serving entities may find it advantageous to encourage demand responsiveness from their customers. They could offer prices which hedge customers yet let them save money by responding to real-time prices. Such pricing would ensure that if their load-profile is no more costly than the average load profile they will pay no more than fully hedged customers. In short, a more volatile spot market for demand can encourage a greater demand response immediately and improve demand responsiveness in the long run.

Allowing the demand-side price to exceed the supply-side price means more revenue will be collected than dispersed—the system operator will make a profit. This is not different, except in sign, from the fact that the system operator normally runs a deficit. To compensate, extra costs beyond the cost of energy, such as the fixed cost of wires, are typically collected through an "uplift" spread over all load on a per-MWh basis. Extra revenues could be returned the same way. In fact they would tend to cancel the uplift and remove a small taxation inefficiency.

The major unsolved problems associated with this proposal seem political, but the uplift adjustment should be examined carefully and the rules for setting the demand-side price in the absence of demand-side bids deserve attention. The next section addresses this problem.

## 2-10.3   PRICE-ELASTIC DEMAND FOR OPERATING RESERVES

The current pricing of operating reserves is black or white. If reserves are at all short ($OR$ less than required $OR$) then the market price is set at the price cap—at least in theory. If $OR > OR^R$, then extra operating reserves are priced at zero. If the rule is followed in a competitive market such as PJM's, the result is an energy price of no more than \$150 when the market has 1 extra MW of operating reserves and a price of \$1,000 when the market has 1 MW too few. (There are indications from several markets that system operators realize this does not make sense and introduce some price elasticity for reserves on an ad hoc basis.)

There are engineering fables about the risk of blackout taking a sudden jump up when the level of reserves falls below the amount of supply that could be lost by the single largest outage (the single-contingency rule). These are never supported by probabilistic calculation and are seen to be implausible when a few realistic mitigating factors are considered. First if the system is 10% short of 10-minute spinning reserves, this may simply mean that this system will "lean on" the interconnection for 11 minutes instead of 10. This depends on the details of the available spin. Next there is the possibility of voltage reductions, again an uncertain value. Spinning reserve may prove to be noticeably different from its nominal value as may nonspinning reserve. The list goes on. The probability of lost load does not change abruptly with a 1 MW change in operating reserves.

As a consequence, the value of operating reserves changes smoothly, a fact that should be reflected in the system operator's willingness to pay for them. If 4 GW of operating reserves are "required," and if operating reserves have a marginal value of $1,000/MWh when only 3.9 GW have been purchased, then even with 5 GW on hand they probably have a marginal value of at least $500/MWh. With only 1 GW of supply, their marginal value must be much greater than $1,000/MWh.

---

**Result      2-10.1        The Price of Operating Reserves Should Increase When They Are Scarce**
The view that an additional MW of operating reserves is worth any price when they are in short supply and worth nothing when they exceed the required level is inappropriate in a market setting. Like every other demand curve, the demand for reserves should be downward sloping.

---

This should be reflected in an explicit, downward-sloping demand function for operating reserves spanning a range of perhaps 15% of load. The first benefit would be an increased elasticity of demand and a reduction of market power. Second, it would provide a way to set price at times when the supply-side price is at its cap. This would greatly facilitate separate demand-side pricing.

The unsolved problem is how to determine the (elastic) demand function for operating reserves. Valuing them requires a combination of engineering and economics. Their value might be determined from the impact of a 1 MW increase on the expected amount of lost load. But if Assumption 2-3.1 (the basis of the Simple Model of Reliability) is correct, this approach may not make sense. Instead it may be necessary to value operating reserves by a method that explicitly accounts for the demand response to the market prices set by reserve pricing or to reductions in market power from operating-reserve demand elasticity.

---

## 2-10.4   THE PSYCHOLOGY OF SYSTEM-OPERATORS

One objection to price caps, especially to low price caps, is that system operators find adhering to them psychologically difficult. They are used to doing all they can to maintain reliability. When reserves are low, they may not be able to resist the temptation to pay more than the price cap in "out-of-market" purchases. As evidence for this view, out-of-market transactions by the California and PJM ISOs are often cited.

While there is some truth to this view, the out-of-market purchases have other explanations. First, they were required and successful. They were required because there was no regional price cap in effect, and the ISOs were subject to the type of inter-system competition described in Chapter 2-9. This causes a loss of reliability for the system that has the lower price cap. Reliability can be improved by out-of-market operations which in effect raise the low price cap. Second, the ISO rules clearly allowed such breaches of the price cap. Failing to take advantage of this possibility might have brought a reprimand had the system suffered a rotating blackout.

A hard, regional price limit is needed, preferably one that has been implemented by the highest authority. All system operators should be forbidden from using out-of-market operations. This does not preclude the existence of an emergency market with restricted participation.

Assume such a price cap has been implemented, and examine the situation from the system operator's viewpoint. Say reserves are tight and the market price is equal to the price cap. Also assume that the price cap is above the marginal cost of any generator, exceptional generation having been removed to the emergency market. If generators believe the price cap will hold, no generator will withhold output because this can only reduce its profits.

Suppose one generator tried to break through the price cap by approaching the system operator in private with an offer to supply an additional 100 MW of power at a price of \$100/MWh more than the cap. This is tempting to the system operator who needs more power. It may even be forced to shed load without this power. Certainly it is worth the price being asked, which is well below the value of lost load. What should the operator do? It may be a psychologically difficult decision, but it is logically simple.

By accepting the offer, the operator would ensure that the generator would be back next time withholding more and asking a higher price. Moreover, other generators would soon learn to play the same game. By giving in, the system operator encourages more withholding and increases the cost of power. If the system operator continues to accept such offers, the requested price will eventually be raised to an unaffordable level. Some negotiations will end with a high price being paid, but some will end in stalemate with generators withholding capacity to prove their threat is credible. Reliability will suffer. Surely, system operators will understand this and hold firm, especially when breaking the price cap leaves them vulnerable to charges of misconduct.

If system operators hold fast, they have every prospect of soon being free from such difficult situations. Refusing to pay prices above the cap puts the generator in a very difficult situation. First, withholding capacity with price at the cap reduces its profits. Second, as an attempt to exercise market power, it is a violation of the Federal Power Act. If the generator actually does withhold, it does so under emergency conditions and may even cause a blackout. Moreover, the only motive for withholding power once the system operator rejects the offer is to cause a power shortage. It cannot be explained by normal profit maximization of the type that explains market power because the generator has actually reduced its profits.

Given these considerations, it seems reasonable to expect that system operators will overcome any predisposition to violate the price cap. They need only think ahead and realize that maintaining the price cap will, in the not-to-distant future, increase reliability, hold down costs, and prevent ever more difficult dilemmas.

The unsolved problem is how to bring about a hard, regional price cap. This is a political, not an economic, problem. It is made all the more difficult because the relevant region is the entire interconnection. Until this problem is solved, system operators will continue to make secret out-of-market purchases in order to protect their systems from the high price caps and secret purchases by competing markets.[2]

---

2. PJM does not release information on out-of-market prices.

*Part 3*

# Market Architecture

# Chapter *3-1*
# Introduction

*The conclusion today, seventeen years later, is essentially the same . . . industries differ one from the other, and the optimal mix of institutional arrangements for any one of them cannot be decided on the basis of ideology alone. The "central institutional issue of public utility regulation" remains . . . finding the best possible mix of inevitably imperfect regulation and inevitably imperfect competition.*

Alfred E. Kahn
*The Economics of Regulation*
1995

R EAL-TIME TRANSACTIONS REQUIRE CENTRAL COORDINATION; WEEK-AHEAD TRADES DO NOT. Somewhere in between are dividing lines that describe the system operator's diminishing role in forward markets. Where to draw those lines is the central controversy of power-market design. A related controversy, not considered in Part 3, is how finely the system operator should define locational prices. Those who favor a large role for the system operator in one sphere tend to favor it in others. Thus the controversies of market architecture have a certain consistency. Although the rhetoric focuses on how centralized a design is, the litmus test in most of the controversies is the extent of the system operator's role. This too may be a distraction. A larger role for the system operator implies a smaller role for profitable enterprises. One side fears the inefficiency and market-power abuses of private parties playing social roles. The other side fears the inefficiency of nonprofit organizations but also covets the central market roles played by the system operator.

Power markets present unusually acute coordination problems. They are the only markets that can suffer a catastrophic instability that develops in less than a second and involves hundreds of private parties interacting through a shared facility. The extent and speed of the required coordination are unparalleled. Generators 2000 miles apart must be kept synchronized to within a hundredth of a second. Such considerations require a market that in some respects is tightly controlled in real time. Historically, this control has extended to areas far from the precarious real-time interactions. As deregulation brings markets into new areas, it is not surprising to find the proponents of markets reaching beyond their ability and to find the traditional system-control structure attempting to perpetuate now unnecessary roles for itself. This clash of interests has produced much heat and shed little light.

While Part 2 ignores questions of architecture to focus on structure, Part 3 considers alternative designs for the real-time (RT) market and the day-ahead (DA) market, as well as the relationship between the two. Most of the design questions revolve around the extent of the system operator's role.

**Chapter Summary 3-1:** After preliminary definitions of forward, future, real-time, and spot markets, this chapter outlines the controversies over bilateral markets, power exchanges, and power pools. Bilateral markets provide a private coordinating mechanism; exchanges provide a public, centrally determined, market price; and pools provide a price, side payments and instructions on which generators should start up. The final section provides a brief introduction to locational prices, the complexity of which plays a role in assessing the need for central coordination.

**Section 1: Spot Markets, Forward Markets, and Settlements.** Forward markets are financial markets, while the RT market is a physical market. To the extent power sold in the DA market is not provided by the seller, the seller must buy replacement power in the spot market.

**Section 2: Architectural Controversies.** The most basic controversy is over the use of a bilateral DA market as opposed to a centralized market run by the system operator. If the DA market is centralized, the second controversy is over the use of a power exchange with a single price and simple bids as opposed to a power pool with multipart bids and "make-whole" side payments. Multipart bids are used to solve the unit-commitment problem, that is, to decide which generators to commit (start up). This problem and the problem of dispatching around congested transmission lines are the two technical problems that underlie the controversies.

**Section 3: Simplified Locational Pricing.** All markets discussed in Part 3 produce energy prices that are locationally differentiated, but the theory of such prices is not presented until Part 5. The key properties of these prices are (1) they are competitive prices, (2) the locational energy-price difference is the price of transmission, and (3) a single congested line makes the price of energy different at every location. Because they are competitive prices, any perfectly competitive market, whether centralized or bilateral, will determine the same locational prices.

## 3-1.1  SPOT MARKETS, FORWARD MARKETS AND SETTLEMENTS

Trading for the power delivered in any particular minute begins years in advance and continues until real time, the actual time at which the power flows out of a generator and into a load. This is accomplished by a sequence of overlapping markets, the earliest of which are forward markets that trade nonstandard, long-term, **forward contracts**. **Futures contracts** are standardized, exchange-traded, forward contracts. Electricity futures typically cover a month of power delivered during on-peak hours and are sold up to a year or two in advance. Most informal forward trading stops about one day prior to real time. At that point, the system operator holds its DA market. This is often followed by an hour-ahead market and an RT market also conducted by the system operator. All of these markets except the RT market will be classified as **forward markets**.

All except the RT market are financial markets in the sense that the delivery of power is optional and the seller's only real obligation is financial. If power is not delivered, the supplier must purchase replacement power or pay liquidated damages. In many forward markets, including many DA markets, traders need not own a generator to sell power. The RT market is a physical market, as all trades correspond to actual power flows. While the term **spot market** is often used to include the DA and hour-ahead markets, this book will use it to mean only the **RT market**. A customer who buys power in a forward market will receive either electricity delivered by the seller or financial compensation. This financial compensation is called liquidated damages, meaning the damage to the customer has been expressed as a liquid, financial sum. Because customers are virtually never disconnected when their forward contract falls through, power is delivered and they are charged for it. This cost defines the liquidated damages. In most cases, a seller who cannot deliver power from its own generator will purchase replacement power for its customer. In either case the obligation has been met financially.

The most formal arrangement for purchasing replacement power occurs in the system operator's markets. Any power that is sold in the DA market but not delivered in real time is deemed to be purchased in real time at the spot price of energy. This is called a two-settlement system and has a number of useful economic properties which are discussed in Chapter 3-2.

## 3-1.2   ARCHITECTURAL CONTROVERSIES

Three architectural controversies have plagued the design of power markets. All three surfaced early and remain in dispute. Each has a decentralized side (listed first) and a centralized side. These are:

1. Bilateral markets vs. centralized exchanges and pools.
2. Exchanges vs. pools.
3. Zonal pricing vs. nodal pricing.

Because the controversies have often been seen in ideological terms, discussion has been characterized by black and white assertions. In reality, there are many trade-offs and only a few clear-cut answers.

Part 3 does not address the third controversy; it is listed for completeness. But all of the markets discussed in Part 3 are assumed to take place in the context of locational pricing, so nodal pricing, or something similar, must be imagined in the background. The complexity of locational pricing figures in the discussion of market centralization. The first two controversies are addressed repeatedly as different time frames and different problems are considered.

Two technical complexities and one problematic simplicity underlie these controversies. The complexities are (1) transmission limits, and (2) the nonconvex structure of generation costs. The first interacts with the physical laws of power flow to produce different costs of delivered power at every point in the system when even a single line restricts trade. This drives the third controversy, but it also plays an important role in the first. Can a decentralized bilateral market solve the problem of optimizing power flow over a grid in which every trade affects the flow of power on every line?

The second complexity, referred to as the unit-commitment problem, also involves a simultaneous optimization over all of the market's generators. Startup is costly, and the value of committing (starting) a unit of generation depends on the cost of power produced by many other generators. Some believe a power pool is needed to collect all the relevant data and make a centralized calculation in order to determine if the value of starting is worth the cost. Others say a power exchange provides enough centralization by computing a public market-clearing price. Suppliers can use this to solve their own unit-commitment problems individually.

The problematic simplicity is the nature of AC power flow. This is most easily understood by considering the grid at a time when line limits play no role; then the power grid is like a pool of water. Any generator can put power in and any load can take it out; no one knows where their power goes or where it comes from. From a physical perspective it does not matter. But this makes normal private trading arrangements impossible. Unless there is a centralized accounting of all trades, any load or generator can steal power from the pool with impunity.

**Bilateral vs. Central Markets.** The first two controversies concern the role of the system operator, which some wish to minimize on general principles. Chapter 3-4 considers the ancillary services that the system operator must either provide or make sure are provided by appropriate markets. While bilateral markets are reasonably efficient at providing the main service, bulk power supply, Chapter 3-6 argues that they are too slow to provide efficiently the two ancillary services most crucial to reliability: RT balancing and transmission security.

Chapter 3-5 considers centralization of the DA market and finds the answer is less obvious. With more time for trading, the slower bilateral design might perform well enough, but the unit-commitment problem and the need for coordination with regard to transmission limits tip the balance. Though it may not be essential, there is a strong case for at least the minimal central coordination provided by a power exchange run by the system operator.

The decisive factor in all of these decisions is the need for speed. Bilateral markets are slower than centralized markets. Because of the extreme complexity of solving the unit-commitment problem and transmission problem simultaneously, a bilateral market is simply too slow. They are slow mainly because they lack a transparent market price. Price is the coordinating agent of free markets, and bilateral markets make it difficult to discover while centralized markets make it easy. By providing a transparent price, a centralized exchange makes finding an efficient set of trades much easier and much faster. In power markets, that means a great deal.

This does not mean bilateral markets should be suppressed. They should be encouraged as forward markets and allowed to exist beside the centralized DA market.

**Exchanges vs. Pools.**   Integrated utilities have always solved the unit-commitment problem centrally, using many parameters to describe each generating unit. Incorrect unit commitment, starting the wrong set of plants in advance, can lead to two problems: (1) inefficiency, and (2) reduced reliability. Chapters 3-7, 3-8, and 3-9 consider whether it is worth moving beyond a power exchange to a power pool. This would reproduce the old approach but with all of the parameters provided by private parties and with the pool having no direct control over the dispatched generators.

The disadvantages cited for an exchange are inefficiency and lack of reliability due to lack of coordination. The disadvantages attributed to a pool are gaming opportunities, and biases and inefficiencies caused by side payments. The complexity and nontransparency of pools can also lead to design mistakes that are hard to discover and correct.

None of the efficiency or gaming concerns have received serious quantitative assessment, and all seem overrated. Either system should be capable of performing quite well if well designed. Because the startup costs are only about 1% of retail costs, and a simple exchange can already manage them quite efficiently, a small increase in bid flexibility would seem to be sufficient. In other words, a little of the power pool approach may be useful, but elaborate multi-part bids appear to cause more problems than they solve. While startup insurance may give the system operator some useful control over generator ramping, this could be obtained with a more market-oriented approach.

## 3-1.3   SIMPLIFIED LOCATIONAL PRICING

> **Power Exchanges vs. Pools**
>
> Pools are often associated with nodal pricing and exchanges are sometimes thought to require a single price throughout their region. In practice exchanges have been associated with zonal prices, but there is no theoretical reason for this association.
>
> Exchanges can provide nodal prices more easily than pools and pools were always run without nodal pricing before deregulation. Part 3 assumes that the choice of locational pricing will be unaffected by whether a pool or an exchange is selected.

Because several points concerning centralization require a partial understanding of the complexity of locational pricing, this section gives a brief overview. The properties presented here are explained more fully in Chapters 5-3, 5-4, and 5-5.

Energy prices differ by location for the simple reason that energy is cheaper to produce in some locations and transportation (transmission) is limited. When a transmission line reaches its limit, it is said to be congested, and it is this congestion that keeps energy prices different in different locations. For this reason locational pricing of energy is also called "congestion pricing."

Locational prices are just competitive prices, and these are unique. They are determined by supply and demand and have nothing to do with the architecture of the market, provided it is a competitive market. This means a purely bilateral market that is perfectly competitive will trade power at the same locational prices as a perfectly competitive, centralized nodal-pricing market. Of course, a bilateral market is likely to be less precise with its pricing, but on average it should find the full set of competitive nodal prices.

Because there is a unique set of locational prices, there is also a unique set of "congestion" prices, which will also be called transmission prices. Again, these are determined by competition and supply and demand conditions and have nothing to do with market architecture, provided the market is perfectly competitive.

If the competitive energy price at X is \$20/MWh and at Y is \$30/MWh, the price of transmission from X to Y is \$10/MWh. Transmission prices are always equal to the difference between the corresponding locational energy prices. If this were not true, it would pay to buy energy at one location and ship it to the other. In that case arbitrage would change the energy prices until this simple relationship held. The relationship can be expressed as follows:

$$P_{XY} = P_Y - P_X$$

The price of transmission from X to Y equals the price of energy at Y minus the price of energy at X. This relationship is the only one used in Part 3, but a few related facts will provide a broader context.

When power flows from Y to X it exactly cancels (without a trace) an equal amount of power flowing from X to Y thereby making it possible to send that much more power from X to Y. Thus a reverse power flow from Y to X (a counterflow) produces more transmission capacity from X to Y. As a consequence, if the price of transmission from X to Y is positive, then the price from Y to X is the negative of this value. This follows from the above formula, as does another consequence: The cost of transmitting power from X to Y does not depend on the path chosen.

This is not surprising because, although contracts may stipulate a "contract path" for power, there is no way to influence the actual path taken. Locational prices reflect this reality by making sure that $P_{XZ} + P_{ZY} = P_{XY}$ for any intermediate point Z.

Not only is it impossible to select the path of a power flow, power takes every possible path between two points, with more flowing on the easier routes. The consequence for a network with a single congested line is that every location has a unique price. In effect there is a price for using the congested line, and every transaction uses that line to one extent or another. Sending power from X to fifty different locations will use fifty different amounts of the congested line, so there will be fifty different transmission prices and fifty different energy prices (plus the energy price at X). One congested line in PJM produces 2000 different locational prices. A centralized market will compute these so accurately that the true locational differences can be seen. A bilateral market finds them imprecisely, so many observed differences will be mainly due to the haphazard nature of the bilateral process.

# The Two-Settlement System

*We can scarcely avoid the inference that light is the transverse undulations of the same medium which is the cause of electric and magnetic phenomena.*

James Clerk Maxwell
1861

*This velocity is so nearly that of light, that it seems we have strong reasons to conclude that light itself (including radiant heat, and other radiations if any) is an electromagnetic disturbance in the form of waves propagated through the electromagnetic field according to electromagnetic laws.*[1]

1864

$T$HE REAL-TIME PRICE ALWAYS DIFFERS FROM THE DAY-AHEAD PRICE. WHICH IS IN CONTROL? Day-ahead (DA) prices, and especially earlier prices, differ significantly from the corresponding real-time (RT) price. The differences are due to misestimations made before traders know all the details of the RT conditions. In a competitive market the RT prices are true marginal cost prices, and the forward prices are just estimates, sometimes very rough estimates. With most trade occurring in the forward markets, does this imply that only a small proportion of generation is subject to the correct incentives? Not under a proper two-settlement system. The purpose of the RT market is to correct the prediction errors of the past. If the transaction costs in this market are minimized so that profitable trade is maximized, the RT price will be accurate and will control actual production. Past mistakes have financial impacts but will not cause inefficiency which is a purely physical phenomenon.

Contracts for differences (CFDs) insulate bilateral trades from all risks of spot price fluctuations while allowing the inevitable inefficiencies of forward trading to be corrected by accurate RT price signals. Both the two-settlement system and CFDs allow efficient re-contracting—a standard economic solution to the problems of decentralized forward trading. Advocates of bilateral trading have often failed to recognize this point and have opposed the very mechanisms that make decentral-

---

1. Maxwell developed the mathematics of electromagnetic fields, later used to design AC motors, transformers, and power lines. He predicted the possibility of electromagnetic waves and calculated their theoretical velocity from laboratory measurements on electric and magnetic fields. At first his suggestion that light is electromagnetic was dismissed as a "not wholly tenable hypothesis."

ized trading efficient. This chapter assumes the existence of a centralized RT market in which all generators and loads must participate.

**Chapter Summary 3-2:** If a generator sells its output in the DA market, the two-settlement system lets it respond efficiently to the spot price without any risk from the volatility of that price. It can only profit from an unexpected spot price, and never suffer a loss. If a generator sells its power to a load in a bilateral contract months in advance, a CFD will let them profit efficiently from an unexpected spot price. If they trade over lines that may be congested, purchasing an FTR (financial transmission right) will provide the same guarantee with respect to transmission prices. In this way forward trades that prove inefficient in real time because of unexpected circumstances can be corrected without risk to the traders.

**Section 1: The Two-Settlement System.** If the system operator runs a DA and an RT market, generators should be paid for power sold in the DA market at the DA price, regardless of whether or not they produce the power. In addition, any RT deviation from the quantity sold a day ahead should be paid for at the RT price.

A CFD requires the load to pay the generator the difference between the contract price and the spot price whether it is positive or negative. This allows either party to deviate profitably from the contract, when the opportunity arises, without affecting the other. If the spot price differs at the two locations, this hedge is not complete.

**Section 2: Ex-Post Prices: The Trader's Complaint.**   Spot prices that differ by location impose transmission costs on traders. These cannot be avoided by the use of CFDs, and they make trade risky. Some markets in transmission rights exist but are generally limited and illiquid. Design of such markets is continuing. A financial transmission right (FTR) from generator to load can perfectly hedge a bilateral trade that faces congestion charges. Since trade is always allowed in the RT market, an FTR is as good a guarantee as a physical right.

## 3-2.1   THE TWO-SETTLEMENT SYSTEM

If a supplier sells most or all of its power in the forward markets, the RT price may appear to have little chance of affecting the production decisions of suppliers. In a properly implemented two-settlement system the opposite is true. In real time, the supplier will behave as if it were selling its entire output in the RT market, even though, in the forward market, it acts as if that were its final sale.

## Separation of Real-Time from Forward Transactions

Say a supplier sells $Q_1$ to the system operator in the DA market for a price of $P_1$. If this amount of power is delivered to the RT market, the settlement in the DA market will hold without modification. But what if none is delivered, or more than $Q_1$ is delivered? In either case the DA settlement should still hold, but there should be an additional settlement in the RT market. If no power is delivered to the RT market, the supplier is treated as if it had delivered the amount promised in the DA market, $Q_1$, but purchased that amount from the RT market to cover its promised delivery. Consequently the supplier is still paid $P_1$ for $Q_1$ but is also charged $P_0$, the RT price, for the purchase of $Q_1$. In general, if a supplier sells $Q_1$ in the DA market and then delivers $Q_0$ to the RT market, it will be paid:

$$\text{Supplier is paid: } Q_1 \times P_1 + (Q_0 - Q_1) \times P_0 \qquad (3\text{-}2.1)$$

This is called a "two-settlement system." If a customer contracts for $Q_1$ and then takes only $Q_0$ in real time, it is charged exactly the amount that its supplier is paid.

| Result | 3-2.1 | **A Two-Settlement System Preserves Real-Time Incentives** |
|---|---|---|

When the RT market is settled by pricing deviations from forward contracts at the RT price, suppliers and customers each have the same performance incentives in real time as if they had traded all of their power in the RT market.

The incentives of this settlement rule are revealed by rearranging the terms as follows:

$$\text{Supplier is paid: } Q_1 \times (P_1 - P_0) + Q_0 \times P_0 \qquad (3\text{-}2.2)$$

When real time arrives, $P_1$ and $Q_1$ have been determined in the day-ahead (DA) market. Assuming the market is competitive, suppliers will also take $P_0$ as given, so by real time, the entire first term will be viewed as a "sunk" cost or an assured revenue. This leaves the second term as the only one that can provide an RT incentive for generator behavior, and this term pays the generator the RT price for every megawatt produced. Consequently the generator will behave exactly as if it is selling all of its product in the RT market. This can be proven by considering the supplier's profit, which is revenue minus cost, and the profit it would have had if it traded only in the RT market.

**Table 3-2.1** Profit With and Without a Day-Ahead Trade

| | |
|---|---|
| Actual Short-Run Profit: | $SR_{\pi F} = R_F + Q_0 \times P_0 - \text{Cost}(Q_0)$ |
| Only-Real-Time Short-Run Profit: | $SR_{\pi 0} = Q_0 \times P_0 - \text{Cost}(Q_0)$ |

The only difference between the two is the fixed revenue, $R_F = Q_1 \times (P_1 - P_0)$, so the value of $Q_0$ that maximizes one will maximize the other.

This result means that no matter what trades, $Q_1$, have taken place in the DA market, or any other forward market, profit-maximizing suppliers will pursue the same RT strategies as if no prior trades had taken place. Consequently, if the RT market is competitive and therefore efficient, this efficiency will not be undermined by forward contracts. Put another way, if mistakes are made in forward markets they will affect revenues but not efficiency because the RT market will induce least-cost operation regardless of these mistakes. The above argument also applies to loads.

## Separation from Imports and Exports

If a supplier sells and schedules $Q_1$ for export in a forward market for a price of $P_1$, and delivers this much power locally to the RT market, the exporter will owe nothing and be paid nothing by the local DA and RT markets. The exporter will be paid only by the external purchaser. But what if no power is supplied or more than $Q_1$ is supplied? As before, the supplier will be paid $(Q_0 - Q_1) \times P_0$ by the RT market, where $Q_0$ is its RT supply. If $Q_0$ is zero, the supplier is paid $-Q_1 \times P_0$, which means it is charged $Q_1 \times P_0$. This charge assumes that it has purchased in the RT market the power that it exported. Of course the transaction could be cancelled, in which case $Q_1$ would be adjusted to zero, and the RT payment formula would continue to apply.

## Separation from Bilateral Markets

If a trader has arranged to buy $Q_1$ at price $P_G$ from a generator and sell it to a load at price $P_L$, how will the participation of the generator and load in the RT market be handled? This transaction does not make the trader a participant in the RT market, yet the generators and loads must participate. Say the generator injects $Q_1$ and the load withdraws $Q_1$ as the contract demands. The generator will be paid $P_0 \times Q_1$ by the RT market which is different from what was specified in the **bilateral contract** with the traders. To compensate for this, the trader must specify that the generator will pay the trader $P_0 \times Q_1$ and the trader will pay the load the same amount. This is over and above the payments specified for the original purchase and sale of $Q_1$ by the trader. The entire transaction works like this (Table 3-2.2):

**Table 3-2.2** Details of a Bilateral Trade with Adjustments

|  | Trade | RT Market | Adjustment |
|---|---|---|---|
| Generator is paid: | $P_G \times Q_1$ | $+ P_0 \times Q_1$ | $-P_0 \times Q_1$ |
| Load pays: | $P_L \times Q_1$ | $+ P_0 \times Q_1$ | $-P_0 \times Q_1$ |
| Trader's net income: | $(P_L - P_G) \times Q_1$ |  | $+ P_0 \times Q_1 - P_0 \times Q_1 = 0$ |

The first term for both generator and load is the payment specified by the original trade. The second term is the result of each participating in the RT market, and the third term is the adjustment term specified by the trader to keep the bilateral trade separate from the RT market. The adjustment exactly cancels the RT settlements of both generator and load, and for the trader, the two adjustments cancel each other.

**Contracts for Differences.** Combining the trade and the adjustment for the generator defines a bilateral trade of a special type called a **contract for differences**, a CFD. The detailed trades shown above contain two of these plus the trades with the RT market. The following table shows the two contracts for differences used in the above bilateral trade as well as a CFD written directly between a generator and a load.

**Table 3-2.3** Contracts for Differences

| Trader pays generator: | $(P_G - P_0) \times Q_1$ |
| --- | --- |
| Load pays trader: | $(P_L - P_0) \times Q_1$ |
| Direct CFD: Load pays generator | $(P_C - P_0) \times Q_1$ |

Each bilateral trade is the contract quantity times the *difference* between the contract price and the RT price. If the generator contracts directly with the load, there is only one contract price, $P_C$, and the load pays the generator $(P_C - P_0) \times Q_1$.

By writing the bilateral contracts, shown in Table 3-2.3, with the generator and the load, a trader can implement the detailed bilateral contract displayed in Table 3-2.2. The RT market and adjustment terms cancel, and the effect is just as if the RT market did not exist. The same is true for a trade between a generator and a load that is implemented with a CFD. These conclusions assume the trade takes place as specified in the contract.

## Contracts-for-Differences Result #1

**Result     3-2.2**

**A Contract for Differences Insulates Traders from Spot Price Volatility**
Bilateral trades implemented through contracts for differences completely insulate traders from the market price provided (1) the traders produce and consume the amounts contracted for, and (2) the market price is the same at the generator's location as at the load's location.

When CFDs are used, the only effect of the spot market is to provide a convenient remedy for deviations from the contract position. Such deviations only affect the party who deviates.

Remarkably, while insulating the bilateral trade from the spot market, the CFD, together with the two-settlement system, also insulates the traders' use of the spot market from the effects of the bilateral trade. The argument used in the beginning of this section can be used again to show that both customer and supplier will behave as if they were trading in the RT market because their incentives to deviate from the contract quantity are determined by the RT price. Such deviations are desirable and can only make them better off while doing no harm to their trading partners. CFDs give us the best of both worlds. Traders are protected from the volatility of spot prices, and the efficiency of the RT market is protected from the inefficiency of forward bilateral contracting.

## Contracts-for-Differences Result #2

| Result | 3-2.3 | **Contracts for Differences Preserve Real-Time Incentives** |
|---|---|---|

Bilateral traders using contracts for differences feel the full incentive of RT prices. Because they could ignore this incentive, any deviation from their contract can only be profitable.

## Locational Price Differences

So far, spot prices have been assumed equal at all locations in the market so both generator and load see the same price, but this is often not the case. As a consequence, the CFD can be written in different ways. To analyze this situation, assume that generator and load have signed a CFD without the benefit of a middleman. There are two ways to write the CFD:

**Table 3-2.4** Two Possible CFDs

| Effect of CFD | Load's Payment to Generator |
|---|---|
| Load pays for transmission: | $(P_C - P_{G0}) \times Q_1$ |
| Generator pays for transmission: | $(P_C - P_{L0}) \times Q_1$ |

$P_{G0}$ is the spot price at the generator's bus, $P_{L0}$ is the spot price at the load's bus, and $P_C$ is the contract price. In both cases, the CFD specifies the payment by loads to generators. In the first case, because the generator's RT price is used, the generator is insulated from locational price differences, while in the second case, load is insulated. To see this, the full transaction, including the RT market, must be taken into account. Consider the case in which the generator's spot price is used and in which each player produces or consumes as specified in the contract.

**Table 3-2.5** Settlements with the Generation-Centric CFD

| | CFD | Spot Market |
|---|---|---|
| Generator is paid: | $(P_C - P_{G0}) \times Q_1$ | $+ P_{G0} \times Q_1$ |
| Load pays: | $(P_C - P_{G0}) \times Q_1$ | $+ P_{L0} \times Q_1$ |

The first term is the CFD settlement and the second is the spot market settlement. These settlements can be simplified as follows:

**Table 3-2.6** Algebraic Simplification to Reveal Transmission Charge

| | Trade | Transmission Charge |
|---|---|---|
| Generator is paid: | $P_C \times Q_1$ | 0 |
| Load pays: | $P_C \times Q_1$ | $+ (P_{L0} - P_{G0}) \times Q_1$ |

Both will pay or receive the contract price, but the load must also pay the locational price difference times the quantity traded. Typically the spot price will be higher at the load's location, so this charge will be positive. The charge is termed a congestion charge or a transmission charge, and it arises from the scarcity of transmission. It is like a transportation or delivery charge but is usually more volatile.

Congestion charges are covered in Part 5, but for present purposes it is only necessary to understand that they are based entirely on scarcity.[2] They are typically zero because there is plenty of transmission capacity most of the time. When it is scarce, competition for transmission can send the price quite high. To some extent these prices are predictable, but they contain a significant random component that can be problematic for traders (see Section 3-1.3).

The uncertainty in the congestion price ($P_{L0} - P_{G0}$) can be hedged by buying energy forwards in the two locations or by buying transmission rights between the two locations. If the CFD is arranged so that the load pays the congestion charge, the load may want to buy a hedge. If the CFD is set up the other way, the generator may want to buy it. If the trade is arranged by a trader, the trader will probably accept the transmission charge and may want to hedge it.

## 3-2.2   EX-POST PRICES: THE TRADER'S COMPLAINT[3]

Power traders can write CFDs and thereby completely insulate themselves from the spot price (and the DA price). But it is much more difficult for them to insulate themselves from locational price *differences*, that is, transmission prices. If they have used a CFD to execute their bilateral trade and they trade as planned, the spot price, no matter how high, will have no effect. Without a transmission right, the spot *transmission* price must be paid in full.

Why is it so easy to insulate a bilateral trade from the spot *energy* price and so difficult to insulate it from the spot *transmission* price? A buyer and seller, considered as a unit, are unaffected by the energy price because together their net position is zero. As a unit, however, they always take a net position in the transmission market; they consume transmission from generator to load.[4] Because they take a nonzero net position in the transmission market, they are affected by the price of transmission.

The complaint of traders is that the transmission price is "ex-post"—it is established after they commit to a trade instead of being posted ahead of time. They would prefer to check the price of transmission, arrange a trade based on that price, then call up the transmission provider and purchase the desired quantity at the price they were quoted. United Parcel Service works this way, and their prices have two convenient features. First, they are posted in advance, and second, they are good no matter how many packages are shipped. Real-time transmission prices have neither feature. Access charges, used to cover the fixed costs of transmission, do have these features, but they are not the focus of the trader's complaint which

---

2. This ignores the charge for losses, which is almost never above 10% and is far more predictable.

3. These "ex-post prices" are not necessarily those that result from the system of "ex-post pricing" discussed in the preface. True "ex-post pricing" is said to use a quantity optimization procedure that has not been publically specified but is rumored to be central to the PJM and NYISO pools.

4. If they trade against the prevailing flow, they create a counterflow and their consumption of transmission service is negative, but they have still taken a position in the transmission market.

concerns only congestion charges. Two approaches to resolving this problem need consideration: can (1) RT price setting or (2) forward markets for transmission be improved?

## Better Real-time Price Setting?

The system operator could post the RT congestion prices in advance. Because congestion prices fluctuate somewhat predictably, this would be a time-of-use transmission charge. The bulk of bilateral trading takes place months in advance, and these charges would need to be set annually, or preferably even further in advance. But congestion prices, as observed in systems that compute them continuously, have only a small component of predictability. They are susceptible to weather, generation outages, transmission outages, and other factors. When the time-of-use price proved to be too high, valuable transmission would go unused. At other times, demand for transmission would outstrip supply, and traders would find none available at the posted price. Curtailments would occur in real time, an outcome that would be very inefficient and extremely unpopular with traders.

To be of any use, time-of-use pricing must be coupled with a reservation system. FERC's pro-forma tariff provides such a system although it is based on contract paths rather than on actual power flows and thus conflicts with the laws of physics. With a system of interconnection-wide coordination, such as NERC is now devising, a consistent reservation system would be possible. This would avoid most RT curtailment, but traders would still find that the posted price would not guarantee availability of reservations. Without a real market for transmission rights, nonprice rationing would fail to select the highest-value users and lead to an inefficient dispatch. At other times, trades would be curtailed unnecessarily.

Currently, no estimate of the inefficiency of such administratively set prices is available. The benefits of such a system are even more difficult to determine. Facilitating bilateral trading, especially long-term trading, is desirable, but the benefits are hard to quantify. In any case, market-based prices is generally considered preferable to regulated prices.

## Better Forward Markets for Transmission

A second approach to alleviating the trader's complaint is to implement a forward market in transmission rights. This has the same effect as the posted price and reservations, but the price is determined by the supply and demand for transmission instead of being administratively determined. PJM sells such rights in its DA market as described in Chapter 3-3. Although this is a help, by the time traders see the DA price of transmission, it is too late to take advantage of it. Because it is a rough guide to RT congestion prices, it is some help in deciding whether to sign a last-minute bilateral contract for power and take a chance on the RT transmission price. A continuous market is needed with a slowly changing price that traders can observe before they arrange a trade. Afterwards they can purchase transmission at a price close to the observed price. So far, such markets have not been developed because of technical and practical problems, but new proposals are currently being debated.

PJM and the NYISO also sell firm point-to-point financial transmission rights between any two points in their systems. Unfortunately these markets are quite illiquid and a trader may have to wait for a quarterly auction to purchase the desired right, although purchase in a secondary market is sometimes possible.

If well constructed, financial transmission rights provide the security of physical rights without some of the drawbacks. Consider how a financial right is used to hedge a bilateral trade in the RT market. A financial transmission right, FTR, from G to L for $Q_1$ MW is defined to pay[5]

$$\text{FTR from G to L for q pays: } (P_{L0} - P_{G0}) \times Q_1.$$

Using this FTR, reconsider the last bilateral trade of the previous section in which load ended up paying a transmission charge.

Assuming the load has purchased this FTR and the same generation-centric CFD is used, the settlement will be as follows (Table 3-2.7):

**Table 3-2.7** FTR Settlements with the Generation-Centric CFD

|  | CFD | Spot Market | FTR |
|---|---|---|---|
| Generator is paid: | $(P_C - P_{G0}) \times Q_1$ | $+ P_{G0} \times Q_1$ | |
| Load pays: | $(P_C - P_{G0}) \times Q_1$ | $+ P_{L0} \times Q_1$ | $-(P_{L0} - P_{G0}) \times Q_1$ |

Note that the FTR payment is included with a minus sign because "load pays" a negative amount—it is paid this much by the FTR. This time when the settlements are simplified, the load pays $P_C \times Q_1$ to the generator and neither pays a transmission charge. Of course the FTR had to be purchased and that cost is not shown in the settlement, but once it is purchased, the parties are immunized completely against the RT transmission charge. By using both a CFD and the FTR they are protected from fluctuations in the general level of spot prices and from locational differences in the spot price.

If the trade is made, there will be no transmission charge, but what guarantees that the parties will be allowed to trade when they have no physical reservation? Generally, RT locational energy markets allow traders to do as they wish, but they must pay for their injections and withdrawals at the RT price. In this case no special action is needed by the parties; they simply make the trade and accept the charges. Provided their trade matches their FTR, the net charges will be covered exactly. If the RT market requires bids, then the generator submits the lowest possible bid (in PJM this would be − $1,000/MWh), and the load submits the highest possible bid. Unless there is some physical problem with the network (in which case not even a physical reservation can guarantee the trade), they will be dispatched. Because they own the FTR they cannot be harmed by any price that results from their extreme bids. The net result is an assurance that they will be able to complete the trade and be completely unaffected by the RT prices.

---

5. This is the definition of a transmission congestion contract (TCC) as found in Hogan (1992). Many variations on this have been proposed and used.

# Day-Ahead Market Designs

*From a long view of the history of mankind — seen from, say, ten thousand years from now — there can be little doubt that the most significant event of the 19th century will be judged as Maxwell's discovery of the laws of electrodynamics.*

Richard Feynman

*One scientific epoch ended and another began with James Clerk Maxwell.*[1]

Albert Einstein

CⁿENTRAL DAY-AHEAD MARKETS CAN BE DESCRIBED AS AUCTIONS. The most obvious design sets energy prices based on simple energy-price bids. A different approach turns the system operator into a transportation-service provider who knows nothing about the price of energy but sells point-to-point transmission services to energy traders.

Either of these approaches presents generators with a difficult problem. Some generators must engage in a costly startup process (commitment) in order to produce at all. Consequently, when offering to sell power a day in advance, a generator needs to know if it will sell enough power at a price high enough to make commitment worthwhile. Some day-ahead (DA) auctions require complex bids which describe all of a generator's costs and constraints and solve this problem for the generators. If the system operator determines that a unit should commit, it ensures that all its costs will be covered provided the unit commits and produces according to the accepted bid. Such insurance payments are called "side payments," and their effect on long-run investment decisions is considered in Sections 3-8.3 and 3-9.3.

**Chapter Summary 3-3:** Day-ahead markets run by system operators are run as auctions. Although some trade energy, some sell transmission, and some solve the unit-commitment problem, they all use the same philosophy for choosing which bids to accept and for setting prices. Four archetypical markets are summarized: (1) a power exchange, (2) a transmission-rights market, (3) a power pool, and (4) PJM's DA market which mixes all three.

---

1. The universe is governed by four forces, and matter is made of their associated particles. The electron and photon are the carriers of the electromagnetic force, and the other three forces are gravity, the weak force and the nuclear force. Maxwell discovered the mathematics of the electromagnetic force, Einstein the mathematics of the gravitational force, and Feynman the laws of the weak force which he unified with the electromagnetic force.

**Section 1:  Defining Day-Ahead Auctions.**  In DA auctions, bids are selected to maximize total surplus, gross consumer surplus minus the cost of production. Locational energy prices are set equal to the marginal change in total surplus when free power is injected at the various locations. Transmission is priced at the marginal change in total surplus when a counterflow is introduced on a transmission path.

**Section 2:  Four Day-Ahead Market Designs.**  Each auction is specified by three sets of conditions: bidding rules, bid acceptance rules, and settlement rules. Market 1 is a power exchange that uses one-part bids. Market 2 is a "bilateral" transmission-rights market. Market 3 is a power pool which uses multipart bids for unit commitment. Market 4 combines the features of the other three as options.

**Section 3:  Overview of the Day-Ahead Design Controversy.**  Forward markets are bilateral, and real-time (RT) markets are centralized. The DA market can be designed either way, and this causes a great deal of controversy. The "centralized nodal pricing" approach specifies an energy market with potentially different prices at every node (bus) and specifies that the auction should solve the unit-commitment problem as well. This requires complex bids. The bilateral approach specifies that energy trades take place between two private parties and not between the exchange and individual private parties. To trade energy, the private parties require the use of the transmission system, so the system operator is asked to sell transmission.

## 3-3.1   HOW DAY-AHEAD AUCTIONS DETERMINE QUANTITY AND PRICE

Day-ahead markets run by system operators take the form of either exchanges or pools and are operated as auctions. The process of selecting the winning bids is often complicated by transmission and generation constraints which can require the use of enormously complex calculations and sophisticated mathematics. The mathematics is often presented as a way of explaining the auction. This is unnecessary, frequently confusing, and often less precise than an approach that explains the purpose rather than the mechanics.

### A Simplified Description of Auctions

All auctions solve some mathematical problem. Bids are submitted, some function of the bids is maximized or minimized, and the solution determines which bids are accepted. For example, say bids are submitted for the purchase of 100 tulip bulbs. Each bid states a number of bulbs and a total price. The auction problem might be to maximize the sum of the accepted bid prices. The solution defines a set of accepted bids. There is one more step—settlement. The winners must pay or be paid and must provide or receive the goods being auctioned. Settlement is not determined entirely by the solution to the acceptance problem but also uses separate price-determination rules. In this example the rules could specify that each accepted bid would pay as bid and receive the number of bulbs bid for, or it might specify that accepted bids would pay a price per bulb equal to the lowest price per bulb of any accepted bid.

---

**Summary**

### An Auction's Three Stages: Bidding, Acceptance, and Settlement

An auction takes place in three stages

1. Bids are submitted.
2. Some bids are accepted and prices determined.
3. Accepted bids are settled at the determined prices.

The optimization problem that determines which bids are accepted does not automatically determine prices. The pricing rule must be specified separately.

---

The description of DA markets often focuses on presenting the auction problem, which bids to accept and what prices to set, in a format that is convenient for solution using linear programming techniques. The solution technique is of little interest from an economic perspective, so presenting the problem in this way adds little value and often much complexity. For example, the auction problem might be to minimize the total price of accepted bids of generators while purchasing enough power to supply a 1000-MW load at location X and taking into account losses. This problem can be presented as a linear programming problem that approximates power loss equations. But the necessary equations, while invaluable to the computer programmer, add relatively little to understanding the economics.

Much of the complexity of electricity auctions is embodied in the constraints placed on bid acceptance by the physics of the system. Given a source (a generator) and a sink (a load), the path of the power flow is usually uncontrollable. Consequently, to avoid overloading lines, certain combinations of bids cannot be accepted. Losses pose similar restrictions on bid acceptance.

Instead of specifying physical constraints in abstract mathematical detail, the following four market summaries will simply name the constraints. The focus can then shift to the bidding rules, the quantity being optimized in the auction problem, and the settlement rules. Settlement rules in particular often include overlooked inducements and penalties that are crucial to the functioning of a market. The shift in focus to economic aspects, and the standardization of the descriptive format allows easier and more meaningful comparisons of these archetypical designs for DA markets.

## Accepting Bids: Determining Quantities

The auction problem determines what bids are accepted, or partially accepted, and this determines through simple accounting the quantities bought and sold. Prices are determined by a separate set of settlement rules. Quantities determine the efficiency of a particular outcome, but prices provide the incentives that determine what bids are submitted and thereby help control the outcome.

All four auction designs considered in this chapter maximize total surplus as defined by the bids. These designs encourage truthful bidding, and in a competitive market, bids will reflect true costs and benefits, and the auctions will maximize actual total surplus.

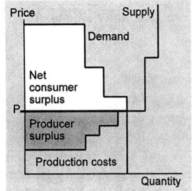

Total surplus is the sum of (net) consumer and producer surplus, but it is also the gross consumer surplus minus production costs. If a consumer offers to buy 100 MWh at up to $5,000/MWh, and the bid is accepted, the gross consumer surplus is $500,000. If the market price is $50/MWh, and the customer's cost is $5,000, the net consumer surplus is $495,000. Similarly, if a generator offers to sell 100 MW at $20/MWh, its cost is presumed to be $2,000. If the market price is again $50/MWh, its producer surplus will be 100 × ($50 − $20), or $3,000. Total surplus is $498,000. Writing this calculation more generally reveals that price played no role in determining total surplus:

$$\text{Total surplus} = Q \times (V - P) + Q \times (P - C) = Q \times (V - C),$$

where $Q$ is the quantity traded, $V$ is the customer's gross surplus, $C$ is the producer's cost, and $P$ is the market price. If many bids are involved, $P$ cancels out of each trade. The problem of maximizing total surplus can be solved independently of price determination.

In an unconstrained system, total surplus can be maximized by turning the demand bids into a demand curve and the supply bids into a supply curve and finding the point of intersection. This gives both the market price and a complete list of the accepted supply and demand bids. Unfortunately transmission constraints and constraints on generator output (e.g., ramp-rate limits) can make this selection of bids infeasible. In this case it is necessary to try other selections until a set of bids is found that maximizes total surplus *and* is feasible. This arduous process is handled by advanced mathematics and quick computers, but all that matters is finding the set of bids that maximizes total surplus, and they can almost always be found.

## The Efficient-Auction Result

| | | |
|---|---|---|
| **Result** | **3-3.1** | **A Single-Price Day-Ahead Auction Is Efficient** |

Ignoring nonconvex costs and market power, single-price DA auctions are designed to maximize the sum of producer and consumer surplus. Having done this they determine locational prices that support this pattern of production and consumption. Efficiency depends on honest bidding which results from single-price settlement and competition between bidders.

## Determining the Market Price

The price determined by supply and demand is the highest of all accepted supply bids or the lowest price of an accepted demand bid. It depends on whether the intersection of the two curves occurs in the middle of a demand bid or in the middle

**Figure 3-3.1**

Changes in total surplus from the addition of 1 kW of free supply will be called the marginal surplus.

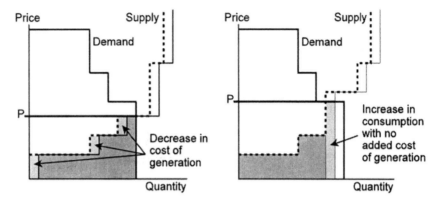

of a supply bid. When the demand curve is vertical, the intersection is always in the middle of a supply bid, and the price is set to the supply-bid price.

Whichever curve is vertical at the point of intersection has an ambiguous marginal cost or value (see Chapter 1-6). If the demand curve has a horizontal segment at $200 that intersects a vertical part of the supply curve that runs from $180 to $220, then the marginal cost of supply is undefined but is in-between the left-hand marginal cost of $180 and the right-hand marginal cost of $220. (See Chapter 1-6.) Consequently it causes no problem to say that the market price equals both the marginal cost of supply and the marginal value of demand.

While the intersection of supply and demand is a convenient method of determining price in an unconstrained market with a single price, it is cumbersome to apply to a constrained market with many prices. An alternative approach sets price equal to marginal surplus. It determines the same price as the intersection of supply and demand and is easier to use in a constrained system.

---

**Definition**

**Marginal Surplus**
Marginal surplus is the change in total surplus with an unit increase in costless supply. (Total surplus is the sum of consumer and producer surplus as well as the difference between the supply and demand curves up to the quantity traded.)

---

First consider the unconstrained markets shown in Figure 3-3.1. The change in total surplus when a kilowatt is added *at no cost* to the total supply of power is the marginal total surplus, which because of its awkward name, will be called simply the **marginal surplus**.[2] In the case depicted on the left of Figure 3-3.1, the free unit of supply shifts the entire supply curve left while the amount produced and consumed remains unchanged. The result is a production cost savings of $P$ times 1 kW, so the marginal surplus is $P$. In the case depicted on the right, consumption is limited by the high cost of the next available generator, but if another kilowatt of free supply were available, consumption would increase by 1 kW. This case is more easily analyzed by adding the free generation out of merit order as shown. The value of consumption increases by $P$ times 1 kW, and the cost of production

---

2. Total surplus should be expressed in $/h. A kilowatt, rather than a megawatt, is used to indicate that only a "marginal" change is being made. Technically one should use calculus, but this is of no practical significance.

**Figure 3-3.2**

Changes in total surplus from introducing a free kilowatt of supply at the consumer and supplier ends of a transmission constraint.

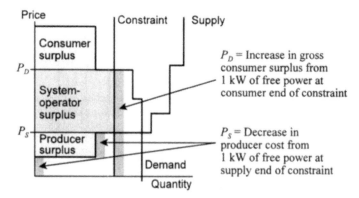

$P_D$ = Increase in gross consumer surplus from 1 kW of free power at consumer end of constraint

$P_S$ = Decrease in producer cost from 1 kW of free power at supply end of constraint

stays constant, so total surplus increases by the same amount. In both cases marginal surplus equals price.

Next consider the constrained market diagrammed in Figure 3-3.2. All suppliers are on one side of the transmission constraint and all consumers are on the other, and the free kilowatt of supply can be introduced at either location. At the consumer end, the kilowatt increases consumption, so the marginal surplus is the marginal value of consumption, $P_D$. At the supplier end, the kilowatt decreases the cost of supply, so the marginal surplus is the marginal cost of supply, $P_S$. Because of the constraint, marginal value and marginal cost are different, and thus the two prices are different. Because the prices are different, the system operator captures part of the total surplus, but in spite of this, total surplus still equals gross consumer surplus minus the cost of production.

If the supply and demand curves were smooth, the results would be the same, but in an unconstrained market, consumer value and production cost would jointly determine marginal surplus. If there were more constraints and more distinct locations, there would be more prices. If demand were vertical, marginal cost would determine marginal surplus at every location. This technique for computing price handles all necessary complexities.

> **Does Price Equal Marginal Cost?**
>
> Prices set by the auctions under consideration are often called marginal-cost prices. When the demand curve is vertical at the intersection of supply and demand, this term is accurate. With the current lack of demand elasticity, it is almost always correct.
>
> When the demand curve intersects a vertical supply curve, a range of marginal costs is determined, as explained in Chapter 1-6. Then price is set to the marginal value of demand which is within the range of ambiguity of marginal cost.
>
> Marginal surplus is defined in both cases and never contradicts marginal cost. Price always equals marginal surplus and is always equal to marginal cost or within its range of ambiguity.

Setting the market price equal to the marginal surplus is justified because it gives the competitive price and thus induces efficient behavior. It also clears the market, which means all accepted bids will voluntarily comply with the settlement, and all rejected bids will suffer no loss given the settlement price. Trade is voluntary at these prices.

A fundamental difference between market-based and auction-based price determination is that in markets marginal cost is set equal to price, while in auctions, price is set equal to marginal cost.[3] (See Section 1-5.1) This causes no problem

---

3. This is not always strictly correct. The logic of auction price determination runs from optimization of trade through determination of marginal costs which then define prices. But the computer programs that solve these problems sometimes make use of the market's logic to find their solution.

as long as marginal costs are correctly determined by the auction, which they are because the auction problem is specified as the maximization of total surplus. This minimizes production costs given consumption and maximizes consumer value given production. This determines the same efficient pattern of production as a competitive market and thus the same marginal costs.

## 3-3.2  SUMMARIES OF FOUR DAY-AHEAD MARKETS

This section summarizes four archetypical DA markets, from simple to complex. Each is a "locational" market, and these locations may be either single buses or zones containing several buses.[4]

**Market 1:  A Power Exchange.**  A power exchange is a centralized market that does not use ("make-whole") side payments. Market 1 is a classic exchange in that it employs one-part bids—its bids consist of only a supply or demand curve. California's Power Exchange was a standard power exchange in its operation but had imposed on it a peculiar relationship to other "scheduling coordinators" and to the CA ISO. Such relationships have nothing to do with the definition of an exchange market. Alberta's power exchange uses two-part bids and serves as the completely centralized RT spot market. Both are exchanges because they do not use generator-specific side payments and both are typical in that their bid formats are simple.

---

**Definition**         **Power Exchange**
An exchange market is a centralized market that does not use side payments. At any given time and location it pays the same price to any generator selling power. It can use multiple rounds of bidding or multipart bids to determine this price and can implement full nodal pricing. Typically, its bids are much simpler than those used by pools, but a centralized market using 20-part bids but not making side payments would still be an exchange.

---

In some respects Market 1 is the simplest DA market. Participants do not search for trading partners and do not have to consider many prices in many locations. Each trader simply trades with the exchange at the trader's location. The system operator's job is simple because it ignores the unit-commitment problem. One difficulty, discussed in Chapter 3-9, is that competitive suppliers will not always find it most profitable to bid their marginal costs.

**Market 2:  A Transmission-Rights Market.**  A transmission market is equally simple for the system operator but requires a complex pre-market step for market participants. Buyers and sellers must find each other and make provisional energy trades that are contingent on the outcome of the DA transmission market, or they must buy transmission on speculation. As with an energy market, most transactions

---

4. If zones are used, the transmission constraints will represent the market less accurately, and a more conservative representation of constraints may be required. This affects only the details of the constraint specification and not the specification of the markets. However, zones and other complications may require settlement rules, not discussed here, to handle special circumstances, for example, intrazonal congestion.

can be handled in longer-term markets. The hallmark of this market is that the system operator does not know the price of energy, only the price of transmission.

**Market 3:  A Power Pool.**  "Power pool" is a term used to describe an organization of regulated utilities that trade power. Such pools did not use nodal pricing, but tight pools did solve the unit-commitment problem centrally. Joskow (2000b) has said of PJM,

> . . . *it is not very different from the power-pool dispatch and operating mechanisms that were used when PJM was a large traditional power pool relying on central economic dispatch based on marginal cost pricing principles.*

"Pool" and "poolco" were initially popular terms to describe this type of market, but proposals with this name were tagged as communist by the California bilateralists; hence those names were dropped in favor of "nodal pricing" and various related terms.[5] Unfortunately these terms refer to the transmission-pricing half of the proposals and not to the unit-commitment part, so there is no common term remaining for a market that mimics an old-fashioned tight power pool. Given the circumstances, it seems best to continue using the old term, *power* pool, in analogy to *power* exchange, with the understanding that in the new context of market design, it refers to a *market* that does what a tight power pool did under regulation. In the present work, "power pool" will mean an auction market that uses side payments and multipart bids to solve the unit-commitment problem centrally; the extent of locational pricing will not be implied. As a corollary, "nodal pricing" will mean nodal pricing and will not imply centralized unit commitment or side payments.

| | |
|---|---|
| **Definition** | **Power Pool** |
| | A pool is a centralized market that uses "make-whole" side payments to, in effect, pay different prices to different suppliers at the same time and location. These payments are only made when an accepted supplier would lose money at its as-bid costs given the pool price. A pool can implement full nodal pricing. It typically uses multipart bids which cover all important aspects of a generator's operating costs and physical constraints, but a centralized market using two-part bids and making side payments is a pool. |

**Market 4:  PJM's Day-Ahead Market (2001).**  Market 4 is modeled on PJM's current market and includes all of the types of bids allowed in the previous three markets. This is the most complex market from the system operator's point of view, but, like market 3, it can be quite simple for suppliers if they simply bid competitively. This requires only that they bid their costs.

---

5. The origins of the English pool can be traced in a memo from Larry Ruff to Stephen Littlechild in June of 1989. It is available with other references at www.stoft.com. The theory of a power pool market is described in Schweppe et al. (1988).

**Table 3-3.1** Notation Used in Auction Market Summaries

| Term or Symbol | Definition |
|---|---|
| $P^S(Q)$, $P^D(Q)$ | Supply and demand bids. Supply bids must be nondecreasing. |
| $Q^D$ | Completely inelastic demand bid. |
| $P^T, Q^T$ | Transmission bid. |
| $P^E, Q^E,$ | Simple energy bid (no supply or demand curve allowed). |
| $Q_1, Q_1^T$ | Day-ahead accepted bid energy and transmission quantities. |
| $P_1, P_1^T$ | Day-ahead locational energy prices and transmission prices determined by the auction. |
| $Q_0, Q_0^T$ | Actual RT transactions. |
| $P_0, P_0^T$ | Real-time locational energy and transmission prices. |
| X, Y | Two different locations. Used to define transmission bids and prices. |
| Uplift | A charge to load to cover unattributed costs, in this case startup insurance. |

## Conventions and Notation for Describing Auctions

**Bidding, Bid Acceptance, and Settlement.**  The market summaries describe market operations in terms of the three auction stages. First, restrictions on *bidding* are described that can have important consequences for the acceptance problem, and for market efficiency. For example, allowing 24 hourly bids instead of a single bid for the day can increase market power. Second, *bid acceptance and price determination* are lumped together because they are computed together, although conceptually they are distinct. Third, *settlement* includes penalties for noncompliance with commitments made in the auction. These typically involve RT market prices and are crucial to the functioning of the market.

**Supply and Demand Curves.**  Usually supply curves are represented by either piecewise linear functions (connect the dots with straight lines) or step functions. Typically they allow the bidder to specify about ten sloped lines or horizontal steps, but all that matters is that bidders can submit a fairly accurate approximation to their actual supply and demand curves. Supply-curve bids must always be nondecreasing.

**Locational Prices.**  Every supply and demand bid at the same location is paid or charged the same price. Transmission prices and quantities are each associated with two locations. The following summaries all assume that demand prices are the same as supply prices and vary by location, but for political reasons, markets must usually charge loads a uniform price.

---

## Market 1: A Power Exchange

---

**Bidding Restrictions:**

| | | |
|---|---|---|
| Supply: | $P^S(Q)$ | 24 hourly bids |
| Demand: | $P^D(Q)$ or $Q^D$ | 24 hourly bids |

**Bid Acceptance and Price Determination:**

| | |
|---|---|
| Acceptance problem: | Maximize total surplus of accepted bids, $\{Q_1\}$ |
| Constraints: | Transmission limits |
| Price determination: | $P_1$ = marginal surplus at each location |

**Settlement Rules:**

| | | |
|---|---|---|
| Supply: | Pay: | $Q_1 \times P_1 + (Q_0 - Q_1) \times P_0$ |
| Demand: | Charge: | $Q_1 \times P_1 + (Q_0 - Q_1) \times P_0$ |

**Notes:** $P_1$ is the DA price appropriate to the location of the accepted bid, $Q_1$. $Q_0$ is the quantity actually produced or consumed, and $P_0$ is the RT locational price.

**Comments:**

Because generators cannot bid their startup costs, it is generally believed they need to submit different price bids in different hours. Loads, whose usage is largely unrelated to price, must do the same. The set $\{Q_1\}$ represents the set of accepted bid quantities, one for each supplier and each demander, a different one in every hour. Acceptances may be for partial quantities.

The auction first finds the set of supply and demand bids which, if accepted, would maximize total surplus to all market participants. Then market price is determined at each location by the marginal surplus of additional supply. This can be computed by making another kilowatt-hour available at no cost, recomputing the optimal dispatch and finding the increase in total surplus. That value is the price per kilowatt-hour assigned to that location. The increase in value can come from either more consumption or from reduced production costs. A kilowatt-hour is used to mimic a "marginal" change.

A DA market is a forward market, and the forward price holds if suppliers deliver and customers take delivery of the DA quantity. Participants may not make or take delivery of the exact quantities accepted, so strategy in the DA auction depends on the penalty for not fulfilling the DA contracts. The NYISO, for instance, confiscates the RT payments specified when $Q_0 - Q_1 > 0$.

## Market 2: A Transmission-Rights Market
### (the "bilateral" approach)

**Bidding Restrictions:**

   Demand:                        $Q^T$ @ up to $P^T$, from X to Y          24 hourly bids

**Bid Acceptance and Price Determination:**

   Acceptance problem:   Maximize total surplus = $\sum (P^T \times Q_1^T)$ with $0 < Q_1^T < Q^T$

   Constraints:          Transmission limits

   Price determination:  $P^T$ from X to Y = the marginal surplus of a 1-kW increase in the transmission limit from X to Y

**Settlement Rules:**

   Demand:          Charge:     $Q_1^T \times P_1^T + (Q_0^T - Q_1^T) \times P_0^T$

**Notes:**   $P_1^T$ is the DA price from X to Y. $Q_0^T$ is the quantity actually produced or consumed, and $P_0^T$ is the RT price from X to Y.

**Comments:**

$P_1^T$ is a price for transmission, not energy. If there are 10 locations there will be 90 pairs of locations and consequently 90 transmission prices. These can be computed by subtracting pairs of energy prices. Adding the same constant to the energy prices leaves the differences and thus the transmission prices unchanged. Consequently the ten energy prices cannot be computed from the 90 transmission prices.

The total surplus from the transmission sold in the auction is the sum of the accepted quantities times the respective bid prices. Bid acceptance is required to maximize total surplus. The price of transmission is set to the marginal surplus of increasing transmission capability on the path in question. If the path is not constrained the price is zero.

The price of transmission, $P_1^T$, is the marginal surplus of increasing the transmission limit from X to Y. This limit may be complicated, but an additional free kilowatt injected at Y and withdrawn at X is always equivalent to raising that limit by 1 kW.

Allowing fixed-quantity bids complicates the auction problem dramatically, so it may be best not to allow them. If a partially accepted bid cannot be used, it could be sold, or it could be returned to the system operator for resale in the RT market. In this case the purchaser would earn $Q_1^T \times P_0^T$ from the RT market. On average, arbitrage between the day ahead and RT markets should keep $P_0^T$ close to $P_0^T$.

---

## Market 3: A Power Pool

### Bidding Restrictions:

Supply:                $P^S(Q)$, startup cost, ramp-rate limit, etc. One bid per day

Demand:                $P^D(Q)$ or $Q^D$     24 hourly bids

### Bid Acceptance and Price Determination:

Acceptance problem:    Maximize total surplus of accepted bids, $\{Q_1\}$

Constraints:           Transmission limits, ramp-rate limits, etc.

Price determination:   $P_1$ = marginal surplus at each location
                       (Computed with accepted generators committed)

### Settlement Rules:

Supply:     Pay:      $Q_1 \times P_1 + (Q_0 - Q_1) \times P_0$

            Pay:      Make-whole side payments for generators with
                      accepted bids.

Demand:     Charge:   $Q_1 \times P_1 + (Q_0 - Q_1) \times P_0 +$ uplift

**Notes:** $P_1$ is the DA price appropriate to the location of the accepted bid, $Q_1$. $Q_0$ is the quantity actually produced or consumed, and $P_0$ is the RT locational price.

### Comments:

In PJM, startup insurance is provided to generators who are scheduled to start up in the DA market and who do start up and "follow PJM's dispatch." Following dispatch amounts to starting up when directed to and keeping output, $Q$, within 10% of the value that would make $P^S(Q)$ equal the RT price. Startup insurance pays for the difference between as-bid costs and the supplier's revenue from DA and RT operations. As-bid costs include energy costs, startup costs, and no-load costs.

Most generators that start up do not receive insurance payments as they make enough short-run profits. The total cost of this insurance is less than 1% of the cost of wholesale power. "Uplift" includes the cost of startup insurance in this simplified market and several other charges in real markets.

## Market 4: PJM's Day-Ahead Market (2001)*

### Bidding Restrictions:

Supply (w/UC):     $P^S(Q)$, startup cost, ramp-rate limit, etc. One bid per day

Demand:     $P^D(Q)$ or $Q^D$.               24 hourly bids

Transmission:     $Q^T$ @ up to $P^T$, from X to Y     24 hourly bids

Pure energy:     $Q^E$ @ up to $P^E$ (supply or demand)   24 hourly bids

### Bid Acceptance and Price Determination:

Acceptance problem:     Maximize total surplus of accepted bids, $\{Q_1\}$, plus the total surplus of transmission, $\sum (P^T \times Q_1^T)$

Constraints:     Transmission limits, ramp-rate limits, etc.

Price determination:     $P_1$ = marginal surplus at each location
(Computed with accepted generators committed)
$P^T$ from X to Y = $(P_1$ at Y$) - (P_1$ at X$)$

### Settlement Rules:

Energy supply:     Pay:     $Q_1 \times P_1 + (Q_0 - Q_1) \times P_0 +$
Make-whole side payments for generators with accepted bids

Energy demand:     Charge:     $Q_1 \times P_1 + (Q_0 - Q_1) \times P_0 +$ uplift

Transmission demand:     Charge:     $Q_1^T \times P_1^T + (Q_0^T - Q_1^T) \times P_0^T$

**Notes:**   See previous notes.

**Comments:**

"Ramp-rate limit" is meant as a proxy for this and various other constraints on the operation of generators, such as minimum down time. Startup cost serves as a proxy for other costs that are not captured in the supply function $P^S(Q)$, such as no-load cost.

The pure energy bids, also called "virtual" bids because they can be made without owning generation or load, are more restricted in format than the energy bids made by actual load and generation.

* This description is still quite simplified as it leaves out PJM's daily capacity market, various other markets, and near-markets for ancillary services and the accompanying uplift charges.

## 3-3.3    OVERVIEW OF THE DAY-AHEAD DESIGN CONTROVERSY

The central debate in power market design continues to be between the advocates of "centralized nodal pricing" and uncoordinated "bilateral" trading. It is generally agreed that RT operation should be centralized and the forward markets beyond a week should be bilateral and decentralized. At some point, as real time approaches, the market structure needs to switch from uncoordinated to coordinated, and at the last second the market is almost completely replaced by the "command and control" of engineers and automatic protective circuitry.

The controversy focuses most intensely on the DA market with one extreme claiming it should be highly centralized and the other claiming it should be completely uncoordinated. Two compromise approaches received little attention until recently. The market can be semicoordinated, or it can contain all possibilities and allow the participants choose to their form of participation. PJM is evolving toward the latter. England and California have compromised by alternating between the extremes. As the English market rejects its centralized approach in favor of minimum coordination, California is headed away from its uncoordinated approach toward maximum centralization.[6]

### Nodal Pricing with Central Unit Commitment

Nodal pricing refers to computing a different price at every node, or bus, of the network. This system computes hundreds or thousands of different prices when one or more of the system's lines is "congested" (meaning they are fully used, and it would be beneficial if they could carry more power). The point is not the many prices as there would be as many in a competitive bilateral market. The point is these prices are computed by an auctioneer.

The centralized-nodal view encompasses a separate and unrelated tenant. It advocates a centralized solution to the unit-commitment problem. More recently this has been advocated as an option instead of a requirement. Centralized unit commitment, but not nodal pricing, requires complex bids that describe in detail the costs and limits of generators. In a power exchange, even one that uses nodal pricing, the auctioneer has no detailed information about generators and makes no attempt to solve this problem. The four summarized markets allow comparison of the unit commitment part of this view but not its nodal pricing component.

### The Bilateral Approach

The bilateral view generally admits that a centralized DA market is needed but contends that it should not trade energy. Instead it should facilitate bilateral trades of energy by selling transmission rights to the traders. This is generally claimed to be a simplification, and for the auctioneer it is a great simplification relative to solving the unit-commitment problem. But it is no simpler than a power exchange,

---

6. In May 2001 the CA ISO  proposed a multi-day centralized unit-commitment pool with the claim that it intends to adopt nodal pricing as soon as practical. This was rejected by FERC. In January 2002, the CA ISO is again contemplating nodal pricing with centralized unit commitment.

and it makes the job of traders far more complex. They must make energy trades with dozens of parties in different locations with different prices, instead of just trading with the exchange at their local price. But the point of the bilateral approach is not to simplify life for traders; it is to minimize the role of the system operator and to maximize the role of traders.

## The Role of the Four Markets in the Controversy

**Market 1:  A Power Exchange.**   The pure energy market is a semi-coordinated market. It compromises by taking a middle path and is simplest in operation for both the exchange and for traders. It prices energy nodally but ignores the unit-commitment problem. This complicates the supplier's bid strategy which must indirectly account for startup costs and generator limitations, but the market itself remains simple. A slight complication in power exchange bidding can help generators solve the unit commitment problem.

**Market 2: A Transmission-Rights Market.**   The transmission-rights market represents a pure bilateral approach. This approach maximizes the role of traders and forces traders and generators to solve the unit-commitment problem on their own. The job of the system operator is no simpler than in a power exchange because it must still account for all transmission constraints and arrive at what is equivalent to a full set of nodal energy prices except that their collective level is indeterminate. Only the differences between locational energy prices are known to the system operator.

**Market 3: A Power Pool.**   Generators report as many details of their costs and limitations as the auctioneer's computer program can handle. The auctioneer then computes which generators should be started ahead of time. It finds the optimal solution, assuming that the bid data and demand forecasts are correct. Because these programs do not find market clearing prices, and the prices they do find are too low to induce a few of the needed generators to start up; these generators must receive "side payments." Once these are guaranteed, the energy prices clear the market.

**Market 4: PJM's DA Market (2001).**   PJM's market combines all the bid types found in the previous three markets. Together all of these bids determine one set of energy prices, and the differences between them determine the transmission prices. Current practice is for almost all generators in PJM to use the unit-commitment bids for bilateral traders within PJM to use the pure energy bids, and for bilateral traders who want to send power through PJM to use the transmission bids. One argument for this approach is that generators should be allowed to let the system operator solve their unit-commitment problem if that is their preference, but this does not answer the more important question of whether the system operator should offer startup insurance to generators.

It should not be concluded that the use of various types of bids in PJM indicate a clear endorsement of one or another type of DA market by any of the market participants. Besides the basic rules, there are still many hard-to-quantify pressures that influence the choices of market participants. Nonetheless, this experiment may someday provide a useful comparison of the various approaches.

*Chapter* **3-4**

# Ancillary Services

*Electric energy thus is the most useful form of energy — and at the same time it is the most useless. [It] is never used as such . . . [but] is the intermediary.*

Charles Proteus Steinmetz
*Harper's*
January 1922

**P**OWER IS THE PRIMARY SERVICE, BUT SIX ANCILLARY SERVICES ARE NEEDED TO ENSURE RELIABLE, HIGH-QUALITY POWER, EFFICIENTLY PRODUCED. Usually ancillary services are defined by how they are provided rather than by the service rendered. This results in a plethora of services and little insight into their relationship to market design. Defining services by the benefit they provide and defining them broadly produces a short but comprehensive list. Services directed at long-term investment are not counted as ancillary to real-time (RT) power delivery. The six listed services require planning by the system operator, but economic dispatch is jointly provided by the system operator and the market. How that task is shared should be the subject of intense debate.

**Chapter Summary 3-4:** Of the six ancillary services, the system operator or its agent must *directly* provide transmission security and trade enforcement, and to some extent economic dispatch. The other services, balancing, voltage stability, and black-start capability, can be purchased from a competitive market, but the system operator must demand and pay for these services.

**Section 1: The List of Ancillary Services.** Services are defined by the benefit they provide to the market and its participants, not by their method of provision. An accurate frequency is required by some motors and particularly by large generators. Many appliances need a fairly accurate voltage, and together these two services define the provision of power from the customer's perspective. Transmission security and occasionally black-start capability are indirect services needed by the market to provide the first two. Economic dispatch can include the solution of the unit commitment problem and often includes efficiently dispatching around congestion constraints. Trade enforcement is required to provide property rights essential for bilateral trading.

**Section 2:  Balancing and Frequency Stability.**  Frequency is determined by the supply–demand balance. Frequency stability cannot be provided by individual system operators, but each is required to balance the real power flows in its control area taking into account a frequency correction. When every system provides this balancing service, they collectively provide frequency stability.

**Section 3:  Voltage Stability.**  Voltage support is provided passively by capacitor banks and actively by generators. Provision is by the supply of "reactive" power which is difficult to transmit. This makes it difficult to purchase under competitive conditions, but long-term contracts that permit competition through entry should be helpful.

**Section 4:  Transmission Security.**  Transmission security can be provided initially through the control of transmission rights or the operation of a day-ahead (DA) energy market with locational prices. In real time, a locational balancing market should be used.

**Section 5:  Economic Dispatch.**  The system operator must assist in the provision of this service by conducting a balancing market and may assist by conducting a DA power exchange or by providing almost the entire service through a power pool.

**Section 6:  Trade Enforcement.**  Once power is injected into the grid it cannot be tracked, so ownership is lost. This makes bilateral trading impossible unless a substitute property right is defined and enforced by the system operator, or in the case of inter-control-area trades, by a higher authority. This property-right system requires the metering of all traders, the recording of all trades, and some power of enforcement to deter or compensate for discrepancies between registered trades and actual power flows.

## 3-4.1  THE LIST OF ANCILLARY SERVICES

There are many lists of ancillary services which differ mainly in how they split and combine the categories of service.[1] The present list combines services whenever they differ only by speed of delivery or method of provision. The purpose is to make the shortest possible comprehensive list that does not combine grossly dissimilar services.

The list is organized very roughly by time scale, starting with real-power balancing, which is of constant concern and sometimes must be provided on the

---

1. As reported in Order 888, when FERC (1996a) asked for help in defining ancillary services it received over a dozen different lists. One list had 38 ancillary services just for transmission. NERC presented a list of 12, though it adopted FERC's final list of six in its Glossary of Terms (NERC 1996). These six are covered by, but do not directly correspond to, the six listed in this chapter.

shortest possible time-scale, and ending with the black-start service which is seldom used.

Ancillary services are those traditionally provided by the system operator in support of the basic service of generating real power and injecting it into the grid. Much more is needed to ensure that the supply of delivered power is reliable and of high quality. Some of these services are indirect, but all ancillary services are concerned with the dispatch, trade, and delivery of power. Looking after investment in generation (discussed in Part 2) and transmission, while they may involve the system operator, are not considered ancillary services. *Because the role of the system operator is at the heart of most architectural controversies, understanding the ancillary services is crucial to evaluating any proposed market designs.*

## Defining the Ancillary Services

This section defines the ancillary services not by how they are provided but by the benefits they provide. The following sections describe the provision of these services.

**Frequency and Voltage.** The two fundamental characteristics of power delivered to a customer are frequency and voltage. As long as these remain correct the customer will have access to the needed power, and it will have the required characteristics.[2] Whenever a customer encounters trouble with the supply of power, either voltage or frequency will have deviated from its allowed range. In the extreme, a power outage is defined by zero voltage.

**Frequency**, measured in Hertz (Hz), is the number of times per second that the voltage goes from its maximum (positive) value to its minimum (negative) value and back to its maximum. This is called a "cycle" and frequency was formerly measured in cycles per second (cps). (When power is used, the frequency of the current flow is the same as that of voltage, and its alternating direction of flow explains the term, AC, alternating current.) Frequency affects the operation of motors, many of which draw more power and run faster at a higher frequency. Synchronous motors, the type used in AC clocks and phonographs, run at a speed directly proportional to frequency. But the greatest service of a stable frequency is provided to large generators—basically synchronous motors operated with a reverse purpose—which suffer less stress when run at a constant speed.

A 10% increase in voltage will cut the life of an incandescent light bulb approximately in half. **Voltage** is electrical pressure, and the more pressure the more electrical current is forced through appliances. Some are built to be insensitive to voltage so they can be used both on 110 volt AC in North America and 220 volt AC in Europe. Low voltage can cause appliances to underperform and, oddly, can burn out some motors.

**Transmission Security.** Compared with other networks, the transmission system is quite fragile.[3] Overuse can cause lines to overheat which can cause them to sag

---

2. The issue of power quality, which includes wave shape and occasional voltage spikes, will be ignored.

3. Too much traffic at one time does not damage a highway, and too many phone calls will not hurt the phone wires.

permanently or even to melt. It can also cause complex electrical problems that interfere with power flow. Because of such problems, most high-voltage power lines have automatic protective circuitry that can take them out of service almost instantly for their own protection. But protecting one line can endanger another and can cut off service to customers. The ancillary service of transmission security keeps the grid operating.

In other lists of ancillary services, this service is buried within the scheduling and dispatch services. Those are the methods of providing transmission security, but they also provide other services.

Voltage support is needed both as a direct customer service and to provide transmission security. This complicates its classification, but within the present classification of ancillary services, reactive power supply is viewed as helping to provide two distinct services: voltage stability for customers, and transmission security. (In the case of voltage support, the method of provision is as useful a method of classification as the benefit of the service, but consistency dictates classification by benefit.)

**Economic Dispatch.**  Economic dispatch refers to using the right generators in the right amounts at the right times in order to minimize the total cost of production.

Like transmission security, this service is provided by what are elsewhere referred to as scheduling and **dispatch** services. From an economic viewpoint, economic dispatch and transmission security are only minimally related. Economic dispatch often has nothing to do with power lines as no line limits are binding, while transmission security has nothing to do with the marginal cost of generators. Rolling both into scheduling and dispatch makes some sense in a regulated world but is of little help when designing a market.

**Trading Enforcement.**  Trading enforcement is an expanded version of part (b) of NERC's sixth ancillary service (NERC, 1996) which is its interpretation of the FERC's first ancillary service required by Order 888 (FERC, 1996a):

> *Scheduling, System Control, and Dispatch Service — Provides for a) scheduling, b) **confirming and implementing an interchange schedule with other Control Areas, including intermediary Control Areas providing transmission service,** and c) ensuring operational security during the interchange transaction.* [Emphasis added]

An expansion of this definition is required because trade is no longer just between control areas but also between private parties within a control area.

Although this service also requires scheduling, its purpose is different from either economic dispatch or the provision of transmission security. This is an accounting and enforcement service. Power trades are inherently insecure as origin and ownership cannot be identified once power has entered the grid. Trading requires property rights which become problematic when ownership cannot be established. As a substitute for ownership identification, all injections and withdrawals from the grid are metered and recorded by system operators. They also record all trades. By matching trades with physical inputs and outputs, the honesty of the traders can be established, and deviations from recorded trading intentions can be dealt with appropriately. Trading enforcement comprises three steps: (1) recording of trades, (2) metering of power flows between all traders and the common grid, and (3) settling accounts with traders who deviate.

Note that this is not physical enforcement of RT bilateral trading whose absence constitutes the second demand-side flaw (see Section 1-1.5). This enforcement occurs much later, but if it were implemented with extremely high penalties (not recommended) it would have almost the same effect as RT enforcement.

**The Black-Start Capability.**   In the worst system failure, either a large part of or an entire interconnection could be shut down. Unfortunately most generators must be "plugged in" to get started; they require power from the grid in order to start producing power. To start the grid, generators possessing black-start capability are needed. Starting a power grid requires the final ancillary service, the black-start service.

---

**Ancillary Services**

1. Real-power balancing (frequency stability)
2. Voltage stability (for customers)
3. Transmission security
4. Economic dispatch
5. Financial trade enforcement
6. Black start

---

**Overview of the List of Services.**   This completes the list of basic tasks that must be carried out in whole or in part by the system operator (or its agents) to support the basic service of RT power production. Frequency and voltage stability are the only two services provided directly to customers. Transmission security (and on rare occasions the black-start service) are provided to the wholesale market to aid in frequency and voltage stability. They are named separately because their role is indirect. Economic dispatch is a service provided to generators, not to end users. Most services must be paid for—this one simply reduces the price consumers pay for power. Trade enforcement is a service to bilateral traders. Without it bilateral trade would be impossible.

## 3-4.2   REAL-POWER BALANCING AND FREQUENCY STABILITY

Frequency stability and power balancing are two sides of the same service. The end-use service is a stable, accurate frequency, but no system operator can provide that because it is mainly controlled by the other systems in the interconnection. Instead all system operators are required to balance their power inflows and outflows according to NERC's ACE formula. When all system operators provide this balancing service to the interconnection, the result is a stable and accurate frequency. The actual service provided by the system operator is balancing, but the service provided to customers as a result of collective balancing efforts is frequency stability. "Real" power is ordinary electrical power, but it is given its full name here to distinguish it from "reactive" power which is used to provide the voltage stability service.

Balancing is provided by a number of different methods which are normally classified as separate ancillary services. Four of the six ancillary services listed by NERC and FERC fall into this category:

1. Regulation and Frequency Response Service
2. Energy Imbalance Service
3. Operating Reserve: Spinning Reserve Service
4. Operating Reserve: Supplemental Reserve Service

Some system operators further classify supplemental reserves into 10-minute, 30-minute, and replacement reserves. Also direct-control load management and interruptible demand, as defined by NERC, would be classified as spinning reserve by some jurisdictions but might legitimately be separated by others.

Frequency instability is caused by a mismatch of supply and demand, as is explained in Section 1-4.3. When total supply in an interconnection is greater than demand the frequency is too high, and when it is less than demand frequency is too low.[4] The comparison must be made for the entire interconnection because frequency is uniform throughout this region as all generators are synchronized. Within an interconnection, one market cannot have a high frequency and another a low frequency.

Supply and demand fall out of balance for three basic reasons. First, load varies in a predictable pattern throughout the day. This requires active "load following" on the part of the system operator and the generators it controls. Second, there are unpredictable, constant small fluctuations in most loads and some generators, such as wind-power generators. Third, there are generator and line outages.

Corresponding to these are three basic approaches to keeping the system balanced. Small random fluctuations are handled by a service called "regulation," which utilizes generators that receive a control signal directly from the system operator. The process is called automatic generation control, AGC. Predictable daily fluctuations, like "the morning ramp," are handled by scheduling generation and, in an unregulated system, by the balancing market. Unexpected generation and transmission outages are handled by operating reserves of various types starting with 10-minute spinning reserves.

**The Market's Role in Provision.** Regulation and operating reserves are generally purchased by the system operator in a market that it organizes and in which it is the sole source of demand. Sometimes there is an attempt to have generators "self-provide" operating reserves as if they might have their own need for such reserves. This makes little sense except as a means of preventing the system operator from scrutinizing the quality of the reserves provided or of capturing the service of administering the market for operating reserves.

Tracking the more predictable and larger-scale fluctuations in demand can best be handled by a DA market, an hour-ahead market, and an RT market. The architecture of this group of markets is the subject of much of Part 3.

---

4. The supply-demand balance for a given control area within the interconnection is defined by ACE, which measures a combination of frequency and **inadvertent interchange** (see Section 1-4.3). This ancillary service is more properly defined as controlling ACE, but this translates into balancing supply and demand locally and results in controlling frequency globally.

## 3-4.3  VOLTAGE STABILITY FOR CUSTOMERS

Voltage tends to drop as power flows from generator to load, and the more power that flows the more the voltage tends to drop. Voltage drop also depends on the type of load. Those involving electromagnetism, like motors and some fluorescent-light ballasts, tend to use a lot of "reactive" power which causes the voltage to drop. If this voltage drop is not corrected, this affects all other loads in the vicinity. Usually the voltage drop on the transmission line is corrected by a transformer that automatically adjusts the voltage before the power is distributed, but this process is not perfect and customer voltage fluctuates as a consequence.

Voltage drop can be counteracted by the injection of reactive power (explained in Part 5). Reactive power is quite different from real power except that its description shares very similar mathematics. For instance, it can be supplied by capacitors which are entirely passive and consume no fuel. It can also be supplied by generators, generally at very low cost. The cost to a generator of supplying reactive power is mainly an opportunity cost because, while supplying reactive power consumes essentially no fuel, it does reduce the generator's ability to produce real power. This relationship is nonlinear. Synchronous condensers are another source of reactive power. These are essential generators that are run electrically. They take real power from the grid and return reactive power.

The stabilization of voltage for consumers interacts in a complex way with voltage support for transmission security. Consequently, the customer service and the voltage-support part of the transmission service should be considered together. However, this book will not examine either service in detail.

**The Market's Role in Provision.**   Some have proposed setting up a market in reactive power in parallel with the market in real power. The price of reactive power would be adjusted to clear the reactive power market.

Because of the externalities associated with the consumption of reactive power, a bilateral market is out of the question. Users of reactive power would purchase far too little. Unfortunately, reactive power is difficult to transmit. In effect it suffers losses that are very roughly ten times greater than the losses of real power. As a result, an RT reactive power exchange would have severe difficulties with market power. The system operator should, however, be able to purchase long-term contracts for the provision of reactive power provided time is allowed for competitive entry. In other words, market power may be exercised over contracts sold before competitors could enter. In the mean time, the system operator may need to exercise monopsony power or regulate reactive power in order to secure it at a reasonable price in the near term.

## 3-4.4  TRANSMISSION SECURITY

Transmission line limits are a complex matter. First there is a thermal limit. If the total real and reactive power on a line exceeds a certain threshold for too many minutes the line will be permanently damaged. The most likely cause of a line exceeding its thermal limit is another line going out of service unexpectedly.[5] Then the power flow on the system re-routes itself instantly, with the most flow being transferred to the nearest lines. In the (almost) worst case, if there are two identical parallel lines between two regions, and these are the only lines between the regions, if one goes out of service the other receives all of its power flow. If each line could handle 100 MW, then to protect each of them, their combined **security limit** would be 100 MW. The only benefit of having two lines instead of one would be the increased reliability.

When security limits are computed accurately, they change frequently. For example, suppose line L1 and L2 both have a 100 MW thermal limit and L1 would dump 50% of its power onto line L2 if L1 were to go out of service. When L1 has 100 MW flowing on it, the security limit on line L2 is its 100-MW thermal limit minus the 50 MW of potential spillover from an outage on L1. The security limit of L1 is then 50 MW. If, however, L1 is carrying only 20 MW, then L2 has a security limit of 90 MW.

Besides thermal limits there are other basic physical limits that may serve as the basis for security limits. These are called voltage limits and stability limits. Both are less stable than thermal limits which vary only a small amount with ambient temperature.

To provide transmission security, the system operator must first compute these limits (and recompute some of them frequently) and then ensure that the dispatch of generation, given the existing load, is a **security constrained dispatch**. This is a set of generation output levels which, given the load, does not cause more power to flow on any line than its security limit allows.

**The Market's Role in Provision.**   Transmission security is a service that must be provided by the system operator. It can be provided by selling transmission rights or by controlling the acceptance of energy bids in a DA energy auction. Neither method is foolproof as both operate at some interval before real time. In the mean time, generators may go out of service, lines may go out, loads may increase or decrease. Any of these can change the flow of power in such a way that a security limit is violated. Fortunately, in most systems, no security limit is threatened in most hours. In PJM, for example, no security limit is binding about $\frac{3}{4}$ of the time.

When a security limit is breached in real time, the system operator needs to induce more generator in one location and less in another. The problem is not total production; it is only a matter of where production occurs. One method of dealing with the problem is to look for a bilateral trade that is causing power flow on the path with the security violation and disallow that trade. The load will then have to find another source of power. This is the current NERC approach to line-loading relief. It is a centralized, nonmarket approach and, as would be expected, is quite inefficient.

NERC is moving toward an RT balancing market. Essentially this amounts to lowering the RT price of power in the region where less generation is wanted and raising the RT price of power in the region where more is wanted. This is a service

---

5. It is the square root of the sum of the squares of real and reactive power that is limited by the thermal limit, but in most cases this is quite similar to a simple limit on real power.

that should be provided by using a market, but the market is only the tool used by the system operator. The demand for security originates with the system operator, and the system operator ensures its provision. The core of the service is a highly complex coordination problem whose solution must be centrally provided.

## 3-4.5   ECONOMIC DISPATCH

Economic dispatch, as defined above, is a matter of using the cheapest generation. There are two parts to this problem: (1) deciding which generators to start up, the unit-commitment problem; and (2) deciding how much to use each generator that is running. However, most generators that are started are used at full output except when they are in transition or when one is avoiding the cost of a night-time shut-down and restart by staying on at a low output level over night.

If the RT dispatch is economic, the DA market doesn't matter. Because RT supply and demand conditions are not known precisely until real time, RT economic dispatch is quite important. Because many generators are slow and expensive to start, the DA market plays an important role in making an efficient RT dispatch possible.

**The Market's Role in Provision.**   There are three basic approaches to economic dispatch: (1) bilateral trading, (2) a centralized DA power exchange, and (3) a centralized DA power pool. In the first approach, suppliers and power traders solve the entire problem. In the second, the system operator helps by running a DA exchange. The exchange provides a public price which is either used by the exchange to select the generators that should start up or could be used by suppliers to self-select during several rounds of bidding. In the third approach, the power pool would require generators to submit complex bids and the system operator would optimize the dispatch using all available bid information. That optimization would be used to control the actual dispatch through prices and start-up insurance. In this approach the system operator provides a full unit-commitment service. In all approaches, the suppliers and the market price jointly determine a dispatch, which if the market is competitive, will be very nearly least cost.

If the system operator relies on a competitive balancing market, the market will provide an economic dispatch. If the system operator chooses such tools as pro rata curtailment or bilateral curtailment for line-loading relief, they will interfere with economic dispatch. There is a debate over how much efficiency can be improved by RT centralized unit commitment.

## 3-4.6   TRADE ENFORCEMENT

When **control area** A sells power across control areas B and C to control area D, how does anyone know whether A really provided the power it sold? Area D cannot measure A's output and does not really care whether A produced the power or not because it can take power out of the grid in any case. The more it uses, the more flows into its area. If the power was not produced, then frequency will decline and all control areas in the interconnect will be responsible for increasing supply to correct the problem.

The origin of this problem is the fact that power cannot be tracked. Once a generator injects power it cannot keep track of it, so it loses ownership of that particular "piece" of power. The solution to the problem is to give the supplier of the power a right to withdraw the same amount of power and let the supplier reassign that right at will, then enforce the rights to withdraw power. For example, when area A sells power to area D, it does so by injecting power and thereby obtaining the right to withdraw power which it then assigns to D. This solves the problem of tracking where the power actually flows, but it requires that injections and withdrawals be metered and compared with the rights to withdraw power that are granted and exercised.

This is an abstract description of the principles on which the actual accounting system is based. In reality flows are measured between adjoining control areas, and each control area has a net scheduled **interchange** that is publically known. *Interchange* is flow on the lines connecting control areas (interties), and every interconnecting line is metered by both control areas. Each gets credit for outflow so they have opposite interests in terms of reporting the flow on the line. One would like to err in one direction, and one would prefer to err in the other direction. This prevents collusion and keeps both honest.

When A sells power to D, this is publically recorded as an increase in A's scheduled net interchange and a decrease in D's scheduled net interchange. The schedules of B and C do not change because any power that flows into them as a result of this trade also flows out. With this system of recording transactions, plus the metering of the interties, bilateral trades between control areas can be enforced. This system creates the appropriate set of property rights.

An analogous system is needed for private bilateral trades within or between control areas. For simplicity consider internal trades. If no one were watching, A could sell power to B within the control area and then fail to produce. B would take the power, and no one would be the wiser. The power taken would be power produced by some other generator somewhere in the interconnection, but there is no physical way to check on who produced the power. Again a system of trade enforcement is needed. No matter how the details of such a system are arranged it requires at least the following three conditions:

1. All trades must be registered with the trade enforcement authority.
2. All traders must have their connection to the grid metered by the authority.
3. The trading authority must have the power to charge traders for discrepancies.

---

**Result   3-4.1**   **Strictly Bilateral Power Trading Requires Centralized Coordination**
Without the central registration and measurement of all bilateral trades and penalties for discrepancies, bilateral power trading would be impossible.

---

**Power Losses.**   One service which is frequently labeled an ancillary service, but has not been mentioned so far, is the replacement of power losses. When the system operator balances the system, it necessarily computes a power flow from which losses can be approximated. Any discrepancy between actual losses and the computed value that is included as part of demand will be reflected in the control-area's ACE which the system operator is obliged to correct. In fact if losses are not computed, ACE measurements could be relied on as a method of determining losses.

Thus losses are just another demand on the system and are handled through normal balancing. The only additional service required of the system operator is to assign the cost of losses to users. The simplest approach is to charge loads for them in the uplift. An economic approach, discussed in Chapters 5-7 and 5-8 is to compute locational loss prices and charge generators. This calculation and its inclusion in billing are the real services associated with losses, but these are simply part of normal economic dispatch.

**The Market's Role in Provision of Trade Enforcement.**   Trade enforcement for bilateral trading must be provided by the system operator. Individual traders have no incentive to provide this service for their own trades and no power to provide it for other trades. No other private party has the power to enforce trades.

# The Day-Ahead Market in Theory

*Economic questions involve thousands of complicated factors which contribute to a certain result. It takes a lot of brain power and a lot of scientific data to solve these questions.*

Thomas Edison
1914

T HE DAY-AHEAD MARKET IS THE FORWARD MARKET WITH THE GREATEST PHYSICAL IMPLICATIONS. By providing financial certainty, it can remove the risk of incurring startup expenses. The more efficient the market, the more accurate the startup decisions and the lower the cost of power. Even without the unit-commitment problem, reducing financial risk would reduce the cost of capital.

As explained in Chapter 3-3, the day-ahead (DA) market can utilize one of three basic architectures or a combination. Bilateral markets, exchanges and pools can each provide hedging and unit commitment. The controversy over the choice of architecture is driven by concerns over the shortcomings of private markets and nonprofit system operators in performing the coordination functions associated with unit commitment. Hedging is also an issue as pools claim to provide it more completely than exchanges.

Some theory of market clearing—when it is possible and when not—helps to provide a framework for evaluating the various designs. "Nonconvex" production costs are the key to this theory, and while conceptually arcane, the focus of current controversy and volumes of market rules attest to their impact on market design.

**Chapter Summary 3-5:** Nonconvex generation costs violate an assumption of perfect competition, but the magnitude of the resulting problems is unknown. The pool approach is designed to minimize these problems, but its pricing ignores investment incentives. The bilateral approach faces formidable coordination problems in the DA market. It may be less efficient and provide less reliability than a centralized approach. Side-payments made by a DA pool do not increase reliability.

**Section 1: Equilibrium Without a Clearing Price.** Startup and no-load costs are "nonconvex" and violate the assumptions of perfect competition. In particular, they often prevent the existence of a market-clearing price. Nonetheless, bilateral markets, exchanges, and pools all have equilibria and equilibrium prices. Bilateral and exchange prices approximate competitive prices as best they can. Consequently, they are approximately right for short-run supply and investment. Pool prices are optimized for short-run supply but ignore investment incentives.

**Section 2: Difficulties with Bilateral Day-Ahead Markets.** In a bilateral market, generators must self commit and, in doing so, they do not consider reliability. When DA load can be served at almost the same price by quick-start or slow-start generators, the market may select too many quick-start generators just by chance. Then the slow-start generators will be unavailable in real time. The uncertainties of the bilateral process cause decreased reliability. Bilateral markets also have more difficulty developing locational prices and can be expected to be less efficient in markets with significant congestion. The magnitudes of these effects are unknown.

**Section 3: Settlement, Hedging, and Reliability.** Selling power in the DA market provides a hedge against the risks of the volatile real-time (RT) price. For the hedge to work, the supplier must commit its generator and be prepared to produce according to its DA contract. Thus DA contracts provide some inducement to commit units as a way of reducing risk. If the DA contract is with a pool and therefore includes a "make-whole" side payment, this leads to no increase in the incentive to commit because that payment is not contingent on RT performance.

**Section 4: Other Design Considerations.** These include transaction costs, facilitation of market monitoring, provision of publicly known prices, and nondiscriminatory access. Penalties for avoiding the DA market should not be used in an attempt to obtain more accurate unit-commitment data.

## 3-5.1   EQUILIBRIUM WITHOUT A CLEARING PRICE

The theory of perfect competition, explained in Parts 1 and 2, assumes that there is a price that "clears the market" when suppliers take that price as given. In this case a competitive market is efficient, but if production costs are **nonconvex**, there may be no market clearing price, and when there is none, economic theory does not guarantee efficiency. Startup costs and no-load costs are "nonconvex," and some constraints on generation output cause similar problems.

Although these nonconvex costs cause deviations from perfect competition, the deviations may be inconsequential. The debate over the power pool approach concerns the magnitude of the resulting inefficiency. This inefficiency can take three forms. Nonconvex costs can cause (1) an inefficient but reliable dispatch, (2) decreased reliability, or (3) financial risks for generators. The justification claimed for the complexity of the pool approach is that it minimizes these three inefficiencies.

## Why There May Be No Clearing Price

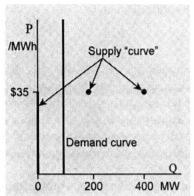

A market clearing price is a single price that causes supply to equal demand. Suppose power is needed for just two hours and exactly 100 MW is needed. Suppose there are a number of competing generators, so they offer to supply at a competitive price. Their marginal cost is $20/MWh up to their capacity of 200 MW, and their startup costs are $30/MW started. Ignoring demand limitations, at a price of $35/MWh a generator can earn a scarcity rent of $15/MWh for two hours on each MW of its capacity for a total scarcity rent of $30/MW.[1] This would exactly cover its start up cost. At a lower price it would not voluntarily agree to sell any power. At a price of $35/MWh or more it would offer to supply its full 200 MW.

Below $35/MWh supply is zero, while at or above $35 supply is 200 MW or greater. There is no price at which supply equals 100 MW. Technically, there is no price at which a supplier could profitably sell 100 MW but could not increase its profit by selling more. Thus there is no price at which supply equals the 100 MW level of demand. There is no market clearing price. (See Section 3-9.2 for another example.) The root of the problem is that production costs are not convex. Convex production costs have the property that twice as much output always costs at least twice as much to produce. In this case 100 MW for two hours would cost $50/MWh, while 200 MW for two hours would cost $35/MWh. Twice as much is cheaper per unit, so the production cost function is not convex.

## A Bilateral Equilibrium

Market clearing is often defined narrowly in the context of perfect competition, but sometimes it is defined more broadly to mean that no profitable trades remain to be made. To avoid confusion this situation will be referred to as an equilibrium but not as market clearing. This definition will be used with bilateral markets. In the above example, a bilateral market could reach an equilibrium by arranging a trade in which 100 MW of power was sold for two hours at a price of $35/MWh. Having made that trade, demand would be satisfied and no further trade could profitably take place. In this context "profitably" takes account of the consumer's value as well as the supplier's profit.

Because of bargaining problems, it is not obvious that a bilateral market would reach this equilibrium. In a more complex and realistic market, the frictions of bilateral trade inherent in the costly process of gathering information and arranging

---

1. Scarcity rent (see Section 1-6.6) is revenue minus costs that vary with output. After startup and no-load costs are subtracted, short-run profits remain.

trades could also prevent the market from reaching equilibrium, especially given the limited trading time of a DA market.

Even if equilibrium is reached there is no guarantee that it will be efficient (least cost) in a market with nonconvex costs. But even if inefficient, in a large market, this inefficiency may prove to be too small to matter. The problems caused by nonconvex costs in a real power market with a peak load of 10 GW and a 1 GW interconnection to a larger outside market may be entirely negligible. This is a matter for empirical research or detailed theoretical calculations.

## An Exchange Equilibrium

A bilateral market can reach an equilibrium by trading at more than one price, but an exchange is limited to a single price. With nonconvex costs, there may be no price that clears an exchange market. In spite of this the exchange will always have an equilibrium. Exchanges are auctions and, as described in Chapter 3, have a definite set of rules that determine what bids will be accepted and how they will be settled. The Nash theorem states that such games always have at least one Nash equilibrium.[2]

An equilibrium of an exchange market will be defined to be a **Nash equilibrium**, a situation in which no player could do better by bidding differently, provided other players maintain their equilibrium bids. Using this definition, the Nash theorem guarantees the market will have at least one equilibrium. Examples of this are given in Chapter 9.

Although the exchange's equilibrium price cannot clear the market, it will tend to come as close as possible in the sense of minimizing the gap between competitive supply and competitive demand. For example, the market may clear except for a single generator that sells less than its full output while wishing it could sell it all at the market price.

## A Pool Equilibrium

Like an exchange, the equilibrium of a power pool is defined as any Nash equilibrium of the pool auction. Like an exchange and a bilateral market, a pool determines an equilibrium set of quantities traded. Unlike an exchange, a pool does not determine a single market price. Instead, it determines a nominal market price, the pool price, and a set of supplier-specific side payments which, in effect, create a different price for each supplier. In practice many suppliers will receive no side payment and so this group will be paid the same price, the nominal market price. But, by design, the pool allows every supplier to be paid a different price.

The average of the individual prices in a pool approximates the price in an exchange, but the nominal pool price is lower and could be much lower. In practice it appears to be lower by (very roughly) 3%. As an example consider a market with generators having marginal costs of \$20.00/MWh, \$20.10/MWh, \$20.20/MWh

---

2. Auction rules specify what bids are acceptable including the fact that prices must be rounded to the penny and quantities to the MW. These rules and limitations imply the auction is a finite game (players have a finite though large number of strategies available). In solving such games, it appears that the finiteness of the strategy set is not required to produce a Nash equilibrium.

and so on. Assume they have no startup costs but have a no-load cost that averages $10/MWh when they produce at full capacity. Assume the generators are small relative to the market size so that the integer (lumpiness) problem is of little concern. A pool price is defined by system marginal cost which is $25/MWh if the $51^{st}$ generator is needed. This generator would receive a side payment of $10/MWh to cover its no-load cost. An exchange would need to set a single market price of $35/MWh to induce this supplier to produce. Thus the nominal pool price can be dramatically lower than the price that comes closest to clearing the market and dramatically lower than the market price of a power exchange. The difference is mainly, but not entirely, made up by the side payments.

### The Efficiency Of Equilibrium Prices

Bilateral prices and exchange prices approximate competitive prices in the sense that they come about as close to clearing the market as possible. Pool prices are based on a different philosophy. They are designed to solve one problem optimally while ignoring two others. Competitive prices, when they exist, play three useful roles: (1) they induce least cost supply given demand, (2) they induce efficient consumption, and (3) they induce efficient investment in generation capacity. Bilateral markets and exchanges provide prices that approximately achieve all three efficiencies while pool prices succeed perfectly at (1) while ignoring (2) and (3).[3] The demand-side efficiency of pool prices is discussed in Chapter 3-8, while long-run efficiency is discussed in Chapters 3-8 and 3-9.

Notice that, in the above example, the generator with a $20/MWh marginal cost would lose money if paid only the nominal pool price of $25/MWh, so it would be given a side payment of $5/MWh. Its total cost at full output, $30/MWh, would then just be covered. Had this been an exchange or bilateral market, the "almost-clearing" price would have been $35/MWh. With regard to short-run supply, this makes no difference since, given demand, the pool always induces an efficient dispatch. But with $5/MWh less revenue to cover fixed costs, long-run incentives will be quite different. A pool does not concern itself with these incentives when setting price and its philosophy of price setting does not produce efficient long-run incentives as a side effect.

Again, the real-world implication may be minimal and empirical research or detailed theoretical analysis is needed. The more difficult question is whether the modest increase in efficiency attributed to optimizing short-run supply exceeds the modest loss of efficiency from ignoring the long run.

## 3-5.2   DIFFICULTIES WITH BILATERAL DAY-AHEAD MARKETS

Markets should be encouraged as long as they do not inhibit the provision of ancillary services by the system operator. These services are needed in real time,

---

3. Pool prices claim to send efficient signals to consumers but ignore some implications of nonconvex production costs. When the optimal dispatch occurs at a production level for which marginal cost is not defined, pool prices send the wrong signal to consumers. This circumstance occurs with a significant probability. Chapter 3-8 gives an example which proves this point.

not a day ahead, so the DA market can be problematic only to the extent it interferes with RT operations. Trades made in the DA market can be rearranged in real time as needed—with one exception. If a generator has not been started and is slow to start, it may be unavailable when needed in real time. Thus, proper unit commitment is sometimes required for *balancing* and calls for special attention when evaluating DA market designs. The ancillary service of *transmission security* also impinges on the design of the DA market by adding complexity to its operation.

## Balancing and Unit Commitment

Real power balancing mainly refers to small, but often quick, adjustments of supply to keep it equal to demand. If demand cannot be balanced because of a shortage of supply, load must be shed. This is the most extreme result of imbalance and the most important problem that could be caused by the DA market. The possibility of this failure is greatest in a bilateral market.

**Example of a Unit-Commitment Failure.**   Consider a bilateral market with 20 identical slow-start, midload generators, each with a different owner, only 10 of which are needed on a particular day. With 20 identical generators and only 10 needed, competition will be stiff and expected profits will be zero, but depending on the number that commit, actual profits will be positive or negative. Before trading begins, each generator knows that it has a 50/50 chance of selling its power in the DA market, but it is indifferent as to whether it does or not because in either case its expected profit is zero.

On the first round of trading, because there is no coordination in a bilateral market, an average of 10 generators will sell power, but because of randomness, 5.8% of the time six or fewer will be committed.[4] When this happens there is a good chance that an unexpected RT event will cause a shortage and prices will be very high. There is even a chance of a backout because there may not be enough quick-start generators to make up for the lack of committed slow-start generators.

Normally, a bilateral market would fix this type of problem. Some customers would realize they might be the ones demanding extra power in a tight market, or some generators would realize they might have a forced outage and need to buy extremely expensive replacement power. But the re-contracting process also takes place without coordination. Just as the outcome of the initial round is partially random, so is the outcome of re-contracting.

With enough time and with low-enough re-contracting costs, a bilateral market will reach the optimal solution. But with only a few hours of trading it may not even learn what collective mistakes were made in the first round. The problem is not that bilateral markets have an inefficient equilibrium; the problem is that without coordination and with a very limited trading time, they have a great deal of trouble *finding* the equilibrium. This is especially true when the optimal set of trades depends on the distribution of generation technologies selected and not just on the character of individual trades. As a consequence, bilateral trading in a constricted time frame produces results that are more random with regard to reliability than

---

4. This assumes that there are enough quick-start generators to substitute for most of the slow-start generators that have nearly the same average costs including startup costs.

does a centralized market. The optimal solution to the reliability problem is not a randomized level of commitment, so bilateral markets provide less reliability.

The magnitude of the problem deserves investigation, but for the present, randomness of commitment in a bilateral market must be taken seriously. Power markets that rely on bilateral DA markets usually do take it seriously and provide some non-market arrangement to help ensure sufficient commitment. One mitigating factor is that days with serious reliability problems are usually extremely profitable, and any hint of such a problem is likely to produce ample commitment.

---

**Result    3-5.1        A Bilateral DA Market Decreases Reliability**

Because of information problems, a bilateral market, while getting essentially the right answer on average, will produce a more random outcome than a centralized market with respect to the level generating units committed in advance. This randomness reduces the efficiency of the market and its reliability. The magnitude of this effect has not been determined.

---

## Transmission Security

The DA market cannot easily interfere with transmission security, but transmission limits complicate the DA market. If locational pricing is needed because of transmission congestion, it will complicate all three types of market. The power pool and power exchange will be complicated by having to compute them directly. The bilateral market will be complicated by having locational prices induced by the cost of purchasing transmission rights. The latter is the more complex process.

To ensure transmission security either physical rights must be centrally allocated and sold, or security must be provided through RT pricing of transmission. The latter produces volatile congestion prices that the bilateral trader will wish to hedge with financial transmission rights. Financial rights are simpler in practice, but physical rights will more easily illustrate the trading difficulties of a bilateral market. The nature of the difficulty is the same under the two systems.

Consider a remote generator that wishes to sell its power. With physical rights, the generator must buy a transmission right in order to implement a trade. Thus, two steps are required for a complete trade. If the power is traded first, then the generator must guess the cost of the transmission right before trading the power in order to know how much to ask for the power. If the transmission right is purchased first, the generator must guess the price at which power will be sold in order to know how much to offer for the transmission right. In either case the generator is forced to guess market conditions in order to set an appropriate offer price in one market or the other.

In a centralized energy market, the system operator, in effect, sells transmission rights along with any trades it arranges. There is no risk to traders because they only end up purchasing these implicit transmission rights when they are worth purchasing. In such a market, the competitive generator need not consider market conditions but only needs to bid its marginal cost. If the market is tight, it will receive the appropriately high price and earn a profit; if demand is slack, the price will be low and it will neither sell its power nor accidentally buy a transmission right it does not need. In a centralized market the generator needs only to understand its own cost and has no need to forecast the weather.

Selling transmission rights is a more complex indirect method of determining the same locational prices that are determined directly by a power exchange or power pool. It increases the transaction costs of the market and the randomness of its outcome. The result is not disastrous, but it is not beneficial.

### Economic Dispatch

The last relevant ancillary service is economic dispatch. Although the bilateral market may introduce more randomness in advance of real time than the other two designs, most of this should be taken care of by re-contracting in the RT market. Ideally the RT market should facilitate the rearrangement of prior bilateral contracts at as low a cost as possible. If this is done, then with the exception of commitment failures, the right dispatch should occur regardless of mistakes in prior markets. Penalties for trading in the RT market interfere with re-contracting and so are most harmful when applied to bilateral markets.

## 3-5.3   SETTLEMENT, HEDGING, AND RELIABILITY

Traders use forward markets, including the DA market, for hedging. The system operator is interested in it for reliability. By scheduling generation in advance, it hopes to guarantee its availability when needed in real time. Settlement accounting provides the basis for analyzing the relevant incentives.

When power is sold in the DA market, the outcome is a quantity sold, $Q_1$, and payment, $R_1$. The payment, $R_1$, would be $Q_1 \times P_1$ in a power exchange and that plus a side payment in a power pool. It is useful to define DA short-run profits, $SR_{\pi 1}$, to be the profits that would result from fulfilling the DA contract exactly.

$$SR_{\pi 1} = R_1 - MC \times Q_1 - SC$$

DA short-run profit equals DA revenue less marginal cost and startup cost (which should be thought to represent all nonmarginal costs). If the supplier performs as specified in the RT market, it will earn exactly the DA profit regardless of the RT price (See Chapter 3-2).

Depending on the RT price, the supplier may be able to increase its profit by producing at a level different from the one specified in the DA contract. Assuming that $Q_1$ is the supplier's full capacity, it can only earn more by producing less and then only when the RT price is low. This is investigated with the help of final short-run profits. These are given by

$$SR_{\pi F} = R_1 - MC \times Q_0 - SC_0 - P_0 \times (Q_1 - Q_0)$$

Final short-run profit includes DA payments, actual production costs and an adjustment for any shortfall in actual production relative to the DA contract. Solving for $R_1$ from the DA short-run profit equation and substituting into the above equation gives

$$SR_{\pi F} = SR_{\pi 1} + (MC - P_0) \times (Q_1 - Q_0) + (SC - SC_0)$$

**Figure 3-5.1**

Final profit including DA contract revenue.

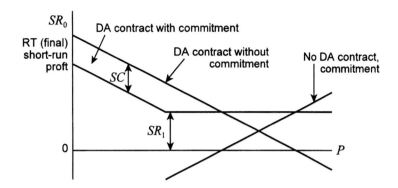

Final profit is DA profit plus any profit that can be made by buying RT power for less than marginal cost plus any savings from not starting up. $SC_0$ is actual startup cost which will be either $SC$ or zero.

As shown in Figure 3-5.1, a supplier with a hedging contract in the DA market can guarantee itself at least $SR_{\pi 1}$ if it starts up. If after starting, it finds the RT price below its marginal cost, it can buy power instead of producing it to satisfy its DA contract and thereby increase its profit. In this case it will wish it had not incurred the cost of starting up, $SC$. But without incurring this cost it cannot protect itself against the danger of high RT prices. This is shown by the RT short-run profit function that goes negative for high RT prices.

Because startup is necessary to achieve the risk-reduction of a DA market hedging contract, suppliers will be motivated to start up. But if they become convinced RT prices will be very low, they may decide to save the cost of starting up. It is unlikely that this will cause a reliability problem because whenever such problems are likely, there is a good chance of high RT prices. Then suppliers with DA contracts will start because they do not wish to risk not starting.

Notice that the pool's side payment plays no role in this analysis. A DA contract with a power exchange provides exactly the same incentives to start up as does a DA contract with a power pool, provided the two contracts pay the same amount for the same accepted quantity of power. This is discussed further in Section 3-7.3. As shown in Example 2D in Section 3-9.2, a power exchange, if its bid structure is too simple, may cause bidding to be somewhat random, which may result in inappropriate bids being accepted. These inappropriately accepted bids are more likely to fail to lead to physical commitment. In example 2D, this improves efficiency and does not decrease reliability.

With flexible bidding a power exchange will accept bids with essentially the same accuracy as a pool, and so exchange and pool contracts will pay the same. Because DA side payments do not induce commitment, the two approaches will have the same reliability properties. The consequences for reliability of less flexible exchange bids are unknown.

**Result    3-5.2    Side-Payments in a Day-Ahead Pool Do Not Increase Reliability**
Receiving a side payment, as opposed to being paid the same total amount entirely through the market price, does not increase a generator's incentive to commit. Thus it does not increase reliability.

## 3-5.4    OTHER DESIGN CONSIDERATIONS

Ancillary services do not give rise to all of the factors that need consideration. Other factors include transaction costs, facilitation of market monitoring, provision of publicly known prices, and nondiscriminatory access (see Section 1-8.1). Although these have been given less consideration both here and in the public debate, some may be equally important.

### Penalties To Promote Centralized Day-Ahead Trading

One design consideration that makes a frequent appearance is a penalty for not trading, and thus not scheduling, in the DA market. Three reasons have been given for this proposal: (1) the limitation of market power, (2) increased reliability, and (3) increased efficiency. PJM requires its generators to bid in the DA market but sets no limit on the price they can bid, so some bid extremely high thereby circumventing the requirement.

About 11% of PJM's generation does not enter the required multi-part bids with a low enough price to be accepted in the DA market, even though they generate in real time. In spite of this, because of arbitrage bids, more power is sold day ahead than in real time. Most of these arbitrage bids may be speculative bids because hedgers would use multipart bids.

Because of the bidding inaccuracy in the DA market, PJM does not rely on its outcome for reliability purposes. The same inaccuracies call into question the efficiency claims of a pool.

To make an effective penalty, generators must be penalized for the difference between their contract position and their RT production. There is some cost to forcing trades into the DA market, but this is difficult to determine. There may also be costs to last-minute scheduling. If these can be determined, imposing them on last-minute traders is sensible. Market power considerations are more difficult to evaluate.

Given PJM's success at providing reliability without relying on the DA market and given the doubtfulness of efficiency claims for a pool, penalties do not seem to be warranted in order to increase the accuracy of the DA pool. Even if generators are forced into the DA market, penalties cannot guarantee the accuracy of their bids.

## Complications and Simplification

Improvements in the basic designs should also be considered. As discussed in Chapter 3-9, a slight complication of the power exchange's one-part bids may capture most of the benefit of a power-pool's multipart bids. Similarly, greater control over ramping might usefully be purchased by a system operator with a power exchange.

Finally, simplicity and transparency of design are great virtues, not so much because of the direct cost of operating under complex designs, which can be substantial, but because of the flaws complex designs conceal.

# The Real-Time Market in Theory

*It must be done like lighting.*

Ben Johnson
*Every Man in his Humour*
1598

**U**NLIKE A DAY-AHEAD EXCHANGE, A REAL-TIME EXCHANGE CANNOT USE BIDS. The real-time (RT) market consists of trades that are not under contract—power that just shows up, or is taken, in real time and accepts the spot price. An RT exchange works like a classical Walrasian auction. A price is announced and suppliers and customers respond. If the market does not clear, a new price is announced. The difference is that in a power market trade takes place all the time; there is no waiting to trade until the right price is discovered. Like a Walrasian auction and unlike a day-ahead (DA) exchange, an RT exchange may find there is no price that balances supply and demand.[1] Consequently, if an exchange is used, it must be supplemented with another exchange or perhaps an operating-reserve market in the form of a pool. There are many possibilities, and little is known about their relative merits.[2]

**Chapter Summary 3-6:** Pure bilateral markets are too slow to handle RT balancing and transmission security. A centralized market is needed which can take the pool approach, the exchange approach, or something in between. Pools have an easier time achieving a supply-demand balance than do exchanges because they utilize different prices for different generators. This allows them to offer an option in a forward market that depends on the real-time market price. An RT exchange can achieve a similar effect but only by employing one or more additional exchanges. These could be for detrimental generation or operating reserves. Alternatively an operating-reserve pool could supplement an RT exchange.

---

1. Although the lack of a market clearing price is indicated by a static analysis, the dynamics of a power market which are limited by ramping constraints may reverse this conclusion.

2. See Rothwell and Gomez (2002) for descriptions of Norwegian, Spanish, and Argentinian spot markets.

**Section 1: Which Trades Are Part of the Real-Time Market?**  RT trade is trade that takes place at the RT price and not at a contract price. Thus, deviations from quantities specified in forward contracts comprise the transactions of the RT market.

**Section 2: Equilibrium Without a Market-Clearing Price.**  As in the day-ahead market, nonconvex production costs often prevent the RT market from having a clearing price. Bilateral markets and pools, with their multiple prices automatically have an equilibrium, but an RT exchange will often find it impossible to exactly balance supply and demand. The assistance of a second or third market may be required. This could be an RT market for decrementing generation below forward contract levels, or it could simply be the market for operating reserves which is needed in any case.

**Section 3: Why Balancing Markets Are Not Purely Bilateral.**  Bilateral markets are slower than centralized markets because traders must find a partner. This makes it difficult for them to maintain a precise system balance while ensuring transmission security. A centralized market takes advantage of electricity's greatest trading virtue: Electricity does not need bilateral delivery. When a load needs power, power from any combination of generators will do. This flexibility speeds the balancing process in a centralized market.

## 3-6.1   WHICH TRADES ARE PART OF THE REAL-TIME MARKET?

A clear understanding of what constitutes the RT market is required before the problem of nonconvex costs can be considered. This question is best approached by first identifying the amount of power traded in real time. For simplicity it will be assumed that the RT price is set every five minutes and that settlements are performed on a five minute basis.

### Real-time Sales Are Not Under Contract.

If the RT price were increased by $1/MWh from 12:00 till 12:05, how much would that increase the revenue of suppliers? Equation (3-2.1) provides the answer and is repeated here for convenience.

$$\text{Supplier is paid: } Q_1 \times P_1 + (Q_0 - Q_1) \times P_0 \qquad [\text{3-2.1}]$$

This shows that the increase in revenue would be the total amount by which RT supply exceeded power sold in the DA market times the RT price. Thus only the deviations from day-ahead contracts are sold at the RT price. If power is also sold in an hour-ahead market, these contracts would be settled according to the same principle and again the RT price would only be paid for deviations from the forward quantity.

All power supplied in real time can be classified as sold under a forward contract or not sold under a forward contract. If not sold forward, it is sold in the RT market at the RT price. Thus power sold in the RT market is not sold under financially binding bids but is instead the power that is not covered by any contract. Instead it is simply produced and sold at the market price prevailing at the time of production. Real-time trades are not based on accepted bids.

### Side Payments and Incentives

When a generator realizes that it will receive a "make-whole" side payment because the RT price will have too low an average, the RT price loses its incentive properties. As shown in Chapter 3-2, the two settlement system preserves the incentives of the RT price under normal forward contracts.

Usually this incentive problem is corrected by a substitute incentive: If the generator does not "follow the dispatch," i.e. follow the RT price, its make-whole payment will be taken away. Typically such regulatory incentives have hidden loopholes that do not occur with a price incentive.

Ramp-rate limits are said to be under-reported in multipart bids. This would allow a generator to ramp slowly without being seen as "not following dispatch." Under a price incentive, the generator might well find it profitable to ramp more quickly. Side-payments may dampen the responsiveness of generators to the RT price.

Fortunately, side payments play a small role and their interference with the price incentive is probably not a serious matter.

**First Illustration.** A complex illustration of this conclusion occurs in PJM's RT pool. PJM accepts bids in the operating-reserve market that takes place just after the close of the DA market, but no quantity is associated with the accepted bid so there is no binding commitment on the part of the generators. But PJM does make a binding financial commitment. If a generator performs according to its bid and still loses money at the RT price, it will receive a "make-whole" side payment. This is what makes PJM's RT market a pool instead of an exchange market. Power sold under such an agreement is sold at the RT price if the average price is high enough so that no side-payment is needed. But if the RT price is low, a generator's payments are determined by its bid. In other words, the generator has been given an option to sell at the RT price or under a forward pay-as-bid price. If the RT price is high, the generator sells the power in the RT market rather than under a forward contract; while if the RT price is low, it sells the power under a forward contract accepted in the operating-reserve market. Oddly, the generator does not always know which market it is selling power in until the end of the day, after the power has been produced.

**Result    3-6.1    Real-Time Power Is Not Bought or Sold Under Contract**
Power sold in the RT market is power sold at the RT price. Power sold under contract is sold at the contract price and its cost cannot be affected by the RT price. All contracts are forward contracts. Deviations in supply or demand from forward contract specifications, intentional or unintentional, comprise the forward market trades.

**Second Illustration.** A second illustration may further clarify the distinction between the RT market and forward markets. Suppose a generator has offered to sell 50 MW for $60/MWh for two hours and the system operator needs the power. Say it accepts the contract at 12:00, sets the RT price to $60/MWh and the supplier begins to flow power immediately. It appears that power is being traded under contract in the RT market. But by 12:05 supply and demand conditions will have changed and the system operator may well feel it needs to raise price to attract more supply or reduce it to attract less. Suppose it raises the RT price to $65/MWh. The power flowing under the 12:00 contract continues to flow but is paid for at the contract price of $60/MWh, not at the RT price. Five minutes after the start of the contract, the power sold under the contract is no longer part of the RT market. In fact it never was; the fact that it was traded at the RT price was merely a coincidence and, as will be explained, probably a mistake. In some later five-minute interval the RT price might again equal the contract price but that would again be a coincidence and not an indication that for five minutes the 12:00 contract is again part of the RT market. The litmus test is this: Would raising the RT price raise the cost of the power? The answer is *no*, hence the contract power is not part of the RT market.

The 12:00 contract is actually a forward contract with an extremely short lead time. Usually such contracts are agreed to several hours in advance, but sometimes they are agreed to only a few minutes in advance. In any case, the power delivery takes place in the future, not at the time of agreement. Consequently the system operator should evaluate the purchase over that time horizon. If the RT price will be $60/MWh at the start of the contract but will likely increase to $200/MWh during its two-hour duration, the system operator should be willing to pay more than the RT price for the contract power. If the RT price is expected to decline it should pay less than the RT price at the start of the contract. It is only by coincidence that the price of power sold in a contract equals the RT price during the first five minutes of the contract, or during the last five minutes, or at any other time. In any case, by the time the power starts to flow, the payment for that power is not affected by the RT price and so the power is not part of the RT market.

**Third Illustration.**   Suppose a contract is written that stipulates a supply of 100 MW for two hours to be paid for at the RT price. Is this an exception? Is this power under contract and also part of the RT market? There are two possibilities: (1) the contract may stipulate the usual settlement rule, or (2) it may stipulate a penalty for deviations from the 100 MW flow. With the usual settlement rule, deviations are paid for at the cost of the deviation, that is, at the RT price. In this case if only 50 MW is delivered the supplier is first paid for the 100 MW at the RT price and then it must buy 50 MW of replacement power at the RT price. The net effect is simply that it is paid only for the amount it delivers and is paid at the RT price. This is exactly what would happen without a contract. This is a contract without effect and will be deemed not to be a contract.

If the contract specifies a penalty for deviations that is different from buying replacement power at the RT price, this constitutes a genuine exception. There is little reason for such a contract, but power sold in this way is part of the RT market and is also sold under contract.

## What Determines the Real-Time Price

The RT price is determined by total actual (RT) supply and demand. This includes power traded under forward contracts and power traded in the RT market. Equation 3-2.2, repeated here, makes this clear.

$$\text{Supplier is paid: } Q_1 \times (P_1 - P_0) + Q_0 \times P_0 \qquad (3\text{-}2.2)$$

Because suppliers cannot alter their forward quantity, $Q_1$, and, assuming a lack of market power, they cannot alter $(P_1 - P_0)$, they optimize their choice of RT output, $Q_0$, solely on the basis of the RT price, $P_0$. (The quantity sold in the RT market is $(Q_0 - Q_1)$, not $Q_0$ which is total RT production.)

When the system operator sets an RT price, it must balance total supply and demand. Suppose more power is needed and there is a generator with a day-ahead contract and a marginal cost of \$65/MWh while the RT price is \$60/MWh. In a competitive pool, that generator would not produce even if it had sold its power in the day-ahead market. It would earn more by buying RT replacement power than by generating. Setting the market price to \$66/MWh will call forth that supply and help to balance the market. Thus raising the RT price can balance the market by increasing the amount delivered under forward contracts, but that does not change the amount sold under forward contracts; that amount is fixed. So any change in the amount delivered is a change in the amount sold in the RT market. This is why the RT price controls all RT power flows even though most of these flows are forward trades and not RT trades.

Changing the RT price changes the quantities delivered under some forward contracts as well as quantities delivered in excess of forward contracts. For example, if some load it taking more power than it has contracted for, raising the RT price may reduce that demand. This is a purely RT transaction.

The RT price should be set by taking into account the full supply and demand response. If this is not done, it will be necessary for the system operator to circumvent the market in some way to balance supply and demand.

## 3-6.2    EQUILIBRIUM WITHOUT A MARKET-CLEARING PRICE

Although the unit-commitment problem is less severe in real time than it is a day ahead, the startup cost problem still complicates the dispatch. Even relatively quick-start generation has startup costs and these make production costs nonconvex. Also, gas turbines are usually **block loaded**, meaning they are run at full capacity once started. Again this causes a nonconvexity in the production-cost function. As discussed in Chapter 3-5, such cost functions violate an assumption of perfect competition and can prevent the existence of a market clearing price. As in the day-ahead market, the lack of a market-clearing price does not indicate the market lacks an equilibrium, but the RT character of the market changes the analysis.

### Equilibrium Prices for Pools and Bilateral Markets

Nonconvex costs make the task of balancing supply and demand more difficult. In this regard, a power pool has an advantage because it can control output with more than the pool price. The possibility of side payments can help with fine tuning. Thus if it needs a generator to start and produce at half of full output, it can set the pool price to the generator's marginal cost. If the generator has a forward contract that gives it an RT option on a make-whole side payment, it will know that its startup cost will be covered even when the pool price is insufficient. Setting the pool price to the generator's marginal cost is a signal that it will be selling power under the forward contract with the make-whole side payment. The power is not sold just for the RT price. By taking advantage of its forward contracts in this way, a power pool can achieve an equilibrium in spite of the lack of a market-clearing price. Essentially an RT pool is a combination of two markets, a forward market and a real-time market.

A pool has one more degree of latitude not available to an exchange. It can call up a generator with a forward contract and tell how much to produce. If the generator obeys, then it retains the protection of it's side-payment guarantee. If it does not, then it loses the guarantee. There are cases when an RT pool must make such phone calls if it is to achieve an efficient dispatch. For example suppose one of two generators is no longer needed for reliability. If one has a marginal cost of $25, while the other has a marginal cost of $20, the one with the lower marginal cost would be more expensive if its no-load cost were $10/MWh at full output. In this case the efficient dispatch requires it to shut down, but it will not do so until the pool price is below $20/MWh, its marginal cost. At that price both generators would shut down and that would reduce production too much. The solution is to phone the low-marginal-cost generator and tell it to shut down while keeping the pool price at $25/MWh.

| Result | 3-6.2 | **Real-Time Pools Sometimes Require Direct Control of Generation** <br> To achieve an efficient dispatch, it is sometimes necessary for a pool to directly control a generator's output because the pool price does not provide the necessary signal. |
| --- | --- | --- |

Like power pools, bilateral markets have the advantage of trading at many different prices. Although these tend towards a single market-clearing price, when that does not exist, there is nothing to prevent deals from being made at whatever price is required. Given enough time and information, all profitable trades should be made and the market will be in equilibrium.

## The Problem of a Real-Time Exchange

A power exchange has the most difficulty balancing supply and demand because in real time there is no formal bid acceptance process. True bidding is impossible. In effect, an RT power exchange is a Walrasian auction. Without some elaboration of this mechanism, and in the presence of nonconvex costs, the supply-demand balance may be impossible to achieve. There are several possible responses to this problem.

First, the problem may not need to be solved. The magnitude of the problem has not been evaluated and a properly run exchange may come extremely close to balancing supply and demand. All control areas experience some balancing error and the result is "inadvertent interchange" between control areas. For the Interconnection as a whole these approximately cancel due to the coordination provided by NERC rules (which rely on the area control error, ACE, as an indicator). To balance the market with a single price, the system operator would need to control the RT price very cleverly. This could mean computing and announcing the RT price somewhat in advance in order to assure generators that if they started, they would be guaranteed a high price for at least some minimum amount of time. Such pricing dynamics are difficult and presently not understood.

A second approach to balancing is to run an additional exchange which provides more flexibility. The second exchange could be for decrementing generation.[3] Suppose a generator has a marginal cost of $20/MWh and a no-load cost of $5/MWh at full output. That generator will bid into an exchange at $25/MWh, but at that price it will produce at full output because that is the only level at which it recovers its full no-load cost. However it will be willing to buy back power at a price of $20/MWh, its marginal cost. This can be accomplished in a second exchange that accepts **decremental bids**. If bids are accepted, they must have an associated duration and the exchange takes the form of a forward market. It would also be possible to have an RT exchange for decremented power. This would apply to all forward contracts. Any supplier who produced less than specified by its forward contract would buy power at the RT decremental price, while any supplier producing more would sell power at the RT incremental price.

## A Hybrid Solution

Power pools will inevitably do much of their RT trading in the exchange mode because imports, load and some native generation will have no option of receiving side payments. These traders must trade at the pool price. Cost nonconvexities on either side of the market may make finding the supply-demand balance difficult because large loads may turn off when a price threshold is reached and generators may start and immediately ramp to full output. These problems are similar to the traditional load-following problems faced by system operators. The solution to such imbalances has always been operating reserves.

PJM runs a pool after the close of the DA pool and considers it to be an operating reserve market. Generators accepted in this market do not automatically sell any power under contract, but if they follow the PJM dispatch they are guaranteed make-whole side payments if they are needed. The make-whole guarantee provides sufficient inducement for them to allow PJM to control their output. In PJM such contracts cover a fairly small number of generators. In effect PJM has a small operating reserve pool that operates alongside a larger RT power exchange.

In general, operating reserves need to be, and are, under the control of the dispatcher. Consequently an operating reserve market seems to be the appropriate tool for handling the small discrepancies in balancing caused by the use of an exchange. This does not imply that the use of both incremental and decremental RT exchanges is inappropriate.

---

3. This is just one possibility. Any additional exchanges will provide added flexibility and improve the supply-demand balance. The design problem is to do the best job with the fewest number of markets. An additional exchange could be an hour-ahead market or a market for ramping services or operating reserves.

## The Efficiency Of Equilibrium Prices

Real-time pool prices suffer from the same defects as day-ahead pool prices. They are not designed to be optimal signals to short-run demand or long-run supply. See Sections 3-5.1 and 3-8.3.

## 3-6.3   WHY REAL-TIME MARKETS ARE NOT PURELY BILATERAL

Most markets operate under a regime of pure bilateral trade, so it is reasonable to ask why RT electricity markets do not. The answer is related to the provision of the ancillary services described in Chapter 3-4. Some of these must be provided by the RT energy market, and a bilateral market is not well suited to their provision. The critical issue is speed. Bilateral markets are slow to make complex trades. Centralized markets are quick largely because they can take full advantage of the homogeneity of electric power.

One ancillary service plays a special role with respect to bilateral trading. Trade enforcement is required for the very existence of the bilateral market. It provides the necessary property rights (see Section 3-4.6). This means that the bilateral market is dependent on centralized accounting, metering, and enforcement. Still, this does not preclude the bilateral market from handling all real power trades.

Voltage support is generally purchased in long-term markets and any interaction with the real-power market is minimal, so this ancillary service plays a minor role in RT trading and will be ignored. The black-start service is procured far in advance and is irrelevant. This leaves three ancillary services to consider.

1. Real-power balancing.
2. Transmission security.
3. Economic dispatch.

The system operator must ensure that these services are provided but may do so by utilizing a market. The question for an RT bilateral market is whether, by stringently enforcing balanced trades, the system operator can induce a bilateral market to provide these RT services. The first two of these are critical as they are the basis for reliability. In combination they are extremely difficult for a bilateral market to provide, which explains why all RT power markets are centralized. In addition, the bilateral market would be so inefficient at providing frequency stability (balancing) while complying with transmission security constraints that it would fail to provide an economic dispatch.

## The Balancing Market and Bilateral Trade

Although the primary goal of the balancing market is to maintain the real power balance, it must do so while respecting transmission constraints and it should do so at least cost. Real-power balancing means keeping the area-control error, ACE, near zero. This is done by keeping local supply and demand, net of scheduled interchange, in balance. NERC's requirement for ACE is quite stringent.[4]

---

4. See Hirst (2001) for a clear explanation of the NERC requirements.

Electricity's greatest virtue as a tradable commodity is that it does not need bilateral delivery. There are no different grades of electric power, and often the point of supply matters very little. If a customer receives generator B's power instead of generator A's power, no one cares. In a power market, this advantage is amplified by the need for strict RT balancing. When one supplier falters, or one load takes more than planned, any other supplier or group of suppliers can make up the difference; there is no need for bilateral trades to stay physically balanced. Forcing a power market to trade only bilaterally would limit the advantage that can be gained from this interchangeability of supply. While many commodity markets have interchangeable products, no other market needs to balance physical supply and demand minute by minute to prevent physical damage to, or disruption of, the marketplace.

**Random Fluctuations.**   Load and intermittent generators, such as wind generators, fluctuate constantly. This makes it difficult to keep a bilateral contract in balance. A centralized market has a natural advantage in this respect. Consider a market at a time of day when load is not changing systematically, and suppose there are 25 bilateral contracts. Say the 25 loads each change randomly up or down by 40 MW. The corresponding generators will need an equal adjustment, for a total adjustment of 1000 MW. In a centralized market, because random fluctuations tend to cancel, the total fluctuation in load would, on average, be only $\frac{1}{5}$ as much $(1/\sqrt{25})$. Thus only 200 MW of adjustment in generation would be needed. Moreover, the adjustment would be made by whichever generators in the market could adjust most cheaply.

**Forced Outages.**   An unexpected forced outage of a line or generator causes frequency-control problems that must be corrected within a matter of minutes. Such outages cause imbalances in specific bilateral contracts, but sufficiently high penalties could cause traders to contract for their own private spinning reserves. This would impose unnecessary transaction costs and complexity on the market.

## Transmission Security and Bilateral Trade

When the transmission network is unconstrained (as has been tacitly assumed so far), the problems of the balancing market are *relatively* simple. If a bilateral contract is out of balance due to a lack of generation, the trader responsible for the imbalance may contract with any other generator in the market to compensate. When there is a transmission constraint, this is not the case. Then it is necessary to check with the system operator before completing the new agreement. Every unscheduled adjustment must be checked in a purely bilateral market.

Consider a line that is operating at its security limit with power flowing from A to B. If a particular bilateral trade has generation at B and load at A, it creates a **counterflow** on the line. This trade decreases the physical flow on the congested line. If the load is unexpectedly reduced and its generator follows, this will increase the actual power flow on line A–B. In this case the trader responsible for the trade must find another generator at the A end of the line and arrange for it to back down instead of the generator that is part of its bilateral trade and located at the B end.

This type of complexity has proven too great for bilateral markets and as a consequence the RT markets in fully deregulated systems are all centralized. Because the system operators know the transmission constraints and can trade with any generator in the system, they are able to keep the system in balance with a great deal of accuracy, except in extraordinary circumstances.

# The Day-Ahead Market in Practice

*Predicting is pretty risky business, especially about the future.*

Mark Twain
(1835–1910)

**I**S THE DAY-AHEAD POOL PRICE DETERMINED BY ARBITRAGE OR COMPUTATION? Forward prices are usually determined by arbitrage between the forward market and the real-time (RT) market. Day-ahead (DA) markets are forward markets, but DA pools with multipart bids have been promoted for their ability to determine through computation the optimal dispatch and the efficient price. Both theories might prove true, or half true, but more likely one is essentially right and the other wrong.

There is no question that the computation takes place accurately and in a mechanical sense determines the DA pool price. The result of the computation is also determined by its inputs. Because the computation itself is a fixed procedure, while the input changes daily, it may be best to view the pool price as determined by inputs. Then the question becomes: Do the pool's input data accurately reflect the producer's reality, or do they deliberately misrepresent that data in order to take advantage of arbitrage opportunities. In the first case, the pool calculation makes use of good input data to produce a price that reflects the true details of generation costs. In the second case, the calculation is not a sensible unit-commitment calculation because its inputs are false. The bidders have manipulated the pool's computation, and the outcome may be thought of as being determined by arbitrage.

Unfortunately, it takes only a small percentage of arbitragers to dominate the outcome. In PJM's DA market, 11% of the generators that are needed, and eventually produce power, in the RT market are rejected by the unit commitment calculation (PJM, 2001, 30).[1] This alone proves the calculation is not highly accurate. Morever, while 3,260 MW of multipart bids representing needed generation are rejected, 6,169 MW of supply-side arbitrage bids (one-part bids) are accepted on

---

1. Only 26,771 MW submit DA multipart bids, while 30,031 MW proves to be necessary in real time.

These data suggest even more strongly that the hypothesis should be rejected, at least in the 95% of hours with price below $80/MWh. Since the true arbitrage price, as defined, would be more accurate than the very simplistic predictor used for this test, the hypothesis would be even more strongly rejected if tested more rigorously.

Another piece of evidence comes from Hirst (2001) who reports that the correlation between DA and RT prices was 63% in PJM, 62% in California's market, and 36% in NYISO. Note that the California market and the PJM market, both of which allow arbitrage, score the same, while the NYISO's DA pool predicts the RT price less well than California's arbitrage-based market. This may be because arbitrage in the NYISO market was more difficult during this period. That would indicate that arbitrage improves the accuracy of pool pricing and is thus the determining factor.

Both sets of evidence are inconclusive, but both suggest that the possibility that pool prices are determined by arbitrage rather than computation (as defined above) should be taken seriously. If arbitrage is determining the pool's outcome, then the computation is largely a charade. This question deserves investigation.

## 3-7.2  EFFICIENCY

Three primary arguments have been made for the use of DA pools: (1) efficiency, (2) reliability and control, and (3) risk management for generators. The efficiency argument asserts that a DA pool will increase the efficiency of the dispatch over that obtained with a DA power exchange. The case for this has never been made in writing, but it is still widely believed among those who support the pool approach.

The theory of the efficiency of DA pools begins with the observation that nonconvex generation costs prevent the existence of a market clearing competitive price and so invalidates the Efficient-Competition Result (Result 1-5.1). This proves a power exchange has no claim to perfection, so a pool might do better. The pool approach is explicitly designed to handle nonconvex costs, but these occur only at the generating unit level and affect only a portion of the generation costs.

Because the cost nonconvexities are very small, accurate data are required if they are to be properly taken into account and the dispatch is to be optimized. With no generating unit ever exceeding 5% of the market and the typical unit closer to 0.5% of the market, it would be surprising if incorrect data for 11% of the generation did not significantly degrade the ability of a pool to find an optimal dispatch. In fact, more than 11% of the data in PJM is incorrect from the point of view of a unit-commitment program because of the large number of arbitrage bids, and the result is a dispatch that is 11% short of committed units. This must negate the effect of the detailed optimizing calculations of a pool. Before the efficiency argument can be given credence, some theoretical or empirical support for it must be advanced.

average in the DA market. Thus more power is sold in the DA market than is needed in real time. Many of the accepted arbitrage bids probably represent speculation as hedgers would be inclined to use multipart bids in order to take advantage of the insurance provided to accepted bids by "make-whole" side payments. This circumstantial evidence does not prove that the arbitrage bids are controlling the DA price, but it does suggest that this is a serious possibility.

Because the accepted DA bids commit 11% less generation on average than will be needed in real time, PJM re-solves the unit-commitment problem with better data immediately after the DA market closes. The outcome of this second computation is used to commit generators for reliability purposes. No power is traded in this market and PJM views it as a method of providing operating reserves.

**Chapter Summary 3-7:**  Limited empirical evidence from the ISO markets suggests that DA pool prices may well be determined primarily by arbitrage. DA pool prices are not obviously more efficient than exchange prices, and they are not needed for reliability as is demonstrated by PJM's lack of reliance on them for this purpose. The side payments of the DA pool seem useful for reducing the DA market risk of generators, but this can be accomplished with far simpler bids and without side payments.

**Section 1: Arbitrage vs. Computation.**  The DA arbitrage price is the expected RT price. The DA computational price is the pool price that would result from the most accurate possible set of bids. These two prices determine whether the actual pool price is primarily the result of computation or arbitrage.

**Section 2: Efficiency.**  Given the extent of missing data in PJM's DA pool and the small magnitude of the errors that the pool computation is designed to address, it seems likely that the computation is ineffective. There is currently no evidence that DA pools provide a more efficient dispatch than DA exchanges.

**Section 3: Reliability.**  PJM's DA market does not serve directly as the basis for reliability. A post-DA-pool calculation plays that role.

**Section 4: Risk Management.**  The DA market provides insurance for some generators with large startup costs and unpredictable profits. But the risks seem modest, and a less-elaborate insurance scheme may be appropriate.

## 3-7.1  ARBITRAGE VS. COMPUTATION

The introduction to this chapter discussed two possibilities: The DA pool price could be controlled by (1) arbitrage with the RT market, or (2) accurate bids and computation. This section defines the conceptual difference more precisely and suggests one observable consequence that might be used to distinguish between the two possibilities.

## Operationalizing the Conceptual Difference

If all generators and loads in the PJM market submitted their bids as accurately as possible, the DA pool calculation would produce a market price and a dispatch. This price, $P_{Comp}$, is the benchmark for a "computationally" determined pool price.

Perfect arbitrage would result in the DA price equaling the expected RT price. A good statistician, which an arbitrager would hire, can determine this price which is the benchmark for a pool price determined by arbitrage, $P_{Arb}$.

If the actual DA pool price is $P_1$, the question of whether the pool price is determined by arbitrage or computation can be operationalized as: Is $P_1$ closer to $P_{Arb}$ or to $P_{Comp}$? The most difficult step in answering this question would be the computation of $P_{Comp}$, but PJM has apparently done most of the necessary work. It has made its best judgement (which is probably excellent) of an accurate bid set. These estimates could be run through the DA market's algorithm to determine $P_{Comp}$. In fact this may already have been done. (The actual computation used for unit commitment after the close of the DA market is not public information.)

## Why Computational and Arbitrage Prices May Differ

If $P_{Comp}$ and $P_{Arb}$ were the same, there would be no meaning to the question, and perhaps they are very close. But if they are the same, then there is little to be gained from computation. A standard market would find the same price and that would imply much the same dispatch. The main selling point for the pool approach has been the improved accuracy of the DA dispatch, which is said to improve efficiency and reliability. This might occur as follows.

Suppose midload plants have a startup cost of $60/MW and a variable cost of $15/MWh. If peakers have no startup costs and a marginal cost of $30/MWh, the breakeven point is four hours. If the plant is needed for less time the peaker is cheaper, but for more time, the midload plant is cheaper. Suppose that on half the days power is needed for just less than four hours and on the other half for just a little more. If the price is computationally determined, the accurate input data will provide better load and generator-availability data than is publicly available. Also the computation will correctly make complex unit-commitment decisions that the market cannot predict. The result will be that the DA pool will price power at $15/MWh on half the days and $30/MW on the other half, and it will usually be right.

If the pool is usually right, the DA pool is providing a valuable service by dispatching accurately. The arbitrage price, as defined, is the best feasible public prediction of the RT price. Since the DA pool price is assumed to anticipate the RT price, if the RT price could be predicted, $P_{Arb}$ would equal $P_0$ and thus $P_{Comp}$. But the only way to predict the RT price is by replicating the pool's computation with good data and this is not feasible. So the arbitrage price, $P_{Arb}$, will get the RT price right on average but will be wrong every day. The arbitrage price will always equal the *average* RT price, $22.50. In this example, if the pool really works, its price will be a more accurate predictor of the RT price than the arbitrage price.

## A Simpler Test

Although PJM has the necessary inputs to compute $P_{Comp}$, without those inputs, it is hard to estimate. But, as seen in the present example, if the DA pool were very accurate, the pool price would come quite close to the RT price, while the arbitrage price would be much further away. The average arbitrage price is just as close to the average RT price, but on any particular day, the actual pool price exactly equals the RT price. The arbitrage price is always off by $7.50 one way or the other.

**Hypothesis**

### The DA Pool Price is More Accurate

Because the DA pool has access to much accurate data that is not publicly available, and because of its superior computational abilities, the DA pool price more accurately predicts the RT price than does a DA estimate based only on public information.

If the DA pool has better information and better computing abilities than market at large, then the pool price should be closer to the RT price than predictions based only on public information. This hypothesis could be test follows.

The actual DA pool price, $P_1$, could be compared with the RT price ho hour for a year and the mean absolute deviation computed. Then the arbitrag could be approximated compared in the same way. (Root-mean-square de could also be used.) If $P_1$ were found to be significantly closer to the RT pr to the estimated arbitrage price, the hypothesis of a computationally det DA pool prices would be accepted.

If $P_1$ is found not to predict the RT price any better than the estimated price, then the possibility that the DA pool price is determined by arbitra open.

## A Glance at the Data

A little calculation shows that using only the hour of the day and no c tion, the RT price can be predicted with an mean absolute error of $ with the $12 error achieved by the DA pool price, $P_1$. It seems plaus into account day-of-week, season, and weather forecast might re to $12. This would favor rejection of the above hypothesis.

If only the 95% of hours with RT prices below $80/MWh ar simple prediction, based on a constant and the hour of the day when predicting the RT price (as measured by standard errors). of $P_1$ is $15.99 while the standard error of the estimated arbitra Measured by average absolute errors, the DA price is a sligh

---

2. To test whether these results were due to some difficulty with DA bid "physical" marginal costs, all hours with RT prices above $80 were remov error was then $9.25, while the simple prediction absolute error was $10.50. error was $29.47 and the standard error of the simple prediction was $45.

average in the DA market. Thus more power is sold in the DA market than is needed in real time. Many of the accepted arbitrage bids probably represent speculation as hedgers would be inclined to use multipart bids in order to take advantage of the insurance provided to accepted bids by "make-whole" side payments. This circumstantial evidence does not prove that the arbitrage bids are controlling the DA price, but it does suggest that this is a serious possibility.

Because the accepted DA bids commit 11% less generation on average than will be needed in real time, PJM re-solves the unit-commitment problem with better data immediately after the DA market closes. The outcome of this second computation is used to commit generators for reliability purposes. No power is traded in this market and PJM views it as a method of providing operating reserves.

**Chapter Summary 3-7:** Limited empirical evidence from the ISO markets suggests that DA pool prices may well be determined primarily by arbitrage. DA pool prices are not obviously more efficient than exchange prices, and they are not needed for reliability as is demonstrated by PJM's lack of reliance on them for this purpose. The side payments of the DA pool seem useful for reducing the DA market risk of generators, but this can be accomplished with far simpler bids and without side payments.

**Section 1: Arbitrage vs. Computation.** The DA arbitrage price is the expected RT price. The DA computational price is the pool price that would result from the most accurate possible set of bids. These two prices determine whether the actual pool price is primarily the result of computation or arbitrage.

**Section 2: Efficiency.** Given the extent of missing data in PJM's DA pool and the small magnitude of the errors that the pool computation is designed to address, it seems likely that the computation is ineffective. There is currently no evidence that DA pools provide a more efficient dispatch than DA exchanges.

**Section 3: Reliability.** PJM's DA market does not serve directly as the basis for reliability. A post-DA-pool calculation plays that role.

**Section 4: Risk Management.** The DA market provides insurance for some generators with large startup costs and unpredictable profits. But the risks seem modest, and a less-elaborate insurance scheme may be appropriate.

## 3-7.1 ARBITRAGE VS. COMPUTATION

The introduction to this chapter discussed two possibilities: The DA pool price could be controlled by (1) arbitrage with the RT market, or (2) accurate bids and computation. This section defines the conceptual difference more precisely and suggests one observable consequence that might be used to distinguish between the two possibilities.

## Operationalizing the Conceptual Difference

If all generators and loads in the PJM market submitted their bids as accurately as possible, the DA pool calculation would produce a market price and a dispatch. This price, $P_{Comp}$, is the benchmark for a "computationally" determined pool price.

Perfect arbitrage would result in the DA price equaling the expected RT price. A good statistician, which an arbitrager would hire, can determine this price which is the benchmark for a pool price determined by arbitrage, $P_{Arb}$.

If the actual DA pool price is $P_1$, the question of whether the pool price is determined by arbitrage or computation can be operationalized as: Is $P_1$ closer to $P_{Arb}$ or to $P_{Comp}$? The most difficult step in answering this question would be the computation of $P_{Comp}$, but PJM has apparently done most of the necessary work. It has made its best judgement (which is probably excellent) of an accurate bid set. These estimates could be run through the DA market's algorithm to determine $P_{Comp}$. In fact this may already have been done. (The actual computation used for unit commitment after the close of the DA market is not public information.)

## Why Computational and Arbitrage Prices May Differ

If $P_{Comp}$ and $P_{Arb}$ were the same, there would be no meaning to the question, and perhaps they are very close. But if they are the same, then there is little to be gained from computation. A standard market would find the same price and that would imply much the same dispatch. The main selling point for the pool approach has been the improved accuracy of the DA dispatch, which is said to improve efficiency and reliability. This might occur as follows.

Suppose midload plants have a startup cost of $60/MW and a variable cost of $15/MWh. If peakers have no startup costs and a marginal cost of $30/MWh, the breakeven point is four hours. If the plant is needed for less time the peaker is cheaper, but for more time, the midload plant is cheaper. Suppose that on half the days power is needed for just less than four hours and on the other half for just a little more. If the price is computationally determined, the accurate input data will provide better load and generator-availability data than is publicly available. Also the computation will correctly make complex unit-commitment decisions that the market cannot predict. The result will be that the DA pool will price power at $15/MWh on half the days and $30/MW on the other half, and it will usually be right.

If the pool is usually right, the DA pool is providing a valuable service by dispatching accurately. The arbitrage price, as defined, is the best feasible public prediction of the RT price. Since the DA pool price is assumed to anticipate the RT price, if the RT price could be predicted, $P_{Arb}$ would equal $P_0$ and thus $P_{Comp}$. But the only way to predict the RT price is by replicating the pool's computation with good data and this is not feasible. So the arbitrage price, $P_{Arb}$, will get the RT price right on average but will be wrong every day. The arbitrage price will always equal the *average* RT price, $22.50. In this example, if the pool really works, its price will be a more accurate predictor of the RT price than the arbitrage price.

## A Simpler Test

Although PJM has the necessary inputs to compute $P_{Comp}$, without those inputs, it is hard to estimate. But, as seen in the present example, if the DA pool were very accurate, the pool price would come quite close to the RT price, while the arbitrage price would be much further away. The average arbitrage price is just as close to the average RT price, but on any particular day, the actual pool price exactly equals the RT price. The arbitrage price is always off by $7.50 one way or the other.

| Hypothesis | **The DA Pool Price is More Accurate**<br>Because the DA pool has access to much accurate data that is not publicly available, and because of its superior computational abilities, the DA pool price more accurately predicts the RT price than does a DA estimate based only on public information. |
| --- | --- |

If the DA pool has better information and better computing abilities than the market at large, then the pool price should be closer to the RT price than DA predictions based only on public information. This hypothesis could be tested as follows.

The actual DA pool price, $P_1$, could be compared with the RT price hour by hour for a year and the mean absolute deviation computed. Then the arbitrage price could be approximated compared in the same way. (Root-mean-square deviations could also be used.) If $P_1$ were found to be significantly closer to the RT price than to the estimated arbitrage price, the hypothesis of a computationally determined DA pool prices would be accepted.

If $P_1$ is found not to predict the RT price any better than the estimated arbitrage price, then the possibility that the DA pool price is determined by arbitrage remains open.

## A Glance at the Data

A little calculation shows that using only the hour of the day and no other information, the RT price can be predicted with an mean absolute error of $17, compared with the $12 error achieved by the DA pool price, $P_1$. It seems plausible that taking into account day-of-week, season, and weather forecast might reduce that error to $12. This would favor rejection of the above hypothesis.

If only the 95% of hours with RT prices below $80/MWh are considered, the simple prediction, based on a constant and the hour of the day, out-performs $P_1$ when predicting the RT price (as measured by standard errors). The standard error of $P_1$ is $15.99 while the standard error of the estimated arbitrage price is $14.57. Measured by average absolute errors, the DA price is a slightly better predictor.[2]

---

2. To test whether these results were due to some difficulty with DA bids when RT prices exceeded "physical" marginal costs, all hours with RT prices above $80 were removed. The DA average absolute error was then $9.25, while the simple prediction absolute error was $10.50. For all hours, the DA standard error was $29.47 and the standard error of the simple prediction was $45.16.

These data suggest even more strongly that the hypothesis should be rejected, at least in the 95% of hours with price below $80/MWh. Since the true arbitrage price, as defined, would be more accurate than the very simplistic predictor used for this test, the hypothesis would be even more strongly rejected if tested more rigorously.

Another piece of evidence comes from Hirst (2001) who reports that the correlation between DA and RT prices was 63% in PJM, 62% in California's market, and 36% in NYISO. Note that the California market and the PJM market, both of which allow arbitrage, score the same, while the NYISO's DA pool predicts the RT price less well than California's arbitrage-based market. This may be because arbitrage in the NYISO market was more difficult during this period. That would indicate that arbitrage improves the accuracy of pool pricing and is thus the determining factor.

Both sets of evidence are inconclusive, but both suggest that the possibility that pool prices are determined by arbitrage rather than computation (as defined above) should be taken seriously. If arbitrage is determining the pool's outcome, then the computation is largely a charade. This question deserves investigation.

## 3-7.2   EFFICIENCY

Three primary arguments have been made for the use of DA pools: (1) efficiency, (2) reliability and control, and (3) risk management for generators. The efficiency argument asserts that a DA pool will increase the efficiency of the dispatch over that obtained with a DA power exchange. The case for this has never been made in writing, but it is still widely believed among those who support the pool approach.

The theory of the efficiency of DA pools begins with the observation that nonconvex generation costs prevent the existence of a market clearing competitive price and so invalidates the Efficient-Competition Result (Result 1-5.1). This proves a power exchange has no claim to perfection, so a pool might do better. The pool approach is explicitly designed to handle nonconvex costs, but these occur only at the generating unit level and affect only a portion of the generation costs.

Because the cost nonconvexities are very small, accurate data are required if they are to be properly taken into account and the dispatch is to be optimized. With no generating unit ever exceeding 5% of the market and the typical unit closer to 0.5% of the market, it would be surprising if incorrect data for 11% of the generation did not significantly degrade the ability of a pool to find an optimal dispatch. In fact, more than 11% of the data in PJM is incorrect from the point of view of a unit-commitment program because of the large number of arbitrage bids, and the result is a dispatch that is 11% short of committed units. This must negate the effect of the detailed optimizing calculations of a pool. Before the efficiency argument can be given credence, some theoretical or empirical support for it must be advanced.

## 3-7.3    RELIABILITY AND CONTROL

Reliability has become the primary justification for a DA power pool. The basic argument has two steps. First a DA power pool will find the efficient dispatch more accurately. Second, its side payments will assure the selected generators start up. The first step can be questioned for the reasons explained in the previous section.

At least in PJM, the second step is incorrect. As explained in Section 3-5.3, because PJM relies strictly on a two-settlement system, DA market side payments are not contingent on RT performance. Consequently they have no influence on a generator's decision to commit or not. DA markets of any variety encourage commitment because, once a hedging contract is signed, the signing party must perform in order to be hedged (see Section 3-5.3). Consequently, PJM's DA pool has no advantage over a DA power exchange in this regard.

PJM does not use the DA market to assure its reliability. Instead it uses a separate after-market calculation that interacts with the real-time market. This calculation may be quite similar to a unit commitment calculation. It is based on some bids submitted after the close of the DA market as well as on DA market bids. It could just as well rely entirely on bids submitted outside of the DA market. Thus PJM's market demonstrates that a DA pool is not needed for reliability. Perhaps a multi-part bid calculation with side-payments is needed, but this can be done outside of the DA market.

---

**Result     3-7.1**     **A Day-Ahead Power Pool Is Not Required for Reliability**
PJM achieves all of the reliability benefit normally attributed to the pool approach without relying on its DA pool for this purpose.

---

PJM's post-market calculation is not a power market—no power is traded—but it is a type of market. Essentially it is a market in which PJM trades profit insurance for an implicit guarantee to start up. Generators submit bids which for simplicity can be taken to consist of a startup cost ($SC$) and a marginal cost ($MC$). If PJM accepts the bid, it guarantees that the generator's profit, as computed by

$$Q_0 \times (P_0 - MC) - SC + \text{Side Payment}$$

will be nonnegative where $Q_0$ is actual production as a function of time and $P_0$ is the real-time price. The side payment is an insurance payment because it is made only when the generator has the misfortune to earn a negative profit. Presumably, PJM minimizes its expected insurance payments subject to the constraint that enough generators must be started to satisfy reliability requirements. This arrangement benefits both the generators and PJM. The implicit guarantee of the generator to start up is believable because it cannot lose money if it starts and has some chance of earning a positive profit. To claim the insurance it must "follow the PJM dispatch" which means keeping its output within about 10% of the quantity indicated by $P_0$ and its marginal-cost bid, $MC$. Ramp-rate limits, as indicated by its bid, are also taken into account.

Alternatively this market may be viewed as a market for spinning reserves. In fact this is essentially PJM's view of it. For the accepted generators, it has the effect of making side payments contingent on RT performance thereby turning part of PJM's RT market into an RT pool.

## 3-7.4   RISK MANAGEMENT

The third justification for a DA pool is that it helps generators manage their risk. As seen in the previous section, the insurance market also provides risk management, but the DA market plays a similar role.

In a power exchange, it is quite possible to bid in such a way that a profit is expected yet find that money has been lost if the bid is accepted. For example, a generator with a $30/MWh marginal cost may bid that cost on the belief that the market price will be above $30 by enough and for long enough that it will recover its $40/MW startup cost. But its bid might be accepted for four hours at a price of only $35/MWh. In this case it recovers only half of its startup costs. Regardless of what happens in the RT market, it has lost money in the DA market. It would have been better off had it not bid. This cannot happen in a pool if the generator bids its true costs. In a pool the generator would bid the $40/MW startup cost as well as its marginal cost, and the bid would probably not be accepted. If it were accepted at a price of $35/MWh for four hours, it would be accepted with a side payment of $20/MW, and the generator would break even.

A primary purpose of forward markets is to hedge suppliers and consumers. When the forward market itself generates risk this partially defeats the purpose. There is a real advantage in the pool's reduction of risk for generators. But if this is the reason for a DA power pool it raises the question of design and complexity.

Unit commitment programs are extremely complex and require many inputs because they are designed to optimize the dispatch. They were not designed to minimize market risk. Had they been, their design would have been far simpler. Two part bids, as are used in the Alberta power exchange would do the job. The bids in that market include a minimum run time. The actual risk involved in committing for a day is quite small, both because a day is a small fraction of a year and because generators can predict prices well enough that they will rarely start and then find they earn no scarcity rent at all. Adding a second part to the bid should make a small risk negligible. The big market risks for generators come from annual changes in load and supply, not from small daily errors. So the effect of a small remaining DA market risk will be undetectable.[3]

While risk management argues for two-part bids it does not argue for a pool. The signature of a pool is side payments. Side payments are not required to remove the DA market risk. Alberta's power exchange does not make side payments and yet removes most of the risk with its two part bid. An exchange sets the market price high enough to cover the costs of all accepted bids. No bid is accepted that loses money on its own terms. If two or three part bids are used, the generator can

---

3.  Risks are subadditive because they are proportional to standard deviations. Variances of uncorrelated events are additive and standard deviations are proportional to the square root of the variance.

bid so that any bid that does not lose money on its own terms will, at worst, breakeven if accepted and fulfilled.

# The Real-Time Market in Practice

*Where there is much light, the shadows are deepest.*

Goethe
Wilhelm Meister's Apprenticeship
1771

**W**HEN MARGINAL-COST PRICES WON'T CLEAR THE MARKET, ARE THEY STILL THE RIGHT PRICES? When production costs are "nonconvex," as are startup and no-load costs, competitive market theory predicts the market may not clear, marginal-cost prices may not be optimal, and the market may not be efficient. The sole purpose of the multipart bids used by power pools is to overcome problems caused by nonconvex costs. The pool approach recommends setting the market price equal to marginal cost exactly as if there were no problem and then making side payments to generators who are needed for the optimal dispatch. These payments cover only the costs that marginal-cost prices fail to cover.

The pool approach recognizes the first failure of marginal-cost pricing and corrects it with side payments. But true competitive prices do more than minimize production costs; they send the right signals (1) to consumers and (2) to investors in new generation. Can a power pool's combination of marginal-cost prices and side payments replicate these benefits of competitive prices? Using an example from a dispute over NYISO's pricing, this chapter shows that pool pricing fails both of these tests. Surprisingly, it was NYISO's position that standard marginal-cost pricing—pool pricing—was inefficient for at least five reasons.

Pool prices are neither the prices of Adam Smith, nor those of competitive economics. They are not right for the demand side and they are not right for long-run investment. Pool prices are right for the centralized solution of the problem of minimizing short-run production-costs, given an output level that is incorrectly determined when demand is elastic.

This does not mean the pool approach is a bad idea; it simply means that adopting it because it gets the prices right would be naive. The problem of nonconvex costs is difficult and a complex market design, such as a pool, could be needed.[1] No market design is likely to get the prices exactly right even in theory, so getting them wrong proves little.

A simpler approach is used as widely and deserves equal attention—the power exchange approach. Unlike pools, exchanges do not make side payments. As a consequence exchange prices tend to be slightly higher than marginal cost. The change in pricing introduced by the NYISO and approved by FERC raised the market price above marginal cost just to the point where it covers the costs of a generator that might otherwise have needed a side payment. This is exactly the philosophy of power exchange pricing. The problems with pool pricing listed by the NYISO correspond well with the theoretical problems of pool prices, and the advantages they sought are those offered by a power exchange.

**Chapter Summary 3-8:** The real-time (RT) market can use a pool approach or an exchange approach. Pools use marginal cost prices and side payments. These prices are too low in cases where another unit of demand would cause a jump in total cost. Instead, price should be raised to curb demand until demand increases to the point where the cost increase is warranted. Pool prices also send the wrong signals for investment. Exchange prices are also suboptimal, but they are simpler and potentially more accurate with regard to demand and investment.

**Section 1: Two Approaches to Balancing-Market Design.** The power-pool approach collects data on generators through multipart bids, computes an optimal dispatch, and sets the market price equal to marginal cost. This does not clear the market, so it makes side payments to keep some required generators in the market. The power-exchange approach typically uses one- to three-part bids, but its distinguishing feature is a lack of side payments. It relies on a single price, while a pool uses prices that are effectively tailored to individual generators.

**Section 2: The Marginal-Cost Question As Decided by FERC.** FERC was asked to direct the NYISO to use marginal-cost pricing as specified in its tariff, and on July 26, 2000 it did. The NYISO objected and proposed to keep its above-marginal-cost prices. FERC accepted NYISO's old pricing scheme with little modification. Non-marginal-cost pricing will be applied in real time while the same situation will call for marginal-cost pricing in the day-ahead (DA) market.

**Section 3: Making Sense of the Marginal-Cost Pricing Charade.** Although NYISO's prices are not marginal-cost prices, they are probably better. Theory indicates that marginal-cost prices can be too low for all purposes. Side-payments

---

1. Because nonconvex costs violate a basic assumption on which the Efficient-Competition Result is based, they invalidate many of the conclusions previously reached in this book. Fortunately the problems actually caused by nonconvex costs are small in magnitude, as is explained in Chapters 3-8. So, while there may be no market clearing price, a power market will have a (Nash) equilibrium price that is extremely efficient.

fix the short-run supply-side problem but not the demand-side or investment problems. NYISO's above-marginal-cost pricing moves toward the power exchange approach in this circumstance.

**Section 4: The Power Exchange Approach.** An RT power exchange could set a market price different from any bid and let the non-bid supply and demand responses, along with accepted bids, clear the market. Because of the nonconvex-cost problems that pools focus on, an additional market for ramping services or decrementing generation is beneficial.

## 3-8.1  TWO APPROACHES TO BALANCING-MARKET DESIGN

A centralized, RT, energy market must provide three ancillary services: balancing, transmission security, and efficient dispatch. When system operators worked for regulated monopolies, they had direct physical control, but in a market, they must rely on RT prices and contracts for control. This increases the difficulty of providing the first two services and introduces the problem of market power, which complicates the job of providing an efficient dispatch.

Having ruled out the purely bilateral approach to the RT market, two basic approaches remain: the **exchange** approach and the **pool** approach. The exchange approach uses simple bids and relies on a single price which is set to clear the market as well as possible. The pool approach is based on the observation that **cost nonconvexities** make clearing the market impossible. It compensates for this flaw with **multipart bids**, optimal bid-acceptance procedures, and **side payments**. The main cost problems (nonconvexities), are **low-operating limits**, no-load costs, and startup costs. The litmus test for a pool is the existence of side payments, which are not used by exchanges.

### The Block-Loading Example

The different natures of the two approaches are best explained by considering how each would handle the same nonconvex cost problem. Consider a system with only low-variable-cost steam units ($20/MWh) and high variable cost gas turbines (GT's) ($40/MWh). The GT's must be block loaded; that is, they must be dispatched at their full 100-MW output capacity or not at all. For simplicity, they are assumed to have no startup costs, and steam units are assumed to have constant variable costs. Suppose that the total maximum output capacity of all steam units is 20,000 MW, that load has just increased to 20,020 MW, and that the system operator is required to keep the system perfectly balanced.

### The Pool Approach

This problem will first be analyzed according to the pool approach which has three steps:

1. Find and implement the optimal dispatch.[2]
2. Set market price equal to system marginal cost.
3. Make side payments as needed to compensate dispatched generators for negative profits.

One GT must be dispatched, but it provides too much supply so a steam unit must be backed off by 80 MW. Then total supply equals load at 20,020 MW. System marginal cost is popularly defined as "the cost to supply the next increment of load." This requires clarification at a point where all generators are fully loaded and another must be started. But the present case is free of this ambiguity as the GT has been started and there is an unconstrained generator. One more megawatt of load, or two more, or 20 more would all be supplied at a cost of $20/MWh hour by the steam generator that has been backed down. Similarly, one less would save $20/h.

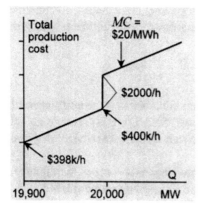

**Miscalculation of Marginal Cost.** Because both FERC and NYISO miscalculate marginal cost in this example it is necessary to consider their misinterpretation in some detail. The basic notion of marginal cost is that it is the slope of the total cost curve at a particular operating point. Thus it has nothing to do with the actual chronology of load arrivals or generator startups. In the present example the total cost curve has a constant slope starting well below the operating point and continuing for 80 MW above that point. Its slope is unambiguous at the operating point and can be calculated in the usual way as the cost of serving the next increment of load. This is shown at the left and explained by Figure 3-8.1.

The misinterpretation assumes that the relevant "increment" of load is the change in load that caused the GT to be started. But this is not the "next" increment; it is the prior increment. Moreover it misinterprets "next" as having to do with time, while the phrase "next increment" in the marginal cost definition actually means "an additional increment." (Note that when the 20-MW increment of load caused the GT to start, the incremental cost was actually $120/MWh, not $40/MWh, because that 20-MW increment raised system cost by $2,400/h.) The system's marginal cost when $20/MWh generation is available is $20/MWh.

**Side payments.** The third step in the pool approach is to pay generators that have been dispatched but have lost money. In this example the GT lost $20/MWh and needs a side payment of $2,000/h.

**The Rationale.** The rationale for an RT pool is first that it achieves an efficient dispatch. The required power could not be produced at a lower cost. Second, consumers are sent the right price signals in the sense that if they choose to consume five more megawatts, that will cost $100/h to produce and consumers will be

---

2. As will be demonstrated, the standard pool approach does not find the optimal dispatch when demand is elastic. Instead it finds a dispatch corresponding to a price that clears the market at marginal cost with the help of side payments (when this is possible).

**Figure 3-8.1**

Marginal cost is $20/MWh because marginal power is provided by a steam unit.

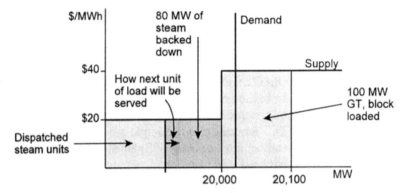

charged $100/h to consume it.[3] As a final consideration, the market should send the correct signals for long-run investment. Marginal-cost prices do this when costs are convex. When they are nonconvex and side payments are included, economics makes no guarantees. The hope is that any long-run damage will be less than the short-run gains from efficient dispatch.

## The Exchange Approach

The exchange approach is not as well specified in theory as is the pool approach, though in practice it may have less variability. Its first rule is to make no side payments. This means that to induce the GT to start up it must set the market price to $40/MWh, or higher.[4]

To prevent oversupply, the exchange approach needs some way to induce a steam unit to back off. This might be done by accepting a **decremental bid**. In this case it would be an offer by the steam unit to pay the system operator perhaps $19.90/MWh to allow the steam unit to reduce its output while still being paid according to all prior contracts as if it were producing its full output. If it has sold its power for $22/MWh and the system operator bought a decremental contract at $19.90, the steam unit would be paid $22 and would in turn pay the system operator $19.90 for the megawatt-hours that it sold but did not produce.

Many designs are compatible with the exchange approach, and some will work better than others. In an exchange approach, there might be several parallel markets, such as one for incremental energy, one for decremental energy, one for reserves, and another for ramping or load-following. Each would attempt to set price to clear the market. Pools use a similar set of markets but handle the problem of backing down units differently.

---

3.  This assumes that the consumer sees the RT price, which is unlikely, but the job of the RT market is to produce the right signal; passing it on to consumers is a separate problem.

4.  While the market has no "market-clearing price" in the classic economic sense, an exchange will have an equilibrium price which typically comes much closer to clearing the market than the pool price.

## 3-8.2   THE MARGINAL-COST QUESTION AS DECIDED BY FERC

On April 24, 2000, the New York State Electric and Gas Corporation (NYSEG) filed a complaint with FERC that challenged NYISO's fixed block (block-loaded GT) pricing methodology on the grounds that it violated NYISO's (FERC-approved) Tariff and theory of locational-based market prices (LBMPs). In effect NYSEG claimed (among many other things) that NYISO should set price to **system marginal cost (SMC)** at all times including when block-loading a GT caused a steam unit to be backed down.

FERC (2001a) describes its July 26, 2000 decision as follows

> . . . *the Commission concluded that when less expensive generation resources are dispatched down for the purpose of accommodating more expensive fixed block resources, the marginal cost of supplying the next increment of load should be equal to the bid price of the least expensive unit that has been backed down. The Commission concluded that in these instances **the Services Tariff required the energy price to reflect the marginal cost of the backed-down unit**, and directed NYISO to revise its procedures for setting prices accordingly.*

FERC and NYSEG were right. The NYISO Tariff and NYISO's theory of pool pricing require price to be set to SMC, and the cheap unconstrained generator determines SMC, not the expensive constrained generator.

On August 25, 2000, NYISO filed for a rehearing and proposed a "hybrid" pricing approach which almost entirely rejected the use of SMC pricing in the RT market.[5] In this filing NYISO made two points that are investigated in the next section, and several that go beyond that analysis. All of these points are implicit criticisms of SMC pricing. First it noted that the pool approach (SMC pricing) sends an inefficiently low price signal to loads.

> *The Commission's rule* [SMC pricing] *would discourage the development of price-responsive real-time loads in New York, because the RT prices it would establish would not reflect the incremental cost of meeting load. The Commission's rule would eliminate the incentive of loads to reduce output in real time in many hours when the incremental cost of meeting load is very high, because this **high cost would not be reflected in RT prices**.* (NYISO 2000, 9) (Emphasis added.)

Second it notes:

> . . . *the new rule* [SMC pricing] *might: . . . (ii) frequently understate RT prices when supplies are tight, thereby sending inaccurate price signals to the market and creating **long-term inefficiencies**.* (NYISO 2000, 5) (Emphasis added.)

---

5. SMC pricing was retained for the unusual event that a GT was on because of a minimum run-time constraint but was otherwise not needed by the RT market.

It did not specify the long-term inefficiencies, but the main long-term concern of any market is efficient investment. This would coincide with the finding in the next section that SMC pricing discourages the steam units relative to the GT's when the opposite is desired.

Because their analysis considers the full complexity of the market, the NYISO discovers a number of other problems not covered in the following analysis. SMC pricing would discourage imports.

> *The Commission's fixed block pricing rule* [SMC pricing] *will very likely discourage external suppliers from delivering real-time imports from external resources that are scheduled in the DA market. This disincentive would arise because the real-time imbalance price would be* **artificially driven below the market-clearing level.** (NYISO 2000, 8) (Emphasis added.)

It would discourage loads from participating in the DA market.

> *First, the new rule* [SMC pricing] *can be expected to undermine the willingness of loads to participate in the DA market because of the high cost of meeting incremental load in real-time would not be reflected in RT prices, i.e.* **real-time LBMPs would usually be set by less expensive steam units, rather than fixed block GT's** *that actually were incremental.* (NYISO 2000, 10) (Emphasis added)

Coupled with the use of uplift and side payments, it would distort congestion costs.

> *Second, because the NYCA's* [New York Control Area's] *fixed block GT units are located principally in New York City or on Long Island, they often operate only when there is transmission congestion, and they only set prices in the eastern part of the state. In this situation, the Commission's fixed block pricing rule* [SMC pricing] *will likely lead to lower prices in the east, but not in the remainder of the NYCA. Any reduction in real-time LBMPs below the bid prices of GT's running to meet eastern loads will be recovered in the form of uplift charges from customers all over the state.* **The net effect will be to shift congestion costs from eastern New York into state-wide uplift.** (NYISO 2000, 10)

It would encourage load-serving entities, which would otherwise not want a high market price, to exercise market power.

> *The Commission's fixed block pricing rule* [SMC pricing] **gives certain market participants an incentive to exercise market power** *. . . . such an entity would have an incentive to bid extremely high, knowing that it would receive substantial uplift payments at the expense of customers throughout the NYCA if its resources are called upon, while the price it pays for energy will be set by less expensive resources.* (NYISO 2000, 12)

The above analysis by the NYISO appears to be accurate although decidedly nonrigorous and intuitive. The only significant error in its analysis is required for bureaucratic reasons. It concludes that

> *. . . the Commission could revise its rule to permit fixed-unit GT's
> that must be run to set real-time LBMPs without requiring the
> NYISO to amend its tariff because the GT's bids will truly reflect
> "the marginal cost of supplying one more unit of energy."*
> (NYISO 2000, 8)

This claim avoids the difficult task of amending the most fundamental element of
the tariff's philosophy—price will be set equal to "the cost to supply the next
increment of Load at that location (i.e., short-run marginal cost)" (NYISO Tariff).
Note that it claims their price will "truly reflect" the SMC, not equal it. Their next
reference to the cost of the GT calls it the "actual incremental cost of meeting real-
time load;" it does not refer to it as marginal cost. FERC (2001a) accepts this claim
that marginal cost is not marginal cost, forgetting what it knew in July 2000.

> *We note that the issue being resolved here arose because NYISO's
> tariff did not clearly specify how the LBMP would be calculated
> when lower-priced units were backed down to accommodate fixed
> block units. To avoid such disputes in the future, NYISO's tariff
> should specify how it will treat fixed block units in setting the
> LBMP.*

FERC claims the source of the problem was that NYISO's tariff was not clear
on how LBMP would be calculated. But the tariff was not at fault. It clearly stated
the price would equal "the cost to supply the next increment of Load at that location
(i.e., short-run marginal cost)." In the present context, this is unambiguous. NYSEG
understood it, and FERC understood it the previous July. All of NYISO's substantive arguments were aimed at showing the price *should not* equal marginal cost.
Their argument consists of this:

> *Sometimes our price, X, is a better price than marginal cost and
> sometimes it is not. When it is, then marginal cost must be X
> because marginal cost is always the best price. In the DA market,
> marginal cost will be the best price, so then marginal cost can
> remain marginal cost. But in the RT market, X is better, except
> in certain circumstances when FERC's [correct] interpretation
> of marginal cost is best. Then marginal cost can remain marginal
> cost in real time as well.*

By approving this approach, FERC removed the previous clarity of NYISO's
tariff. NYISO was right that their price was better than marginal cost, and FERC
was right to approve it, but despite the best efforts of regulators and system operators, marginal cost will remain "the cost to supply the next increment of load"
whenever the cost function is smooth.

## 3-8.3   MAKING SENSE OF THE MARGINAL-COST PRICING CHARADE

The New York ISO is supposedly founded on the rock of marginal-cost pricing.
But when one of its members tried to enforce this simple rule, NYISO put up a
vigorous and successful fight against it. In the process it convinced FERC that

marginal cost does not mean what everyone knows it means. The outcome is a rule based on the misinterpretation of the tariff, but which by all appearances is an improvement on SMC pricing.

Could SMC pricing really be as dangerous in this case as claimed by the NYISO? If SMC pricing is wrong how should price be set? Should it always equal the average cost of the last unit started? Even more puzzling is the fact that FERC and the NYISO retained proper SMC pricing for the identical circumstance in the DA market while replacing it in the RT market. Is there some incompatibility between SMC pricing and an RT market?

One clue to this puzzle is that the pricing rule NYISO fought for and won is a step toward the exchange approach. It is designed to eliminate the side payment to the GT, and it attempts to implement a market-clearing price (see emphasis in the third quote above from page 8 of NYISO's filing). But to understand the underlying problem a theoretical analysis is required.

## Analysis of the Block-Loading Example

In the example of Section 3-8.1, when load reaches 20,020 MW the system operator has a problem. At a market price less than $40/MWh, the new generator will produce nothing and demand will exceed supply. At a market price of $40/MWh or more, the next generator will produce 100 MW and supply will exceed demand. There is no price that clears the market. This problem, characteristic of nonconvex costs, makes the exchange approach difficult. When generalized to similar problems caused by the other cost nonconvexities, it is the motivation for the pool approach.

**The Pool Approach.**    As explained in Section 3-8.2, the pool approach dispatches one GT at full output and backs down 80 MW of steam output. It sets the price at system marginal cost, $20/MWh, and it pays the GT $2,000/h as a side payment to make up for dispatching it with a market price below its variable cost.

In competitive markets, economics predicts that a market price equal to marginal cost will provide optimal incentives for suppliers and customers in both the short run and long run alike. These economic conclusions are based on the assumption that costs are convex, so it is not obvious that SMC pricing will prove optimal for solving the problems caused by costs that violate this assumption. It is best to proceed cautiously. Three benefits are desired of the market price. It should:

1. Induce efficient supply in the short run.
2. Induce efficient demand in the short run.
3. Induce efficient investment in the long run.

As is universally acknowledged, marginal cost, $20/MWh in this example, is the wrong price to induce efficient short-run supply, and a side payment is needed. This leaves the questions of short-run demand efficiency and long-run investment. Does setting price equal to SMC induce an efficient demand for power? The marginal condition seems right. Customers will see the true cost of their next unit of consumption. But, this analysis tacitly assumes that demand is inelastic, otherwise demand could not have been known before price was calculated as equal to marginal cost. If demand is inelastic, then all prices "induce" the efficient level of demand

**Figure 3-8.2**

Determining the maximum market price before starting a block-loaded gas turbine.

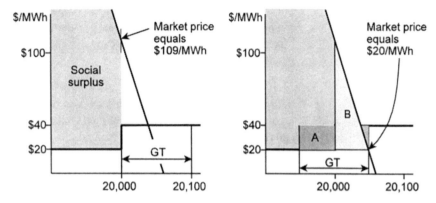

or at least do not interfere with it. Every price is right for demand efficiency. To test SMC pricing on the demand side, demand elasticity must be introduced so that price matters.

Suppose demand is reduced by 1 MW for every $2/MWh increase in price. This is a modest level of elasticity for a market as large as NYISO's.[6] With elastic demand, there is a demand-side question of price. If the market price is raised, the GT will not be needed. This will decrease total surplus by reducing the power consumed and increase total surplus by avoiding the cost of the GT. When demand is 20,001 MW at a price of $40, price need only be raised to $42/MWh to reduce demand to 20,000 MW and balance the system. Because the GT is block loaded, meeting the extra megawatt of demand instead of raising the price would cost $40/h for the megawatt itself and 99 × ($40 − $20)/h for the 99 MW of cheaper power displaced by block-loading the GT. Clearly the price should be raised and the GT not dispatched.

As the demand curve shifts to the right, price can be increased to keep the system balanced, but at some price the demand that is being suppressed becomes so valuable, that it is better to start the GT and serve the load. This point is reached when the net consumer surplus of suppressed demand, which is equal to the area of triangle B in Figure 3-8.2, equals the cost of side payments to the GT, given by the area of rectangle A. This happens at a price of $109.44 when the demand curve has shifted to the point where demand would be 20,045 MW at a price of $20/MWh. From this level of demand until demand reaches 20,100 MW, the market price stays at $20/MWh. The meaning of these price fluctuations will be discussed after investment is analyzed, but in brief, SMC pricing often misses the required price for demand efficiency by a wide margin.

**Marginal-Cost Pricing and Investment.**   The third benefit expected of price is inducement of the efficient mix of generating technologies. To check this, begin with one change of assumption. Suppose the GT's could be flexibly dispatched with a constant marginal cost of $40/MWh. Suppose there are just enough steam units to cover the load-duration curve up to a duration of 50%. Suppose the stock

---

6. If NYISO replaced its zero-elasticity demand for operating reserves with a realistic demand curve, it might well produce this much elasticity in the demand for power.

of GT's leaves room for the price spikes required to cover the fixed costs of GT's.[7] In addition, suppose the fixed costs of steam units are \$10/MWh greater than the fixed costs of GT's. In this case, Result 2-2.2 gives the following market equilibrium condition.

$$FC_{steam} - FC_{GT} = (VC_{GT} - VC_{steam}) \times D^*_{GT}$$

For the given cost values, the market would be in long-run equilibrium. Steam units would earn short-run profits of \$20/MWh half of the time, and this would exactly cover their extra \$10/MWh of fixed cost.

Now imagine that the flexibility of the GT's is removed and they revert to their standard all-or-nothing nature. This decreases their value because it increases the total cost of production relative to what it was when they were flexible. If the market works properly it will send a signal indicating that slightly less GT capacity and slightly more steam-unit capacity would be optimal.

What signal does marginal cost pricing send after the change from flexible to block-loaded GT's? Whenever demand does not exactly match a whole number of GT's, a steam unit must be backed down. And any time a steam unit has been backed down, it can produce a small additional (marginal) increment of power at the marginal cost of \$20/MWh. Thus, marginal cost will almost always be \$20/MWh. Whenever this occurs, it reduces the short-run profits of steam units to zero and the short-run profits of GT's to minus \$20/MWh. But the GT's have their profits restored through side payments and the steam units do not. The few peak hours when GT's and steam units both earn scarcity rents are left undisturbed as is the capacity market if it exists. Thus GT's continue to recover their fixed costs as before while steam units lose nearly all the \$10/MWh of scarcity rents they previously collected when GT's set the market price. This strongly opposes the weak positive signal that should have been sent to steam units. In this example, SMC pricing sends a grossly incorrect investment signal.

## The Mystery of System Marginal Cost[8]

**The Zero-Elasticity Case.**   If demand is totally inelastic then there is no question of what total supply should be—it should equal demand. In this case, the optimal dispatch simply minimizes production cost. In the optimal dispatch, there will always be at least one unconstrained generator (if there is enough generating capacity) except for the infinitely rare circumstance in which the dispatched generators can supply only and exactly the demand. When there are several unconstrained generators they will all have the same marginal cost in an optimal dispatch. Thus, when demand is inelastic, system marginal cost is always well defined and equal to the marginal cost of any unconstrained generators.

In this case, SMC pricing doesn't hurt demand, but it still sends the wrong signal to investors. This is ignored in the hope that the long-run damage is smaller than the short-run advantage of an optimal dispatch. In many cases the same optimal dispatch could be supported by higher prices and no side payments. This would

---

7.  The coverage of fixed costs may or may not include payments from a capacity market.

8.  For a more rigorous treatment of this subject, see Cramton and Wilson (1998, 24).

cause no distortion of demand or distortion of short-run supply, and it might be better for investment.

**The Elastic Case.**  With even a small amount of demand elasticity, price can no longer be set equal to marginal cost without causing demand-side inefficiency. If the demand curve is imagined as gradually moving from left to right, the price spike in the above example occurs entirely while the output is stationary and exactly at the full capacity of the steam units. Without any change in production, price increases from $20/MWh to $109/MWh thereby holding demand constant in spite of the moving demand curve. It is reasonable to say that marginal cost is undefined at this point so none of these values conflict with it. But marginal cost cannot possibly determine all of these prices or even the maximum price. Instead price must be set equal to the marginal value of power to customers. In the elastic case, it is still possible to define SMC pricing and it provides a useful benchmark.

| | |
|---|---|
| **Definition** | **System Marginal Cost (SMC)**<br>SMC is the marginal cost (*MC*) of all unconstrained units when supply equals demand, where demand is computed with $P = MC$. Optimization of the dispatch, given the output level, is assumed. If there is no output level satisfying these conditions, SMC is undefined. |

| | | |
|---|---|---|
| **Result** | **3-8.1** | **SMC Unit Commitment with Elastic Demand Is Inefficient**<br>If production costs are not convex and demand is elastic, SMC prices generally fail to induce efficient short-run supply, short-run demand, and long-run investment. Side payments do not prevent any of these inefficiencies. |

It is useful to define the pricing procedure illustrated in the block-loading example as a second benchmark.

| | |
|---|---|
| **Definition** | **System Marginal Value (SMV)**<br>SMV is the marginal value of demand at the optimal output computed by taking account of demand elasticity. (If marginal cost is defined at the optimal output, then SMV = SMC.) |

Contrary to SMC pricing, SMV pricing is right for demand but wrong for supply. In the example, when price is set to $109/MWh to curb demand, it is much too high from a short-run supply perspective. This is to be expected because whenever costs are nonconvex, there is no assurance of a single efficient price. Also, there has been no demonstration that SMV costs provide the proper long-run incentives when coupled with the necessary side payments. There may well be a theoretically correct pricing scheme and it may require different prices for loads and generators. While interesting as a theoretical problem, more practical approaches should be pursued first.

| Result | 3-8.2 | **System-Marginal-Value Pricing Provides Efficient Demand Incentives** |
|--------|-------|------|
| | | SMV prices induce efficient short-run demand but generally fails to induce efficient short-run supply. |

## How Serious Are the Inefficiencies of SMC Pricing?

In the above example, the maximum short-run inefficiency occurs when the GT is block-loaded to serve 1 MW of load. This decreases quadratically to zero when about 45% of the GT's output is needed. The average over the whole period of increasing need for this GT, provided load increases linearly, is only in the neighborhood of $300/h for a market that might be selling over $300,000/h worth of power. That is, 1 part in 1000, and this circumstance probably happens infrequently. Of course, there are other cost nonconvexities, but it appears that even when handled poorly, they do not cause much trouble.

## Where Does the NYISO's Non-SMC Pricing Lead?

NYISO's pricing innovation leads in two directions. First, it leads toward complexity. It introduces an ad hoc rule for pricing in the RT market. Because NYISO compromised with FERC, adopting FERC's initial request for SMC pricing under some circumstances but not others, it adds complex new distinctions. The NYISO-FERC innovation also introduces different pricing philosophies for the RT and DA markets without any supporting theory. Together these seem to confirm the standard concerns over the arbitrariness and impenetrability of real-world power pools.

Viewed more broadly, the pricing innovation may be seen as leading toward greater simplicity. NYISO saw the benefits to its deviation from SMC pricing as rooted in a single feature. Its price, as opposed to an SMC price, would pay the cost of the GT. In its view, this produces several benefits. First, it avoids side payments and the associated uplift. This prevents one case of market power and the distortion of transmission pricing. Second, it focuses the payment of the cost more closely on those who caused the cost.

It also produced a higher price that is somehow more right on average. This point is intuitively correct but was probably not understood in detail. While SMC prices send the right signals when the marginal-cost curve is continuous, they miss the costly block-loading event. They miss it because price cannot be raised to infinity for a split second thereby tracking marginal cost at the instant the generator is block-loaded. SMC pricing cannot handle this type of discontinuity in marginal cost, so it ignores it. NYISO's cruder pricing scheme does not get tricked by missing any infinitely high and narrow price spikes.

NYISO has moved away from the complex details of standard SMC pricing in this one case, and has moved even further from the details of SMV pricing. Probably, without realizing it, they are giving up on the optimality for which these approaches strive. Instead they are moving toward a simpler but more robust form of pricing, the type of pricing used in every other market. They have moved toward a price that cannot support the optimal dispatch but can pay for the cost of the generators it dispatches—they have moved toward a power-exchange approach to pricing.

## 3-8.4  THE POWER EXCHANGE APPROACH

Like the power-pool approach, the power-exchange approach is still ill-defined and problematic. As the power-pool approach needs some power-exchange features, so a power-exchange approach may need some power-pool features. It may need two- or three-part bids. Perhaps not, but that is a matter that must be determined empirically. Only clever design and numeric evaluation followed by experimentation will yield the answer.

One factor above all others will push markets in this direction. Real-time pricing of demand will eventually produce a large demand-side response that does not operate through bids. Then, if the system operator evaluates the bids it has and determines a market-clearing price and sets it as the market price, the market will refuse to clear. The new price will provoke a significant demand-side response at some of the most awkward and delicate times. This is already happening in PJM but not because of the demand side. That is still incapacitated, but imports and exports play an active role and do not necessarily bid into the PJM market. (The NYISO noted the discouragement of "external suppliers from delivering RT imports.").

When the system operators see load increasing, they look at their bids, which may be few and high, and choose a bid price that is a few hundred dollars above the current price. They often find this is followed by a large influx of external power which causes them to reevaluate and set a new lower price. The imported power then leaves, or the exports resume, and the system operator is again forced to raise the price.

System operators at PJM have learned to take account of un-bid supply elasticity, but the rules still require them to set price at the level of some particular supply bid. This is a contradictory approach. The system operator may choose not to take a $400/MWh bid because it believes this will induce too great an increase in net imports and instead set the price at a $200/MWh bid price. The system operator can adjust the purely bid-determined price down by $200, but it is not allowed to adjust it by a mere $100 if there is no supply bid at $300. It cannot set the price at $300 even if that would more accurately balance the system. If large price adjustments are made outside of the bidding rules, why can't smaller adjustments be utilized?

The un-bid elasticity of the supply and demand curves causes two problems for the power-pool approach. First it forces the system operator to invent fictitious bids that represent the un-bid portions of these curves. While this can be done scientifically and without bias, it cuts against the grain of the underlying philosophy—enter hard data from market participants and compute the optimum. The second problem is more fundamental. The elasticities are not instantaneous. If price

**Figure 3-8.3**

Market clearing at a bid price and clearing in-between bid prices.

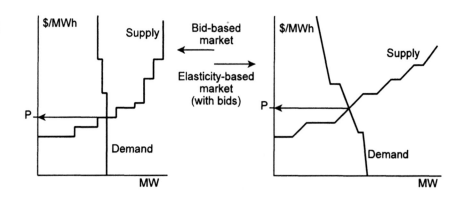

is based on current SMV, it may cause an unwanted supply or demand response half an hour later due to the dynamics of the supply and demand elasticities. Just as supply-side bids account for ramp rate limits, low-load limits, and other lags and constraints, so the fictitious demand-side bids should account for lags and discontinuities in demand if optimization is to be successful.

The result of increasing un-bid elasticity will be an order of magnitude increase in optimization complexity, a new and wide opening for arbitrary decisions on the part of the system operator.

## Improving Prices

The North-East ISOs tend to compute and announce prices every five minutes as if they meant them, but they never use them to pay for power. Instead they replace all 12 prices every hour with a single average price used in every five minute interval. The originally announced prices probably fool only a few. The use of the five-minute price is to instruct generators as to the desired level of output. In other words, these prices do not work through the normal economic channel of increasing and decreasing revenue according to price times quantity. Instead they are regulatory instructions which the suppliers translate according to their bid curves. These instructions are backed by penalties in the form of cancellation of make-whole side payments. In PJM not all generators are subject to control; in fact, real-time side payments may affect only a minority.

**Use Honest 5-Minute Prices.**  Five-minute prices matter most when the price is changing rapidly. At other times, it might be sufficient to compute price every half hour or every hour. Consider an hour with a large price change as shown in Table 3-8.1. When the system operator sets a price of $48, the market participants will guess that the average for the hour will be about $30 and respond to that instead. This will mean a 15-minute delay before many participants begin responding to these 5-minute prices. If five minutes matters, then the delay undermines both control and the precision of the dispatch.

**Use the Best Price.**  The next improvement is to learn to predict un-bid supply and demand responses and use these predictions along with bids to set prices more accurately. Stop restricting RT price to bid values, and instead set the price that

**Table 3-8.1**

| Time | Price per MWh |
|------|------|
| 1:00 | $24 |
| 1:05 | $24 |
| 1:10 | $24 |
| 1:15 | $24 |
| 1:20 | $24 |
| 1:25 | $24 |
| 1:30 | $24 |
| 1:35 | $24 |
| 1:40 | $24 |
| 1:45 | $48 |
| 1:50 | $48 |
| 1:55 | $48 |

will do the best job of balancing supply and demand. Consider the dynamics of the supply and demand response.

**Use a Downward Sloping Demand for Reserves.**   A known elasticity makes control through price adjustments easier. Zero elasticity is inappropriate for operating reserves. Replacing this assumption with a downward-sloping demand function for operating reserves will make balancing easier. When reserves are short, the price automatically rises and this calls forth more bid-in supply and imports and reduces exports and demand. It also helps to reduce market power by increasing the elasticity of demand. If 10% is the current requirement for all types of reserve, then they must be worth nearly $1,000/MWh up to the 10% level. Surely another few percent must be worth something, so the demand curve should go to higher levels of reserves at low prices as well as to lower levels of reserves at high prices. This simple change should dampen prices spikes and make the operator's job easier. The argument that elasticity will jeopardize reliability is without foundation. It depends entirely on what demand curve is selected; some would increase reliability dramatically. Buying more reserves when they are cheaper and fewer when they are dear will reduce the cost of reliability.

**Purchase Balancing Services.**   The system operator buys balancing services directly in two ways: by buying regulation and by buying spinning and nonspinning reserves. Both are limited in the uses to which they may be put and do not cover many normal circumstances that worry system operators. This concern is expressed in many ways, such as tying the startup insurance offered by power pools to following the dispatch. Also, PJM runs a full-scale DA simulation after the DA market to determine what generators to start. Limits are placed on the flexibility of external trades. NYISO confiscates the power of generators that overproduce. While there are many strategies, system operators still find the job difficult, especially at awkward times such as the morning ramp which occurs just after last-minute exports are scheduled.

These difficulties exist partly because the balancing service is thought of in terms of supply and demand instead of in terms of their **derivatives** (rates of change). Ramping is a separate service with a separate cost. This cost includes the wear and tear on machines from changing output as well as startup costs and the costs from dispatching around low-operating limits. One approach to better control would be to pay for ramping, that is, for changes in output between successive five-minute intervals. This price could be either positive or negative.

Sometimes, it is necessary to ask a cheap generator to back down while keeping a more expensive generator on line. There should be a market for the backing-down service, separate from the energy market, and perhaps linked with the reserve markets.

**Accepting Bids.**   For the most part, the previous suggestions ignore bids because an RT exchange does not utilize bids—it is a Walrasian auction (see Chapter 3-6). But accepted supply bids discourage market power. If a large generator has sold all of its output before real time, it will still have the ability to raise the market price, perhaps significantly, but it cannot profit from it. If it raises the market price, the increase does not affect its payment on forward contracts. If it withholds power it must buy it from the market, so when it withholds it wants a lower, not a higher price. For this reason the system operator should encourage as much trading as possible before real time because it is more vulnerable to market power in real time.

This means the RT exchange should be supplemented with an hour-ahead exchange or even with a forward exchange that is capable of accepting bids at any point in time. This will minimize the extent of the RT market without the use of penalties.

Unaccepted forward bids also provide the system operator with information about the supply curve, although a generator may not follow its own bid if it is above marginal cost. In spite of this, bids will help set an accurate RT price. In order to encourage bidding and provide certainty to generators, more complex bids could be allowed. This does not turn an exchange into a pool; only side payments can do that.

# The New Unit-Commitment Problem

*Let us not go over old ground, let us rather prepare for what is to come.*

Marcus Tullius Cicero
(106–43 B.C.)

T HE OLD PROBLEM ASKS WHICH UNITS SHOULD BE COMMITTED; THE
NEW PROBLEM ASKS WHAT MARKET DESIGN WILL BEST SOLVE THE
OLD PROBLEM. The old problem was solved by collecting data on all the generators and applying the techniques of mathematical programming. The new problem
might be solved by a market designed to induce generators to voluntarily and
accurately provide this same data. The market coordinator could then purchase
power from the generators identified by the old algorithm. This is the power-pool
approach. A power exchange is an alternative approach which pretends the old
problem does not exist.

It seems impossible that ignoring the old problem could be the best way to solve
it, but most market architectures ignore just such complex commitment problems.
When the market coordinator ignores the problem, the suppliers take it up, and they
may do a remarkably good job. Although the programming techniques used to solve
the old problem are astoundingly complex, most generators can get the right answer
on most days simply by looking at the calendar. If you have a baseload plant and
it's summer, keep your plant committed. If you have a peaker, don't look at the
calendar, just watch the real (RT) price—day-ahead (DA) forecasts are not needed.
When it really matters, on the hottest days, every supplier knows to commit. But
those who commit units for individual suppliers will do much better. They will
have years of experience and the necessary resources. Moreover, they may have
access to an exchange that uses two-part prices or multiple rounds of bidding. To
beat a good exchange market, a pool must be very good indeed.

Two central concerns have motivated the power-pool approach: efficiency and
reliability. Committing the wrong units costs more, but this problem is limited by
the magnitude of the cost involved and the efficiency of markets without centralized
unit commitment. Reliability is more of a wild card. Perhaps a decentralized market

would be more prone to occasional undercommitment of plants that are too slow to start. This could leave the system operator without the necessary resources and cause a blackout. As with efficiency, this suggestion has not been subject to quantitative analysis, so the new unit-commitment problem remains unsolved.

The old power pools in the Northeast have adopted power-pool markets, while the Western markets have adopted or proposed power exchanges. The choice of design has been historically determined, at least partly because no theoretical or empirical evidence is available. Considering only one well-defined part of the problem, efficiency in the context of a workably competitive market, Professor William Hogan has said:

> I am not prepared to argue for or against the need for unit com-
> mitment without doing more work. It is neither simple nor obvious
> that it is important or not important. If I had to bet, I would go
> 60/40 in favor of it being important.[1]

This chapter provides a framework for thinking about the economics of the unit commitment problem in the context of a competitive market. It suggests that some problems appear small while others are difficult to assess. It confirms Professor Hogan's view that the new unit commitment problem is difficult and our knowledge is sketchy.

**Chapter Summary 3-9:**  The cost of committing units is about 1% of retail costs, and individual generators can come quite close to minimizing this without the help of a central computation—a power pool. Though generators will sometimes renege on commitments made in a DA exchange market, they are least likely to do this when most needed, for that is when the RT price will be highest.

This does not rule out the possibility of improving on the power-exchange design. But when a first-order approximation is accurate to 1%, a twentieth-order approximation may not be required (power pools use bids with at least that many parts). Alberta allows two-part bidding, the second part being a minimum run-time limit. While completely artificial, in simple cases, it can substitute perfectly for bidding a startup cost. The crucial lesson is: Suppliers may not bid the literal truth, but with sufficient competition and a bit of flexibility, they bid in a way that makes their operation and the market efficient.

**Section 1:  How Big Is the Unit-Commitment Problem?**  Startup costs are the main costs of commitment. Typically, these amount to less than 1% of retail costs. Because startup costs are ten times smaller than fixed costs and have first claim on scarcity rents, marginal-cost prices often send the right signals for unit commitment.

---

1. Personal communication, January 5, 2002.

**Section 2: Unit Commitment in a Power Exchange.**  Four market designs are considered in the context of a single market structure in which startup costs are not covered by marginal-cost prices. Case A demonstrates that a power exchange has no competitive equilibrium in the classic economic sense. Case B demonstrates that an auction without startup-cost bids or side payments can have an efficient competitive Nash equilibrium in spite of lacking a classic competitive equilibrium. Case C reconsiders the simple power exchange design and finds the Nash equilibrium which proves to be inefficient but competitive. Case D introduces the possibility of reneging, that is, failing to generate the power sold in the DA market. This leads to greater overcommitment in the DA market followed by reneging to the point of an efficient dispatch.

**Section 3: Investment Under a Power Pool.**  A power pool includes side payments that reduce risk for generators that are accepted in the DA market and that produce according to the supply curve they bid. If all generators bid honestly, the dispatch will be efficient but investment incentives will be distorted. Types of generators that receive more insurance payments will be encouraged to over-invest. Like the short-run inefficiency of the power exchange, the long-run inefficiency of the power pool will be small.

## 3-9.1   HOW BIG IS THE UNIT-COMMITMENT PROBLEM?

A power exchange does not provide a unit-commitment service, but a power pool does. Either can set a single price at any given time, a few prices in different zones, or thousands of prices, one at each electrical bus in the system. This chapter only considers markets without congestion in order to focus on the unit-commitment problem. Exchanges and pools were defined in Section 3-3.1, but key parts of those definitions are repeated here for convenience.

**Definitions**

**Power Exchange**

A centralized market that trades energy at a single price (per location) at any given time and location and does not make side payments. An exchange may use multipart bids.

**Power Pool**

A centralized market that trades energy and uses side payments that depend on the suppliers' bids. Side payments are made only when the pool price is too low to cover the startup or no-load costs of an accepted bidder.

The purpose of power-pool complexity is to solve the unit-commitment problem, but the magnitude of that problem has not been assessed in a market context. Prior to deregulation there was only one unit-commitment problem—the problem of when to commit and de-commit a system's generating units. That problem remains, but it is not the relevant problem for market design. The new problem is: What market design will induce the most efficient commitment of generating units. This problem is different in nature and broader. Sections 3-9.2 and 3-9.3 examine questions of market design, while this section investigates the magnitude of the old problem in the market context.

| | |
|---|---|
| **Definition** | **The (New) Unit-Commitment Market Problem**<br>What market design will induce the most efficient commitment of generating units? "Efficiency" should include the transaction costs of the system operator and suppliers and the reliability consequences of commitment errors. |

## The Magnitude of Startup Costs

Startup costs are usually found in the range between \$20 and \$40/MW.[2] Generators that serve baseload start much less often than once per day, and very few generators start more often. Hydrogenerators have exceptionally low startup costs. As a rough estimate, there may be half a megawatt of startup per megawatt-day of load.

If the average startup cost is \$30/MW, then the cost of startups is \$15/MW for each MW-day of load. That is \$15 per 24 MWh, or about \$0.60/MWh. Retail power costs about \$80/MWh in the United States, so startup costs are $\frac{3}{4}$ of 1% of retail costs. (Preliminary estimates for PJM indicate its startup costs may be closer to half this much.)

---

### Units to Measure Startup Cost

Startup costs are measured in dollars per megawatt of capacity started. For a plant that is started once per day, the cost flow is most conveniently stated in \$/MWday. When used in formulas, startup costs are converted to \$/MWh by dividing by 24.

Duration should be expressed as a pure number when used in formulas but can be converted to hours per day, for convenient interpretation, by multiplying by 1 in the form (24 hours / 1 day). Thus a duration of 0.1 equals 2.4 h/day.

---

A significant amount of this cost is covered by scarcity rents from marginal-cost pricing.[3] If approximately $\frac{1}{3}$ of startup costs are subject to unit-commitment problems, and these problems add 50% to these startup costs, then the total waste of funds would be roughly $\frac{1}{6}$ of the cost of startups. This inefficiency would raise retail rates about $\frac{1}{8}$ of 1%. Because of the assumption that the market increases the costs of the problematic startups by 50%, this should be an upper bound on the magnitude of the startup problem.

Test examples involving two or three generators usually demonstrate inefficiencies on the order of 1% if their startup costs are a realistic fraction of total cost. But startup inefficiencies are inversely proportional to the number of generating units in the market. Together these give a back-of-the-envelope estimate of inefficiency on the order of 1 part in 10,000. So far, advocates of power pools have not produced a contradicting estimate, but the question is still unanswered as no

---

2. See Hirst (2001). **No-load costs** are another source of difficulty and are several times larger than startup costs. They are proportional to the length of time a generator runs and consequently easy for generators to include in their bids in a power exchange. They may present more of a problem in conjunction with startup costs.

3. Scarcity rent (see Section 1-6.6) is revenue minus costs that vary with output. After startup and no-load costs are subtracted, short-run profits remain.

theoretical proof has been given and no calculations have been made on actual power systems.[4]

| **Result** | **3-9.1** | **A Power Exchange's Unit-Commitment Inefficiency Is Less Than 1%.** Startup costs, the principle source of unit-commitment inefficiency, amount to roughly 1% of the retail cost of power or less. A significant portion of this cost is covered by marginal cost pricing, and markets without central unit commitment can, in principle, solve the unit-commitment problem quite efficiently. Consequently the inefficiency caused by nonconvex production-costs may be less than $\frac{1}{100}$ of 1%. |
|---|---|---|

## How Marginal-Cost Bidding Can Cover Startup Costs

One factor mitigating the startup-cost problem is scarcity rents. These must be great enough to cover fixed costs which are more than ten times greater than startup costs. Because startup costs have first claim on scarcity rents, they are quite often covered by them. Example 1 explains how this works and why it sometimes does not.

**Example 1A: Startup Costs Covered Completely.** Consider a market with peakers and midload plants that have the costs shown in Table 3-9.1. In long-run equilibrium, the market will build peakers up to the point where they just cover their fixed costs. This cost recovery must occur when all peakers are in use, and the price is driven above $50/MWh by the system operator's demand for operating reserves in short supply. The price of energy during this period could be moderate or extreme, but in either case, it will be just sufficient to cover the peaker's fixed costs of $120/MWday (see Chapter 2-2).

**Table 3-9.1** Cost Assumptions of Example 1

| Type of Supplier | Variable Cost (per MWh) | Fixed Cost (per MWday) | Startup Cost (per MWday) |
|---|---|---|---|
| Peaker | $50 | $120 | $0 |
| Midload plant | $30 | $240 | $40 |

During this same period, midload plants will also earn $120/MWday from prices above $50/MWh, enough to cover half of their fixed costs. The remaining half plus startup costs total $160/MWday, and this must be recovered from the difference between the $50/MWh variable cost of peakers and the $30/MWh variable cost of midload plants (see Chapter 2-2). This $20/MWh revenue stream is available whenever peakers are running, so peakers must run for 8 hours per day (160/20) if midload plants are to recover these costs. Consequently midload plants will be built up to the point, but not beyond the point, where peakers are needed 8 hours per day on average.

If load had the same pattern every day, high prices set by peaker marginal cost and system operator demands would allow midload plants to recover their full

---

4. The author has spent considerable time checking proposed examples, one of which is posted on www.stoft.com.

$280/MWday of fixed and startup costs by bidding their marginal costs. There would be no need for complex bids in either a power exchange or a power pool.

**Example 1B: Costs Are Still Covered on Some Low-Load Days.**   Load fluctuates from day to day, sometimes spending many hours above $50/MWh and sometimes spending no time at all. On low-load days, this removes $120/MWday of fixed-cost recovery for both peakers and midload plants. The shorter duration of peaker use on such days reduces midload cost recovery still further. When peakers run only half as long as they do on average, midload plants will earn only $80/MWday of scarcity rent. This will still cover all of their startup cost ($40/MWday) and a bit of their fixed cost without any need for complex bidding. As long as scarcity rents cover startup cost and just a little more, the unit will be started, the startup paid for out of scarcity rent, and the remaining rent used to cover fixed costs. In this sense, startup costs have first claim on scarcity rents.

| | | |
|---|---|---|
| **Result** | **3-9.2** | **Marginal Cost Prices Can Solve Some Unit-Commitment Problems**<br>A market in which generators have (nonconvex) startup costs may still have a normal competitive equilibrium that produces an efficient dispatch at all times. |

**What Triggers the Startup Inefficiency?**   Only on days when there is no price spike and peakers run for less than two hours ($\frac{1}{4}$ of normal) will the midload plants be faced with the dilemma caused by startup costs. On these days they must bid above marginal cost in a power exchange, and they must bid marginal cost and startup cost separately in a power pool. (Of course they can and will bid startup costs every day in a power pool, but it does not matter on most days.) Section 3-9.2 considers how well a power exchange performs on such days.

## 3-9.2   UNIT COMMITMENT IN A POWER EXCHANGE

When California adopted a power exchange, advocates of this design suggested that suppliers would find it so simple to bid that they could "sit back and read a book" while the auction proceeded. In spite of subsequent simplification, no optimal bidding strategy was ever presented. To this day, no one has explained the competitive bidding strategy for such an auction. Example 2C shows the startling complexity of such a strategy in a trivially simple market.

Fortunately, as explained above, the inefficiencies that are likely to result from an incorrect solution to the unit-commitment problem are small, and marginal-cost pricing often solves the problem. When it does not, some rule of thumb will most likely produce a reasonable outcome, but this has never been demonstrated. A second problem also causes alarm. Some advocates of power pools see ominous signs in a power exchange's lack of a classic competitive equilibrium. They assume the worst, while advocates of power exchanges assume the best.

This section presents four elementary examples that address the main points of both positions. These analyze the behavior of a market in which load is low and, as a consequence, midload plants are unable to cover their startup-costs by bidding only marginal costs. Each example uses the cost structure of Examples 1A and 1B coupled with a specific low-load condition. The examples differ slightly in auction design and post-auction trading activity.

**Table 3-9.2** Bidding above Marginal Cost in a Power Exchange

| Example | Auction | Bidding | Outcome |
|---------|---------|---------|---------|
| 2A | Walrasian | — | No equilibrium |
| 2B | two-part | Deterministic | Efficient |
| 2C | one-part, no reneging | Random | Inefficient |
| 2D | one-part, with reneging | Random | Efficient |

Example 2A demonstrates the absence of a classic competitive equilibrium. Example 2B shows that allowing bidders to state minimum-run times gives this market an efficient, though not classic, competitive equilibrium. Examples 2C and 2D utilize the simplest standard power exchange design. Example 2C assumes that the outcome of the DA market is enforced. In this case the equilibrium is inefficient due to occasional overcommitment. Example 2D assumes that generators can renege on their DA contracts, but they must buy back the necessary power in the RT market. This results in an efficient competitive equilibrium but one with considerable complexity and randomness.

## Example 2A: Absence of a Classic Competitive Equilibrium

A classic competitive equilibrium is defined by the outcome of a "Walrasian auction."[5] An auctioneer calls out prices. If a certain price elicits more demand than supply, he tries a higher price, and vice versa. To define a competitive equilibrium, economists consider what the outcome would be if both suppliers and demanders assumed they could have no influence on the auctioneer's price. Under this assumption, if there is a price at which supply equals demand, that is the competitive equilibrium.

The assumption on trading behavior is referred to as being a "price taker." If there is a quantity, $Q$, and a (competitive) price, $P$, that satisfy the following two conditions, supply equals demand. Suppliers could not increase profits by selling a quantity different from $Q$ at price $P$. Demanders could not increase their satisfaction (utility) by buying a different quantity $Q$ at price $P$.

| Definition | **Classic Competitive Equilibrium** |
|---|---|
| | A market condition in which supply equals demand and traders are price takers. More precisely, there is some price $P$ and quantity $Q$ such that suppliers cannot increase their profit by selling a different $Q$ at the price $P$, and demanders cannot increase their utility by buying a different $Q$ at the price P. (A competitive equilibrium is often neither a Nash equilibrium nor a likely outcome.) |

5. The competitive equilibrium of economics will be called "classic" to distinguish it from efficient competitive Nash equilibria that are not Walrasian equilibria (see Example 2B).

**Figure 3-9.1**

Load profile for
Example 2 (A, B,
C, and D).

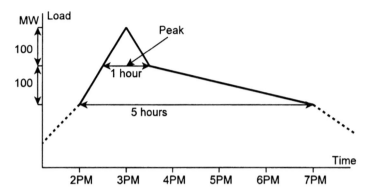

Economics predicts efficiency for markets that have a classic competitive equilibrium and price-taking suppliers and warns of possible inefficiency for markets that do not. The failure of power exchanges to have a classic competitive equilibrium is sometimes seen as a serious flaw.

**Assumptions of Example 2A.**   There are two 100-MW peakers and two 100-MW midload plants. Demand, depicted in Figure 3-9.1, is above baseload, requiring some output from a midload plant for five hours and more than 100 MW above baseload for only one hour. The lower load-slice has an average duration of three hours, and the upper load slice has an average duration of one half hour.

**Table 3-9.3** Assumptions of Example 2 (A, B, C and D)

| Type of Supplier | No. of Suppliers | Capacity (in MW) | Variable Cost (per MWh) | Startup Cost (per MWday) | Avg. Duration of Load Slice |
|---|---|---|---|---|---|
| Peaker | 2 | 100 | $50 | $0 | 0.5 h |
| Midload | 2 | 100 | $30 | $40 | 3.0 h |

To show that the market of Example 2A has no equilibrium, it is only necessary to look for it where it would necessarily be found. Because classic competitive equilibria are always efficient, if one is to be found it must correspond to the most efficient dispatch. That consists of dispatching a midload plant for the bottom load slice and a peaker for the top load slice. The difficulty with this dispatch is that both plants spend time at fractional output levels.

For the peaker to be in equilibrium at a 50-MW output level, the market price must be $50/MWh, its variable cost. Otherwise, if the price were higher, it could increase its profits by producing at full output or, if the price were lower, by producing at zero output. Similarly the midload plant requires a market price of $30/MWh to be in equilibrium with a fractional output level. The only possibility for a classic competitive equilibrium is marginal-cost pricing: a price of $30/MWh between 2 PM and 7 PM, except for the peak hour when it would be $50/MWh. Unfortunately, this allows the midload plant to recover only $20/MW for startup costs (during the peak hour), so it refuses to run. The optimal dispatch is not an equilibrium for any price path, so there is no competitive equilibrium.

| | | |
|---|---|---|
| **Result** | **3-9.3** | **A Power Exchange Lacks a Classic Competitive Equilibrium** |

If price is set equal to marginal cost at all times, as required by a competitive equilibrium, some generators that are part of the optimal dispatch may not cover their startup costs. Frequently this problem prevents a DA power exchange from having a classic competitive equilibrium.

## How a Power Pool Fixes the "Problem"

Power exchanges frequently lack a classic competitive equilibrium, but, as will be demonstrated shortly, they do not lack a **Nash equilibrium**. The classic competitive equilibrium would be *perfectly* efficient, but the Nash equilibrium might be *extremely* efficient.

Using the same example, a power pool would set the market price to variable cost at all times. This means $30/MWh between 2 PM and 7 PM, except for the peak hour when it would be $50/MWh. As noted, the midload plant loses $20/MW, which comes to $2,000. It is offered a side payment of $2,000 on the condition that it start up and "follow the dispatch," that is, ramp up and down as directed. When only part of the midload plant's output is needed, the market price equals its variable cost and so it is indifferent as to what output it produces. With this arrangement, a power pool induces the optimal dispatch in spite of the market's lack of a classic competitive equilibrium.

Notice that the power pool solves two problems: (1) how to get the midload plant to commit, and (2) how to get the midload plant to ramp up and down as directed. The ramping problem may be more serious than the unit-commitment problem, but so far neither has been assessed.

| | | |
|---|---|---|
| **Result** | **3-9.4** | **A Power Pool with Accurate Bids Induces the Optimal Dispatch** |

Power pools lack a classic competitive equilibrium. But if bidding is honest, power-pool side payments, together with marginal-cost prices, will induce the optimal dispatch. With enough competition between suppliers, bidding should be honest. (This result ignores demand elasticity and long-run incentives.)

## Example 2B: Efficiency with Two-Part Bidding.

Walrasian auctions are found in economic texts but rarely, if ever, in commerce.[6] Moreover, the concept of a **Walrasian equilibrium** used to define a classic competitive equilibrium is not used in auction theory—the Nash equilibrium concept is used. The Walrasian equilibrium concept is not intended to predict how a market will actually function, and it does not. A highly monopolistic market may have a competitive Walrasian equilibrium, but that does not indicate it will produce a competitive outcome. The Nash equilibrium *is* a prediction of a market's outcome.

This example finds the Nash equilibrium of a power exchange and shows that, while it has no classic competitive equilibrium, it does have a perfectly efficient

---

6. An RT power exchange resembles a Walrasian auction but does not wait for a market clearing price before allowing trade to take place.

Nash equilibrium. In this case, lacking a classic competitive equilibrium is not a problem.

Alberta uses a simple auction that handles the present example quite nicely. Suppliers bid an energy price and a minimum run time. Example 2B replaces the minimum run time with a minimum quantity of energy for convenience. Bids will only be accepted if at least this much energy is purchased. As with a power pool, the system operator needs a guarantee of performance. The settlement rule states that a generator will only be paid for the quantity accepted if it follows the system operator's dispatch.

The cost-minimizing set of bids is accepted. When there is a tie, the winner is chosen by the toss of a coin. The price paid at every point in time is the highest accepted bid price at that time.

With this auction design, the example market has an equilibrium, which is defined by midload plants bidding $40 and peakers bidding $50. Midload plants will specify, as part 2 of their bid, that their minimum output is 300 MWh. Because of tie bids each generator will be selected half the time, and because they just cover costs, generators will be indifferent as to whether or not they are accepted. No generator can do better by changing its bid, so this is a Nash equilibrium. No generator has exercised market power so it can be termed a competitive equilibrium although it does not fit the classic economic definition of one. The market price produced by this auction is a competitive price, though in the classic economic sense it is not a "market-clearing" price.

Example 2B shows that the lack of a classic competitive equilibrium does not imply a power pool is needed to achieve a perfectly efficient equilibrium. This example is important because it achieves the effect of a power pool without the side-payments of a power pool and without paying any attention to realistic bidding. Generators bid a false minimum energy and do not bid startup costs.

Note that this two-part-bid auction is more complex than a power exchange that uses one-part bids. The two-part bids require a more complex evaluation process, but the result is still transparent. The crucial point is that this two-part-bid auction is not a power pool because it pays all suppliers the same price at every point in time and makes no side payments. The single market price is high enough to cover the costs of all accepted bids.

---

**Result**     **3-9.5**     **A Power Exchange Can Be Efficient Without a Competitive Equilibrium**
In a market that has no competitive equilibrium, a power exchange may still have an efficient and competitive Nash equilibrium.

---

## Example 2C: Inefficiency with One-Part Bids

The question remains: How well does a pure power exchange solve the unit-commitment problem. Example 2C answers this question by computing the Nash equilibrium for a power exchange operating in the market structure under consideration.

In a pure power exchange, suppliers can only state their capacity and a price curve. If suppliers bid the same price in a pure power exchange as in the Alberta auction, the outcome would be very different. The two midload bids would win. One of these suppliers would lose the toss of the coin and be accepted for the top load slice because they cannot specify this as unacceptable.[7] This generator would be paid $40/MWh, which would cover only $5/MW out of its $40/MW startup cost ($10/MWh rent times half a megawatt-hour of output per megawatt of capacity). Without the peaker to set a high price, the generator in the bottom slice would also not recover its startup cost.

Generators will need to change strategy to suit the new auction rules. Peakers will still compete and bid $50, but the midload generators will bid randomly according to a specific set of probabilities. In game theory, a random strategy is termed a *mixed strategy* and a deterministic strategy is termed a *pure strategy*.[8] A mixed strategy specifies a probability for every possible bid that a supplier can make. The equilibrium strategies have been computed for the present auction and are presented in Figure 3-9.3. This pair of strategies forms a **Nash equilibrium** because neither supplier can increase its expected profits by changing its strategy, provided the other supplier holds fast.

Even though the two midload suppliers are identical, they use different strategies—an equilibrium characteristic that will be discussed shortly. Sixty percent of the time supplier A bids just under $50, while 60% of the time supplier B bids any amount over $50. How far over does not matter because any amount ensures that supplier B will not be selected. Supplier A is always selected. When B opts out (by bidding high), A is quite profitable, but when B bids seriously, B underbids A 80% of the time. In this case A loses money and B is profitable. All told, the two suppliers compete away all of their profits, which is not surprising because this is basically a Bertrand model (pure price competition) and such models lead to perfect competition.

Although the equilibrium is competitive, it is not efficient. Forty percent of the time both midload suppliers win. Having a midload supplier serve the top load slice costs $55/MW, while using a peaker costs only $25/MW.[9] Because this inefficiency occurs only 40% of the time, the average excess cost is $1,200 per day ($30 × 100 × 0.4).

The minimum cost of supplying the total load in this example is $2,500 for the top load slice and $13,000 for the bottom load slice, so the inefficiency is nearly 8% of wholesale cost or 3% of retail cost. This is unrealistic for several reasons,

---

7. A one-part-bid auction, though it involves only price and quantity like a Walrasian auction, works differently. The auctioneer does not call out a price and then allow trade at that price. Instead it sets a price and accepts certain bid quantities from each supplier. In this way a set of trades can be determined without a market-clearing price.

8. Mixed strategies are common in the real world, and although difficult to compute, even children learn to employ them without instruction. Any finite game without a pure strategy Nash equilibrium has a mixed strategy Nash equilibrium. This game has no satisfactory pure strategy for a midload plant, so it must have a mixed-strategy Nash equilibrium. All power auctions are finite games because bids are rounded to the penny and to the megawatt.

9. Cost per megawatt is startup cost plus an average of 0.5 MWh of energy times variable cost for each megawatt of the peak load slice.

**Figure 3-9.2**

Randomized bids from two midload plants in a power exchange.

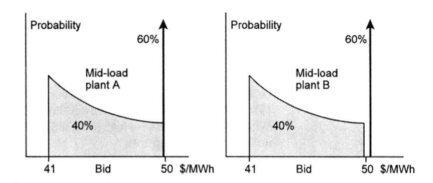

two of which are the exaggeration of startup costs and the underestimation of the number of generators. In an actual market, startup costs are closer to 1% than to 25% and the number of generators is closer to 400 than to 4. Inefficiencies due to problems with startup cost decline in proportion to the number of suppliers in the market. Correcting for these two effects could reduce the estimate to less than $\frac{1}{100}$ of 1%.

This auction has two equilibria, one in which generator B opts out by bidding high 60% of the time, and another in which the roles are reversed. This could lead to further coordination problems, but it seems likely that each will stumble into a particular role and realize it does not pay to switch. Still, when such a simple problem generates such complex behavior, it raises questions about how well a large market will coordinate.

---

**Result 3-9.6**

**A Power Exchange Has a Nearly-Efficient Nash Equilibrium**
When a power exchange has no classic competitive equilibrium, it still has a Nash equilibrium. Suppliers with a commitment problem will use randomized strategies which will cause some inefficiency. In a market with several hundred suppliers the inefficiency may be near $\frac{1}{100}$ of 1% (1 part in 10,000).

---

## Example 2D: Efficiency with One-Part Bids and Reneging

Midload suppliers in Example 2C overcommit 40% of the time. This would tend to increase reliability, but the reliability of a one-part auction cannot be properly investigated without allowing a second round in which generators can renege. To renege means to fail to start up and produce power sold in the DA market. When a generator reneges, it must buy replacement power in the RT market to cover its sales in the DA market. Example 2D extends Example 2C by allowing generators to renege.

When both midload generators are accepted, the one with the higher bid will be given the top load-slice and will fail to cover its startup cost. In this situation the generator might renege and leave the system operator with insufficient resources in real time. To address this concern, a power pool guarantees that suppliers accepted day ahead will not lose money if they produce the amount accepted.

**Figure 3-9.3**

Mixed strategy for midload generators in Example 2D.

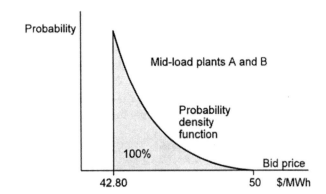

When a midload generator reneges, it will be required to buy replacement power at a cost of $50/MWh. (Competition between the peakers will hold the price down to the competitive level in the RT market.) If the midload generator has sold power day ahead for $49.99, then it will lose only $0.01/MWh by purchasing instead of generating the power it sold. This is far less than what it would lose on startup costs, so it will not commit and will not produce power. Reneging can be profitable, and some reneging must be expected.

The possibility of reneging and purchasing replacement power changes the payoffs to both players in Example 2C. Whenever payoffs change, strategies change, so the auction of Example 2D will have a different equilibrium. This has been computed, and the results are shown in Figure 3-9.4. Because it has become less costly to risk commitment, both suppliers bid under the $50/MWh peaker bids, and both are selected. This produces 100% overcommitment (double the correct number of midload plants are chosen to startup) with the result that the plant with the higher bid always reneges to avoid supplying the top load-slice.

As predicted by those critical of a power exchange, the uncertainty produced by a pure power exchange causes reneging. The net result, however, is an optimal dispatch. Reneging merely reverses the DA overcommitment, so it is difficult to view this behavior as problematic. The resulting dispatch is efficient and the suppliers exercise no market power. This outcome occurs in spite of the absence of a competitive equilibrium and suppliers being unable to submit bids that reflect their true costs.

This model is too simple to tell the whole story, which is still unknown. It demonstrates that if a power exchange uses a bid format that is too simple and to far from generation's actual cost structure, random (mixed strategy) bidding may well be the result. If it

---

### Uncertain Profits in Example 2D

Although the dispatch is the same as in Example 2B, prices and payments are different. In Example 2B, the price was $40 off peak and $50 on peak, so the dispatched midload generator exactly covered its costs. In Example 2D, random bidding causes random fluctuations in short-run profits.

For example, if one plant bids $42.80 and the other bids $43, the low bidder will lose $1.40/day per MW of capacity, while the high bidder will lose $3.50/MWday.

If one bids $49 and the other $49.50, the low bidder will make a profit of $17.50/MWday and the high bidder will lose only $0.25/MWday.

---

is, reneging will probably follow. It also shows that reneging does not necessarily cause a reliability problem. Reneging does not mean turning off a plant that has been started, and when it results from the random bids of a power exchange it may simply cause generators that were incorrectly accepted to correct the error by not starting up.

Possibly, a more realistic model would show that reneging can be problematic, but this must be at least an unusual occurrence because the conditions that cause potential reliability problems are the same conditions that make reneging very unlikely. If a generator is known to be desperately needed in real time, it is unlikely to find it unprofitable to startup.

## Conclusions

Neither power exchanges nor power pools have a classic competitive equilibrium given normal production cost structures. Both have Nash equilibria. If bidding is honest, load forecasts are perfect and demand is inelastic, a power pool's Nash equilibrium should be perfectly efficient in the short run, and a typical power exchange's Nash equilibrium may well be 99.99% efficient. This difference will undoubtedly be swamped by other effects.

If a power exchange has a significant disadvantage, it may be the difficulty of finding the Nash equilibrium. If a one-part-bid power exchange has trouble finding the equilibrium; with reneging, or with the inefficiency of the equilibrium, this can be remedied by using two- or three-part bids. This will not turn it into a power pool because it does not add side-payments. As correctly noted by advocates of power pools the computation of multipart bids adds only a very small cost.[10] Without side payments, the exchange will retain most of its transparency and the beneficial aspects of its prices which more accurately reflect costs (see Chapter 3-8 and the following section).

If a power-pool has a significant disadvantage, it may be its complexity. This imposes some small transactions costs but more importantly may open the door for gaming possibilities. Power-pool auctions are far too complex to check with game theory except with regard to their main features. They are also too complex to be checked by human intuition.

Other considerations of interest are market power, system operator control over ramping, and the effects of reneging on the efficiency of the RT market. None of these areas are currently understood.

---

## 3-9.3   INVESTMENT UNDER A POWER POOL

A power pool is designed to optimize the dispatch, provided that bids are truthful and load forecasts are accurate. If they are, it finds an optimal dispatch. Three questions have been raised as to the actual efficiency of such markets:

1. How compatible are bid formats with actual generation costs?[11]
2. How well do these markets behave in the presence of market power?
3. Do prices plus side payments send efficient long-run signals?

---

10. The main drawback of two- or three-part bids is that the auction becomes less transparent. Still, it will not develop the opacity of power pools in which it is impossible for bidders to evaluate their bids relative to prices because the prices involve side payments which are not public information.

11. See Cameron and Camton (1999, Section III.A).

**Bid Formats.**  The previous section argued that a power exchange can perform remarkably well with a bid format far from representative of physical reality; the generators bid a minimum energy production but had no such physical restriction. If such an unrealistic bid format can work so well, it seems unlikely that the more realistic bid formats of power pools could cause any problems. They might, if generators bid honestly in a literal sense. But generators will certainly compensate for small deviations from the realism of power-pool bidding formats just as they compensate for large deviations from realism in power-exchange bidding.

**Market Power.**  The question of market power remains open. Neither auction design has any obvious claim to an ability to reduce market power. The question is particularly vexing because optimal bidding in the presence of market power often requires a random bidding strategy. Such strategies are notoriously difficult to analyze. In simulations, the author has found cases in which a power pool controlled market power better than did a power exchange but in which the power pool produced a less efficient result. But the opposite situation was also found. Most likely, the choice between a power pool and a power exchange has little impact on the exercise of problematic levels of market power. What may make more difference is whether bids are hourly or daily (see Section 4-4.4). Pools are more likely to do this, but exchanges can if they use two- or three-part bids.

## Long-Run Inefficiency Due to Startup Insurance

Insurance is notorious for dampening economic signals, and side-payments, a form of startup insurance, proves no exception. Example 3 demonstrates that by insuring generators against startup costs, the power pool can encourage overinvestment in generators with high startup costs. Startup insurance prevents the generator from "feeling" the startup cost on days when the insurance is called into play.

| Result | 3-9.7 | **Side Payments in Power Pools Distort Investment in Generation** |
|---|---|---|

**Side Payments in Power Pools Distort Investment in Generation**
Power pools pay the difference between marginal-cost prices and the actual costs, including startup costs, of generators that are required for an optimal dispatch. This encourages too much investment in the types of generation that receive the greatest side payments per megawatt of capacity.

**Assumptions of Example 3.**  There are two daily load profiles, with equally high peaks as shown in Figure 3-9.4, but the second load peak is one third as wide as the first. While this profile is unrealistic, it appears to illustrate a general principle and simplifies the example. Peak-load and midload generators have the cost structures shown in Table 3-9.4. The unrealistically low fixed costs are required by this simple example to ensure that side payments are made on the low load days.

  The first step is to find the minimum-cost equilibrium. Peakers will be built up to the point where their fixed costs are covered, but the exact level at which that happens is not important. What matters is that the high prices which allow peakers to recover their fixed costs allow midload plants to recover an equal amount

**Figure 3-9.4**

Alternating daily load
profiles for Example 3.

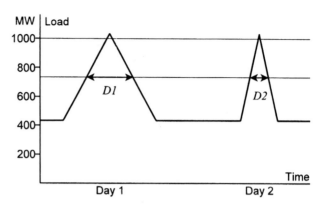

of their fixed costs. This amounts to $48/MW over the two-day load profile, $\frac{3}{4}$
earned on day 1 and $\frac{1}{4}$ on day 2. So midload plants will recover $36/MW on day 1,
and $12/MW on day 2 from prices above $40/MWh. This is two-days' worth of
price-spike revenue.

**Table 3-9.4**  Cost Assumptions of Example 3

| Type of Supply | Variable Cost (per MWh) | Startup Cost (per MWday) | Fixed Cost (per MWh)* |
|---|---|---|---|
| Peaker | $40 | $0 | $1 |
| Midload | $30 | $40 | $1 |

\* The fixed-cost units, of $/MWh, are explained in Chapter 1-3.

Next the optimal level of investment in midload plants must be determined.
This investment level will be characterized by the duration of peaker use, $D_{peaker}$.
The more midload capacity, the shorter the duration of peaker use. Given the load
profile, $D_{peaker}$ determines a particular level of investment. At this duration the
average cost of owning and using capacity, $AC_K$, of a peaker should equal the $AC_K$
of a midload plant. The definition of $AC_K$ given in Equation 1-3.1 must be modified
to include startup cost as follows:

$$\text{Average cost per MW of capacity: } AC_K = FC + D \times VC + SC$$

Equating $AC_K$ of peakers and midload plants and then substituting the example
values with $SC$ converted to $/MWh, and cancelling out the units gives:

$$FC_{peak} + D_{peaker} \times VC_{peak} + SC_{peak} = FC_{mid} + D_{peaker} \times VC_{mid} + SC_{mid}$$
$$1 + D_{peaker} \times 40 + 0 = 1 + D_{peaker} \times 30 + 40/24$$

Solve for $D_{peaker}$ and then convert this duration from a pure number to the convenient
units of hours per day by multiplying by 1 in the form of (24 hours/1 day).

$$D^*_{peaker} = 4/24 \times (24 \text{ h/day})$$
$$D^*_{peaker} = 4 \text{ h/day}.$$

This is the optimal duration because it minimizes the total cost of production
including variable, fixed, and startup costs. This average duration can be achieved

only by a duration of six hours on day 1 and 2 hours on day 2 because of the relative widths of the load profiles.

If the optimal mix of generation has been built and if the power pool sends the right long-run signals, then the economic profit of a midload plant should be zero (short-run profits cover fixed costs), so that there will be no incentive to increase or decrease its share of the mix. Actual profit can be calculated as follows using the values for $D^*_{peaker}$ just computed.

**Table 3-9.5**  Optimal Daily Profit of Midload Plants for Example 3

| Day | $D^*_{peaker}$ | Scarcity Rent | | Non-Energy Cost | | Profit |
|-----|-----------|---------------|---------|---------|---------|--------|
|     |           | P > $40       | P = $40 | SC      | FC      |        |
| 1   | 6 h       | $36           | 6 x $10 | $40     | $24     | $32    |
| 2   | 2 h       | $12           | 2 x $10 | $40     | $24     | -$32   |

Dollar figures indicate $/MWh.

Scarcity rent for midload plants comes from two sources: price-spike revenue (computed above), and prices equal to the peaker's variable cost, $40/MWh. This price only lasts for the duration $D^*_{peaker}$ and then the price falls to $30/MWh. This contribution to scarcity rent is $D^*_{peaker} \times (\$40 - VC_{mid})$/MWh. Notice that economic profit averages zero over the two days, just as it should. Notice also that on day 2, net energy revenue is only $32 so it does not cover startup costs. Consequently, midload generators receive a side payment of $8/MW on day 2. This makes midload plants profitable and encourages additional investment.

The additional investment in midload capacity will move the system away from the optimal mix of generation and increase the total cost of production. In an actual power market this effect should be small because the side payments will be small, but it may be large enough to cancel any positive effect of side payments. Without quantitative analysis little can be said about the efficiency of pool prices relative to the efficiency of power exchange prices.

# The Market for Operating Reserves

**A Telegrapher's Valentine,** by James Clerk Maxwell, 1860.

*The tendrils of my soul are twined*
*With thine, though many a mile apart.*
*And thine in close coiled circuits wind*
*Around the needle of my heart.*
*Constant as Daniel, strong as Grove.*
*Ebullient throughout its depths like Smee,*
*My heart puts forth its tide of love,*
*And all its circuits close in thee.*

*O tell me, when along the line*
*From my full heart the message flows,*
*What currents are induced in thine?*
*One click from thee will end my woes.*
*Through many a volt the weber flew,*
*And clicked this answer back to me;*
*I am thy farad staunch and true,*
*Charged to a volt with love for thee.*

$\mathbf{A}$ MARKET FOR OPERATING RESERVES PAYS GENERATORS TO BE-
HAVE DIFFERENTLY FROM HOW THE ENERGY MARKET SAYS THEY
SHOULD. If generators are cheap and will produce at full output, the market might
tell them to produce less. If they are too expensive to produce at all, it may tell them
to start "spinning," and this may require them to produce at a substantial level. Its
purpose is to increase reliability and moderate price spikes.

**Chapter Summary 3-10:** Not maximizing profits in the energy market is an
opportunity cost, and generators must be paid for this to secure their cooperation.
There are two philosophies: (1) have the system operator calculate this value from
the real-time price and pay them accordingly; (2) have the generators guess this
value and include it in their bids. The first approach may be quite susceptible to
gaming while the second is optimal in theory but risky for generators in practice
and may increase the randomness of the outcome.

**Section 1: Types of Operating Reserve.** Operating reserves come in several
qualities classified by how quickly the generator can respond. "Regulation" keeps
the system in balance minute by minute. Ten-minute spinning reserve can start
responding almost instantly and deliver its full response within ten minutes. This
type of reserve will serve as a model for considering market designs. The problem
of linking the different reserve markets is not considered.

**Section 2: Scoring by Expected Cost.** One approach to conducting a market
for spin is to have suppliers submit two-part bids, a capacity price, $CC_{bid}$, and an
energy price, $VC_{bid}$. An obvious way to evaluate such bids is to score them by their

expected cost $CC_{bid} + \hat{h} \times VC_{bid}$, where $\hat{h}$ is the fraction of the time that energy is expected to be required from a supplier of spin. Unfortunately there is no single correct $h$, because the amount of energy used depends on the supplier's energy price. Evaluating the bids by using the wrong $\hat{h}$ leads to gaming, which, depending on the structure of the auction, can be either extreme or moderate. Although a reasonably efficient auction based on expected-cost scoring seems possible, this has not yet been demonstrated.

**Section 3:  Scoring Based on Capacity Price Only.**  An alternative scoring approach calls for the same two-part bids ($CC_{bid}$, $VC_{bid}$), but evaluates them by picking those with the lowest capacity price, $CC_{bid}$.[1] Remarkably, this is sufficient, provided the bidders are exceptionally well informed.

Information requirements are problematic under capacity-bid scoring. To bid efficiently, bidders must know how the probability of being called on to provide energy will depend on their energy-cost bid, but this probability function depends on who wins the auction and what the winners have bid.

**Section 4:  Opportunity-Cost Pricing.**  Auctions in which suppliers offer both a capacity and an energy bid require that they guess what their opportunity cost will be in the real-time market. For instance, if they believe that the spot market price will be $50/MWh, and their marginal cost is $49/MWh, they may offer spin capacity for $1/MWh. If the market price turns out to be $80 and they are not called on to provide energy, they will have missed a significant opportunity. Of course with a low spot price they would have won their gamble. The problem is not with the averages but with the randomness such guesswork will introduce into the bidding process. As a remedy for this problem, suppliers can be paid their opportunity cost, whatever that turns out to be.

# 3-10.1   TYPES OF OPERATING RESERVE

Spinning reserve (spin) is the most expensive type of reserve because a generator must be operating (spinning) to provide it. Spin is typically defined as the increase in output that a generator can provide in ten minutes.[2] Steam units can typically ramp up (increase output) at a rate of 1% per minute which allows them to provide spin equal to 10% of their capacity.[3] Spin can also be provided by load that can reliably back down by a certain amount in ten minutes.

---

1. This approach was developed by Robert Wilson for the California ISO and is explained along with the problems of expected-cost bidding in Chao and Wilson (1999b).

2. Australia often defines spin as the five-minute increase in output.

3. This value can be improved by the generator owner, and markets may lead to such improvements. Some reports indicate this value may be more an economic than a physical limit.

The next lower qualities of operating reserves are ten-minute nonspinning reserves, typically provided by gas (combustion) turbines, and 30-minute nonspinning reserves. The highest quality of operating reserve is called "regulation" and is used continuously, not just for contingencies. It is bought separately and is often not included when operating reserves are discussed. This chapter will consider only spinning reserve because it is the most critical and illustrates many important design problems.

Spin can range in price from free to expensive. "Incidental spin" is provided by marginal generators whose maximum output is not required but more often by those in the process of ramping down. Sometimes generators are given credit for spin when they are ramping up at full speed to keep up with the "morning ramp." While these may meet the letter of the definition, they do not meet its spirit because they could not help to meet a contingency such as another generator dropping off line. Typically the spinning reserve requirement of a system is roughly equal to the largest loss of power that could occur due to a single line or generator failure, a "single contingency."

Providing spin from generators that would not otherwise run is costly for several reasons. Most importantly, generators usually have a minimum generation limit below which they cannot operate and remain stable. If this limit requires a generator to produce at least 60 MW, and its marginal cost is $10/MWh above the market price, it will lose $600/h if it provides any spin. If it can provide 30 MW of spin when operating at 60 MW, the spin will cost $20/MWh. In addition, there will be a **no-load cost** due to power usage by the generator that is unrelated to its output. Startup costs should also be included.

Providing spin from inframarginal generators, ones with marginal costs below the market price, is also expensive. If a cheap generator has been backed down slightly from full output, its marginal cost may be only $20/MWh while the competitive price is $30/MWh. In this case, backing it down 1 MW will save $20 of production cost but will require that an extra megawatt be produced at $30/MWh. The provided MW of spin costs $10/MWh. Sometimes it is necessary to provide spin in this manner because too little is available from marginal and extra-marginal generators. This is typically the case when the market price is above the variable cost of most generating capacity.

The three operating reserve markets are tightly linked to each other and to the energy market. California demonstrated the folly of pretending differently and managed to pay $9,999/MWh for a class of reserves lower than 30-minute nonspin at times when the highest quality reserves were selling for under $50/MWh.[4] This chapter will not consider the problem of how the markets should be linked, although the most straightforward suggestion is to clear all three simultaneously using a single set of bids that can be applied to any of the markets.

---

4. The root of this problem was a "market separation" ideology, although several peculiar rules played a role as did FERC.

## 3-10.2  SCORING BY EXPECTED COST

The expected cost of spinning reserve depends on the cost ($CC$) of the reserve capacity, the cost of the energy provided ($VC$) when reserves are called on, and the chance they will be called ($h$). The cost of providing capacity is always associated with a megawatt value of the capacity and the length of time for which it is provided, so $CC$ is measured in dollars per megawatt-hour as is $VC$. (See Chapter 1-3 for a discussion of units.)

Because the expected cost of spin is $CC + h \times VC$, it seems natural to allow two-part bidding and to score the bids using this expected-cost formula. If $h$ has a single value known to the system operator, this is a reasonable scoring procedure. But if the system operator is mistaken about $h$ and the bidders know $h$, the procedure is susceptible to a classic form of gaming.[5] Say $h$ is the correct probability but the system operator believes it is $\hat{h}$, and the bidder bids $CC_{bid}$ for capacity cost and $VC_{bid}$ for energy cost. The bid's score, $S$, the payment to a winning bid, and the net profit of a winner are shown in Table 3-10.1.

**Table 3-10.1**  A Bidder's Accounting in an Expected-Cost Auction

| | |
|---|---|
| True expected costs: | $CC + h \times VC$ |
| Bids: | $CC_{bid}, VC_{bid}$ |
| Score: | $S = CC_{bid} + \hat{h} \times VC_{bid}$ |
| Expected auction payoff: | $CC_{bid} + h \times VC_{bid}$ |
| Expected profit: | $CC_{bid} + h \times VC_{bid} - (CC + h \times VC)$ |

The lowest score wins. Say the bidder wants to achieve a score of $S$. It must choose $CC_{bid} = S - \hat{h} \times VC_{bid}$, where it is free to choose any energy bid, $VC_{bid}$. With this chosen and $CC_{bid}$ determined, expected profit will be

$$\begin{aligned}
\text{Expected profit} &= CC_{bid} + h \times VC_{bid} - (CC + h \times VC) \\
&= (S - \hat{h} \times VC_{bid}) + h \times VC_{bid} - (CC + h \times VC), \\
&= [S - (CC + h \times VC)] + [(h - \hat{h}) \times VC_{bid}]
\end{aligned}$$

$S - (CC + h \times VC)$ is unaffected by the bidder's choice of $VC_{bid}$, so profit is controlled by the term $(h - \hat{h}) \times VC_{bid}$. The bidder can first choose any score, $S$, and then achieve any profit level by choosing the correct $VC_{bid}$! The choice of $VC_{bid}$ will determine the bidder's choice for $CC_{bid}$, as described, but together $VC_{bid}$ and $CC_{bid}$ will produce any desired level of profit and any desired score. This depends on the bidder knowing $h$, and the system operator choosing $\hat{h} \neq h$. If $\hat{h} < h$, then the bidder should bid an extremely high cost for energy and a low cost for capacity. If $\hat{h} > h$, the bidder should bid an extremely low (negative) $VC_{bid}$. (California's 1993 series of Biennial Resource Planning Update auction contained this flaw, and the winning bids proposed huge up-front capacity payments and negative energy

---

5. The discussion of this point is based on Chao and Wilson (1999b).

payments.[6]) As long as the system operator gets $\hat{h}$ wrong and the bidder knows it, the bidder can achieve any score and any level of profit simultaneously.

While this would seem to make the expected-cost auction entirely useless, a closer look is required. The probability, $h$, of selling energy from the accepted reserve capacity depends on the energy price of the accepted bid. Suppose the system operator uses an $\hat{h}$ that is intermediate. Those who believe their $h$ will be lower, will enter very low energy bids. This will make their probability of use, $h$, higher than $\hat{h}$, in contradiction of their assumption, and their strategy will prove ineffective. This self-limiting effect does not eliminate the problems of gaming but provides some hope for expected-cost auction design.

## 3-10.3   SCORING BASED ON THE CAPACITY BID ONLY

The best theoretical work on spinning-reserve auctions (Chao and Wilson 1999b) suggests the energy-price part of the bid should be ignored when accepting bids, and winning bids should be those with the lowest capacity-cost values regardless of their energy-price bids. This surprising result is based on the observation that suppliers of spin with a low cost of energy will profit from the spin energy market when more expensive spinning reserve is called. This gives them a strong incentive to be selected in the spin auction and they will submit a capacity bid, $CC_{bid}$, lower than their capacity cost, $CC$, in order to win the auction. In this way, the relatively low cost of their energy bid, $VC_{bid}$, is indirectly accounted for when the auctioneer evaluates only their capacity bids.

An example will make clear that precise mathematics lies behind this qualitative argument. Consider two types of suppliers, the second of which has the higher energy price. Assume that all suppliers bid competitively, and that enough spin is needed so that some spin must be bought from each type.[7] If capacity scoring works, the cheaper type will win and all of them will be selected before any of the more expensive types is taken. If capacity scoring fails, it will do the reverse.

In this example, when spin is called on to supply energy, all available spin is called into service, and the more expensive energy supplier will set the market price for energy. The costs and bids of these two types are indicated in Table 3-10.2.

**Table 3-10.2** Capacity Scoring Example 1, Costs and Bids

| Supplier Type | Expected Cost | Capacity Bid, $CC_{bid}$ | Energy Bid, $VC_{bid}$ |
|---|---|---|---|
| 1 | $CC_1 + h \times VC_1$ | $CC_1 - h \times (VC_2 - VC_1)$ | $VC_1$ |
| 2 | $CC_2 + h \times VC_2$ | $CC_2$ | $VC_2 \ (> VC_1)$ |

A key advantage of capacity scoring is that it makes the energy part of the bid irrelevant for winning acceptance as spinning reserve. Competitive bidders will consider only the spin energy market when they set their energy bids and so will

---

6. This was anticipated by Bushnell and Oren (1994) and described by Gribik (1995).

7. Although cheaper suppliers could bid above the competitive level, this chapter is only concerned with the efficiency of spinning reserve auctions under *competitive* assumptions.

bid their true variable costs (in a competitive market). This is reflected in Table 3-10.2.

Since competitive bids are always break-even, both bidders subtract from their capacity bids any profit they expect to make in the spin energy market. To accomplish this, they must know what energy bids will set the spin energy market price. In this case the problem is simple. Some bidders of each type will win, so $VC_2$ will set the market price. Type 1 bidders will earn $h \times (VC_2 - VC_1)$ from this relatively high price so they subtract this from their capacity cost when bidding. Type 2 bidders make no money in the spin energy market and so bid their true capacity costs.

If this auction is efficient, the type of supplier that can provide power most cheaply must win the auction. The winner is selected purely on the basis of its capacity bid, so type 1 will win if and only if

$$CC_1 - h \times (VC_2 - VC_1) < CC_2$$

Type 1 can provide spin more cheaply if and only if its cost is less:

$$CC_1 + h \times VC_1 < CC_2 + h \times VC_2.$$

These conditions are algebraically equivalent, so the cheaper supplier will always win the auction (see Chao and Wilson (1999b) for a rigorous treatment). The bidders have adjusted their capacity bids to take account of their difference in marginal cost, so the system operator need only evaluate the capacity part of the bid.

---

**Result      3-10.1**        **Capacity-Bid Scoring for Spinning Reserves Is Optimal**
With perfect information, capacity-bid scoring is efficient. Suppliers should bid a capacity cost and an energy cost. Bidders with the lowest capacity bids should be selected to supply spin, and from these, the ones with the lowest energy bids should be selected to run when energy is needed. Accepted bids are paid the market-clearing capacity price and energy price.

---

## Inframarginal Reserve Capacity

A second example will help demonstrate the robustness of the Chao-Wilson efficiency result. Spinning reserves can be provided by generators that back down from their maximum output even though their cost of production is less than the average price of spot energy, $P$. By producing at only 90% of capacity, a steam unit can convert the last 10% of its capacity into spinning reserves. When extra-marginal sources of spin are too expensive, such inframarginal sources are preferred. When reserves are in short supply, extra-marginal sources prove insufficient, and inframarginal sources are required.

Redefine the type 1 supplier in the previous example to be an inframarginal steam unit. This changes both the cost of using this generator and the generator's bidding strategy. First consider its strategy. Instead of comparing its profit in the spin market to zero profit, the inframarginal generator compares it with its profit in the spot energy market. If accepted in the spin auction, it would be paid its capacity bid, $CC_{bid-1}$, plus $(VC_2 - VC_1)$ with probability $h$. In the spot market it would earn its normal scarcity rent at all times. These profits are shown in Table 3-10.3. Competitive type-1 generators will bid to break even in the two markets thereby ensuring they will at least cover their opportunity cost. To find their bid, $CC_{bid-1}$, equate the profits in the two markets and solve for $CC_{bid-1}$.

**Table 3-10.3** Equating Profits Determines the Capacity Bid Price

| Type | Spin-Market Profit | Spot-Market Profit |
|------|--------------------|--------------------|
| 1 | $CC_{bid-1} + h \times (VC_2 - VC_1)$ | $P - VC_1$ |

The expected *cost* of a type-1 generator providing spin also must be found in order to determine the efficiency of this auction. This time there is no physical capacity cost, $CC_1$, but there is a cost to replacing the energy that would have been provided by this inframarginal generator had it not withdrawn some of its output to provide spin. The cost of providing a small amount of additional energy is the spot price, $P$, so the net replacement cost is $(P - VC_1)$. This is the actual increase in production cost in a competitive market, so it is the real cost, not the opportunity cost, of providing spinning reserve with an inframarginal generator. In addition, there is the possibility, $h$, that spin energy will be needed; this increases the production cost of type-1 spin by $h \times VC_1$. This total as well as the value of $CC_{bid-1}$ found by equating the two profit levels in Table 3-10.3 are shown in Table 3-10.4 for the type-1 generators. The entries for the type-2 generators remain unchanged from the previous example.

**Table 3-10.4** Expected Costs and Capacity Bids

| Type | Expected Cost | Capacity Bid, $CC_{bid-1}$ |
|------|---------------|----------------------------|
| 1 | $(P - VC_1) + h \times VC_1$ | $(P - VC_1) - h \times (VC_2 - VC_1)$ |
| 2 | $CC_2 + h \times VC_2$ | $CC_2$ |

This auction is efficient if the type of supplier capable of providing spin most cheaply wins the auction. The winner is selected on the basis of its capacity bid, $CC_{bid}$, so type 1 will win if and only if:

$$(P - VC_1) - h \times (VC_2 - VC_1) < CC_2$$

Type 1 can provide spin more cheaply if and only if

$$(p - VC_1) + h \times VC_1 < CC_2 + h \times VC_2$$

These are algebraically equivalent, so the auction is efficient.

## The Information Burden of Capacity-Bid Scoring

In the standard theory of competitive markets, it is optimal for suppliers to bid their marginal cost curves. This information is readily available. Knowing one's own costs is merely a matter of good accounting and requires no theory of how the market works and no data on other suppliers and market demand conditions. Optimal capacity-only bids cannot be based simply on information about one's own costs. They require knowledge of $h(VC_{bid})$, the probability of being called to provide energy as a function of the energy bid, $VC_{bid}$. The bidder must use this function to compute how much energy it will be called on to provide and the distribution of prices when it is called. This crucial function is determined by the outcome of the auction. If many low bids for $VC_{bid}$ are accepted, $h(VC_{bid})$ will be low, the profit from supplying energy will be low, and the bidder will need to submit a higher capacity bid.

In a very stable market, learning $h(VC_{bid})$ would be relatively easy. Bidders would observe yesterday's $h(VC_{bid})$, base their bids on it, and that would determine a new $h(VC_{bid})$ the next day. This would be observed, and, as the process repeated, $h(VC_{bid})$ would most likely converge to some stable function which would be known to all. But the conditions in the spinning reserve market vary with weather, time of year, and the state of repair of generators. The market is also unpredictable from year to year due to load growth, rainfall, and new investment in generation. Consequently, the function $h(VC_{bid})$ is likely to be difficult to predict. If not correctly predicted, suppliers will bid incorrectly and the dispatch will be inefficient. Because the bidding problem is so complex, larger suppliers will probably have an advantage unless a market develops to supply estimates of $h(VC_{bid})$. In this case, part of the inefficiency will become visible in the form of revenues to the suppliers of information, but presumably, total inefficiency would decrease.

Many proposals for "simple" auctions prove their efficiency by assuming that bidders have perfect information and unlimited computational abilities. Sometimes this assumption is a reasonable approximation of reality. As a first step toward checking this approximation, those who propose an auction mechanism should be required to provide an explicit method by which bidders could bid optimally, or at least very efficiently, in a competitive market. If the auction designer cannot demonstrate how a bidder could compute its bid under the simplifying assumption of perfect competition, then market participants may find the information and computational requirements of the design overly burdensome. This may lead to inaccurate and conservative bidding.

## 3-10.4   OPPORTUNITY-COST PRICING

One complaint against paying the clearing bid price for spinning reserve capacity is that those selected to provide reserves often find themselves losing money and tempted to cheat. For instance, if the expected spot price is $50, a generator with a marginal cost of $49 might offer to provide reserves for $1/MWh and be accepted. If the real-time price then turns out to be $80/MWh, and no reserves are called on, the generator will perceive that it is losing $30/MWh relative to what it could be making if it were not providing reserves.

Although this situation will lead generators to wish they could cheat, it is not a difficult matter to check on them. The capacity of a generator, which is easily verified, together with its ramp rate change little from day to day. As long as a 200-MW generator is paid for no more than 180 MW of output while it is supposedly providing 20 MW of spin capacity, it cannot cheat.

In spite of this, there may be good reasons for helping generators to avoid such uncertainty. If a generator can base its bids on information about its own costs instead of on estimates of future market outcomes (in a very volatile market), it seems reasonable to assume that bidding will be more accurate, at least in a competitive market. This is the philosophy behind the approach to spinning reserve markets used in NYISO.

## Steps Toward an Opportunity-Cost Pricing Design

In a spin market based on opportunity costs, generators who are required to spin are paid the short-run profit they would make if they were producing energy. This is calculated based on their marginal-cost bid and the market price. For example, suppose inframarginal generators have variable costs ranging from $30 to $40/MWh, and the expected spot price is $40/MWh. In a competitive market, these would bid their true variable costs and a $CC_{bid}$ of zero. Suppose several extra-marginal generators bid their true capacity cost of $8/MWh. The system operators uses the expected spot price to impute opportunity costs in the range of zero to $10/MWh to the inframarginal generators. The imputed opportunity costs are then added to their capacity bids.

If some, but not all, of the extra-marginal generators are needed, they set the clearing price for capacity at $8/MWh. All inframarginal generators with variable costs above $32/MWh are accepted, and all are paid $8/MWh provided the spot price is $40/MWh. If the market price proves to be $50/MWh, then they are all paid a capacity price of $18/MWh. Because they are paid a market clearing price and treated as if they had been the most expensive accepted bid, they have no reason to regret their bids and they will bid honestly.[8]

This market, based on opportunity costs, uses a capacity-bid design. This is unusual. Opportunity-cost markets tend to be unit-commitment markets with multi-part bids and optimization programs. In this setting it is not standard to "score" the bids by a simple rule, but in effect an optimization program is just a complex scoring rule. As a result, there are winners and losers. To decide if a bid with a high marginal cost is nonetheless a winner in the spin auction, the program assigns some probability to the chance of the supplier being called on to provide energy—it assigns an estimate of $h$. Such markets are closely related to the two-part-bid-evaluation, expected-cost designs described in Section 3-10.2.

Unfortunately there is little if any literature on the gaming possibilities in this kind of market. Section 3-10.2 suggested they may be serious, but paying opportunity costs gives generators with market power in the ancillary service market a reason to lower their marginal-cost bids. Since their marginal-cost bids are used as well in the energy auctions, this reason to bid low should partly cancel a tendency to exercise market power in the energy market.

---

8. This point was suggested by David Mead in a personal communication, September 18, 2001.

*Part  4*
# Market Power

# Defining Market Power

*Sixty minutes of thinking of any kind is bound to lead to confusion and unhappiness.*

James Thurber
(1894-1961)

Market power, a central topic in economics, has been defined carefully. The standard economic definition is a central concept of industrial organization. Market power is the ability to alter profitably prices away from competitive levels. This definition, with slight variations, has probably been in use for more than a hundred years and is supported by a large body of empirical and theoretical work. It is terse and carefully worded as a good technical definition should be. Frequently, regulators ignore it and attempt their own definition. FERC announced a new one in its report, State of the Markets 2000.

> *Market power is defined as the ability to withhold capacity or services, to foreclose input markets, or to raise rival firms' costs in order to increase prices to consumers on a sustained basis without related increases in cost or value.* (FERC 2000a)

By the end of 2001 it had been discarded and a new one was under design.

Market power is a three step process: (1) an exercise, (2) an effect on price and quantity, and (3) an impact on market participants. The first step can take on many forms and appearances. For clarity, the economic definition considers only price (step 2) and profit (step 3). FERC's definition focuses on step 1 and misses some methods of exercise such as raising the offer price of a supply bid. Because it omits "profitably," all baseload plants would have market power whenever they are needed even though they would lose money if they exercised it. It ignores the concept of *competitive price*. It adds the clause "without related increases in cost"—any time a supplier exercises market power it is "trying to recover" some "related increase in cost." The notion of "sustained basis" may exclude peakers and is vague. No authority is cited for this definition.

Both the FERC definition just cited and the revisions to it under consideration in January 2002 include the concept of raising competitors' costs. In a nontechnical,

Internet posting, "Maintaining or Creating a Monopoly" (FTC, 2000), the FTC discusses this possibility as follows.

> *While it is not illegal to have a monopoly position in a market, the antitrust laws make it unlawful to maintain or attempt to create a monopoly through tactics that either unreasonably exclude firms from the market or significantly impair their ability to compete.*

In other words it is unlawful to raise a rival's cost, but it is lawful, under anti-trust law, to exercise monopoly power. (The FTC and DOJ define market power in essentially the same way as economics does.) FERC combines the concept of raising competitor's costs—shifting competitive supply curves up—with the concept of market power—moving the market away from these supply curves. This cuts against the grain of both the economic and anti-trust distinctions and can only create confusion and make communication more difficult. As Paul Joskow noted, FERC effectively had no definition of market power in 2001.[1]

**Chapter Summary 4-1:** Market power it usually exercised by asking a higher price than marginal cost or by withholding output that could be produced profitably at the market price. When either strategy is successful, the result is a higher market price, higher profits, and withheld output. These are two different paths to the same result.

**Section 1: Defining Market Power.** Market power is the ability to profit by moving the market price away from the competitive level. According to economics, any ability to do this, no matter how fleeting or minimal, is still market power. Most firms have some market power and this causes no significant problems, provided the amount is small.

**Section 2: Monopoly Power in a Power Auction.** Power auctions typically set the market price at the price of the last bid accepted. In some circumstances, this rule keeps the market price below the competitive price. In this case, a supplier may profitably raise the market price toward the competitive price, an action sometimes misdiagnosed as the exercise of market power.

**Section 3: Demand-Side Market Power.** Monopsony power is market power exercised on the demand side with the intention of lowering the market price. An ISO can exercise it by interrupting load at a cost that is greater than the market price or by curtailing exports. Monopsony power can be an effective and beneficial method of combating monopoly power, but it can also be abused.

---

1. Joskow (2001b, 7) has described to the U.S. Senate the shortcomings of FERC's market-power definition as follows : "FERC does not appear to have a clear definition of market power, has not identified the empirical indicia it will use to measure the presence and extent of market power, . . . [and] , has not defined how much market power is too much market power to satisfy its obligations to ensure that wholesale electricity prices are just and reasonable, . . ."

## 4-1.1   DEFINING MARKET POWER

Market power is the ability to affect the market price even a little and even for a few minutes. This definition may seem harsh, but it is not. It is simply a definition without punitive implications. Two more qualifications are needed to complete the formal definition: The effect must be profitable, and the price must be moved away from the competitive level. In economics (and under most U.S. laws), market power is not viewed as antisocial behavior; it is simply a rational form of market behavior that usually leads to inefficient outcomes. To analyze such behavior scientifically, it needs a nonsubjective definition. The one provided by economics has proven useful partly because it allows a clear-cut, noncontroversial determination once the facts are known. It also defines a concept that has proven extremely useful in theory and application.

| | |
|---|---|
| **Definition**<br>(economic) | **Market Power**<br>"The ability to alter profitably prices away from competitive levels." (Mas-Collel et al. 1995, 383) |

| | |
|---|---|
| **Definition**<br>(regulatory) | **Market Power**<br>"Market power to a seller is the ability profitably to maintain prices above competitive levels for a significant period of time." (DOJ, 1997) |

---

### Having vs. Exercising

Regulators carefully distinguish between having and exercising market power in order not to pre-judge market participants. Long-run considerations sometimes make this a useful distinction. Without such considerations there is little difference.

The fundamental assumption of economics is *rational* economic behavior. Because market power is profitable when exercised, according to both economics and DOJ, the only *rational* course for a participant with market power is to exercise it. Within the economic paradigm, all available market power will be exercised.

The profits referred to in the market power definition are usually taken to be short-run profits; hence it is common to speak of not exercising market power for fear of long-run consequences (see Chapter 4-2).

---

In their Horizontal Merger Guidelines, the U.S. Department of Justice (DOJ) and the Federal Trade Commission (FTC) have redefined the term "market power." Their definition rules out cases that economics considers to be market power but that regulators consider to be insignificant. This narrows the definition and makes it vague. The regulatory definition will be discussed in Chapter 4-6, while the economic definition is used throughout this book.

**Why Market Power Must Be Profitable.** Both definitions require that the exercise of market power be profitable. Although this is not universally accepted, it is a useful standard. Without this requirement any supplier owning a nuclear power plant would have a very substantial amount of market power even when exercising it would be hugely unprofitable. A plant shut down for any reason would constitute an exercise of market power unless the definition were elaborated to include motive. Although the requirement of profitability does not eliminate every mis-identification of perverse behavior, it is helpful.

More importantly, the profitability requirement is included to aid in theoretical analysis. Suppose one wants to compute the price increase that would be caused by supplier Y's full exercise of market power. Without the profitability requirement, that would mean checking the impact on market price of supplier Y shutting down completely, for that is what would have the greatest impact on price. But perhaps supplier Y can profitably raise the price in only 10% of the hours and then only if it withholds no more than its small gas turbine. This result predicts vastly less impact on price and will be a reasonable estimate of what might happen instead of an implausible, worst-case scenario. To conduct a theoretical discussion of market power using a definition that does not include profitability, one must constantly modify it by saying "profitable exercise of market power." So, for convenience, "profitable" is included in the definition.

## 4-1.2   DEFINING PRICE-QUANTITY OUTCOMES

**Monopoly power** is the market power of sellers, who want higher prices, as opposed to buyers, who want lower prices. Because market power must move price *away from* the competitive equilibrium, the result is always a price that is higher than the competitive price.

The principle method of exercising monopoly power is termed withholding of output, though very often this takes the form of "financial withholding" which simply means charging more than buyers are willing to pay. However, before investigating the complex strategies used to exercise market power, it is best to analyze the effects of market power on price and quantity. To this end consider an extremely simple case in which one supplier with only baseload plants shuts down one plant, thereby withholding the plants output as shown in Figure 4-1.1.

The result is to shift the competitive supply curve to the left by the quantity withheld and to raise the price from $P^*$ to $P^e$, the actual price in this market. (Once shifted, it is no longer the competitive supply curve.) Depending on the amount of generation withheld and the amount of generation remaining in the market, this strategy might or might not be profitable, but for the sake of discussion, assume it is. Three prices and three quantities are of interest, two of each are defined by the competitive equilibrium ($P^*$, $Q^*$) and the actual, monopolistic equilibrium ($P^e$, $Q^e$). The other price and quantity are defined by the competitive supply curve and the actual price-quantity pair as follows.

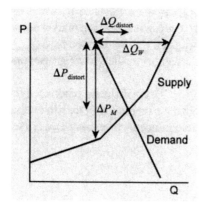

The competitive supply curve can be represented as a supply function $Q^*(p)$ or as an inverse supply function $P^*(q)$. When the supply function is applied to the monopolistic equilibrium price a new quantity is found which is denoted simply by $Q^*(P^e)$. When the inverse supply function is applied to the monopolistic equilibrium quantity a new price is found which is denoted $P^*(Q^e)$. Both values can be read from the competitive supply curve but in different directions as shown in Figure 4-1.1.

**Figure 4-1.1**

The basic strategy of withholding and the price-quantity outcome.

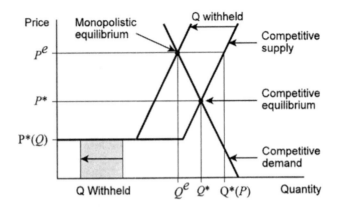

These prices and quantities determine two price differences and two quantity differences that are essential to the understanding of market power. The first is the **quantity withheld**, $\Delta Q_w$, which equals $Q^*(P^e) - Q^e$. This equals the capacity of the generator that was shut down and "withheld" from the market, hence its name. It is particularly important to note that the capacity withheld is <u>not</u> equal to $Q^* - Q^e$. While this is clear in the present example, cases of financial withholding are not as obvious and there is an almost irresistible tendency in power market analysis to misdefine the quantity withheld. The consequence is that $\Delta Q_w$ is overlooked and, $Q^* - Q^e$ which is actually the **quantity distortion** is given the wrong name (Mas-Collel et al. 1995, 385). Notice that the quantity distortion is considerably smaller than the capacity of the generator that is withheld.

---

**Definition**

**Quantity Withheld, $\Delta Q_w$**

Quantity withheld equals $Q^*(P^e) - Q^e$, which is the gap between the total amount that would be produced by competitive suppliers at the monopolistic price, $P^e$, and the amount that actually is produced in the monopolistic equilibrium.

(If the marginal cost curve is flat at the market price, the *lowest* competitive supply at that price [lowest $Q^*(P^e)$] should be used.)

---

The monopoly quantity distortion, $Q^* - Q^e$, is the difference between the competitive output and the actual output. This is the amount by which market power has distorted the competitive equilibrium quantity and will be denoted by $\Delta Q_{distort}$. This quantity is important in determining dead-weight welfare loss (see Section 4-2.4).

The price analog to quantity distortion is $P^e - P^*$, the difference between the actual price and the competitive price. This is called the monopoly **price distortion**, denoted by $\Delta P_{distort}$, and can be used as a reliable indicator of market power (see Section 4-5.2).[2]

---

2. "The monopoly situation for which we observe strong price distortions correspond to those in which demand elasticity is low, . . ." See Tirole (1997, 67).

| | |
|---|---|
| **Definitions** | **Monopoly quantity distortion, $\Delta Q_{distort}$**<br>The decrease in output, $Q^* - Q^e$, below its competitive level caused by the exercise of monopoly power.<br>**Monopoly price distortion, $\Delta P_{distort}$**<br>The increase in price, $P^e - P^*$, above its competitive level caused by the exercise of monopoly power. |

The second price difference is the **markup** over marginal cost, $P^e - P^*(Q^e)$, and is denoted by $\Delta P_M$. Note that $P^*(Q^e)$ is not the competitive marginal cost, $P^*$, but the marginal cost that would result from suppliers producing the monopoly output in a competitive manner. This price difference is completely analogous to the quantity withheld. The markup divided by actual price, $P^e$, is called the Lerner index and is discussed in Section 4-3.1.[3]

| | |
|---|---|
| **Definition** | **The Markup, $\Delta P_M$**<br>Markup equals $P^*(Q) - P^e$, which is the gap between the marginal cost of competitive suppliers supplying market quantity $Q^e$, and the actually price in the monopolistic equilibrium. |

# 4-1.3   THREE STAGES OF MARKET POWER

The market process related to market power will be more clearly understood if conceptualized as occurring in three stages: (1) exercise, (2) price-quantity outcome, and (3) consequences. The four differences ($\Delta Q_W$ etc.) provide a complete description of the price-quantity outcome but tell little about stages 1 and 3. Confusion between stages 1 and 2 occurs because the term "withholding" is used to describe strategies, and the "quantity withheld" is used to describe an outcome of the strategy. Confusion between stages 2 and 3 occurs because the Lerner index, $\Delta P_M/P$, is treated as if it were a stage-3 social consequence, while it is primarily a stage-2 quantitative diagnostic.

## Stage 1: Strategic Exercise of Market Power.

Aberrant definitions of market power, such as the one quoted in this Chapter's introduction, focus on the strategies of exercising market power—"withhold capacity or services . . ." While describing the strategies is informative, the range of possible strategies and the need to qualify them by outcome and consequences leads to complex and imprecise definitions. The standard economic definition ignores stage 1 entirely and uses only price distortion, $\Delta P_{distort}$, from stage 2, and profitability, from stage 3, to define market power.

---

3. See Tirole (1997, p. 66) for definitions of relative markup and the Lerner index.

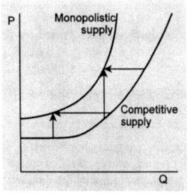

Was the shift leftward or upward?

The two components of market-power strategy are **quantity with-holding** (reducing output) and **financial withholding** (raising the price of output). In most cases, these are equivalent strategies. For example, it is impossible to tell whether a monopolistic shift of a linear supply curve is a leftward shift to reduce output or an upward shift to increase price. Because a monopolist is free to shift different parts of its supply curve by different amounts, it is impossible to distinguish price increases from quantity reductions for any upward sloping supply curve.

The ambiguity of the distinction between price raising and quantity reducing strategies is well recognized by economics. Mas-Collel et al. (1995, 385) first describe a monopolist in terms of setting a high price, then note that "An equivalent formulation in terms of quantity choices can be derived . . . We shall focus our analysis on this quantity formulation of the monopolist's problem." This does not mean that the two strategies will have the same appearance, but that they will have the same outcome and consequences. In stage 1, they look quite distinct, but in stages 2 and 3 they are indistinguishable.

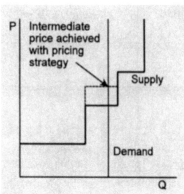

There is one small exception to this conclusion. Because both demand and supply curves may be vertical near the equilibrium, a supplier sometimes can control price more completely with a pricing strategy than with a quantity strategy. By using a withholding strategy, it may be possible only to raise price not at all or to the next highest bid, while a pricing strategy may make it possible to achieve any intermediate value. The only use for such intermediate prices would be to confuse the market monitors as such strategies are always less profitable than pushing the market price to the price of the next highest bid (or the price cap if there is no next highest bid). Mas-Collel et al. can ignore this distinction because pricing strategies cannot ever be more profitable than the best quantity-withholding strategy.

Appearances are important. A strategy of bidding \$40/MWh instead of \$35/MWh may make recognizing or proving the exercise of market power much more difficult than a strategy of withholding half of the generator's output. So even though the same price, quantity and profit are achieved with both strategies, the regulator and market monitor need to be keenly aware of these differences in appearance. But because the definition of market power does not involve stage-1 strategies, these appearances will play a very small role in proving market power or assessing its impact.

| Definition | **Strategy of Withholding** |
| --- | --- |
| | Producing less than would be profitable, assuming all output could be sold at the market price. Not acting as a price taker. This strategy may be executed "financially" by bidding high or "physically" by curtailing output. |

## Stage 2: Price-Quantity Outcomes

Price-quantity outcomes are usually well described by the four price differences discussed in the previous section, but one semantic confusion should be mentioned. Market-power strategies are often described by the verb "withholding." This action can take the form of a high bid, shutting down a plant, or curtailing its output. These actions should not be confused with the quantity difference, $\Delta Q_W$, which is described by the noun phrase "withheld." Sometimes the two are numerically equal, as in the case of the example of Section 4-1.1. When withholding is financial, they are not easily compared.

Again power markets provide a small exception. There are cases in which the stage-2 outcome cannot be fully described by the resulting price and quantity because some generator is curtailed and paid a price different from the market price. This peculiarity relates to the nonconvex costs of generation and the details of the auction design. An example is discussed in Section 4-1.5.

## Stage 3: Social Consequences of Market Power.

**Profit.**  The social consequences of the exercise of market power are fully determined by the stage-2 price-quantity outcomes, so these play the useful role of shielding the stage-3 analysis from the complexities of market power strategies. The first consequence is profit, not just for the exerciser of market power, but for all suppliers. In fact market power may be more profitable for those who do not exercise it because the exercise costs can be significant and the resulting increase in the market price affects all suppliers in proportion to their output. The supplier that exercises the market power gets no special advantage from being the one to do so.

**Wealth Transfer.**  The second consequence is the transfer of wealth from consumers to producers. This is simply the price distortion, $\Delta P_{distort}$, times the total market quantity.

**Dead-weight Welfare Loss.**  The third consequence is called the "deadweight" welfare loss and is the inefficiency resulting from monopoly power. This and wealth transfer are discussed in  Section 4-2.3.

---

## 4-1.4    USING PRICE-QUANTITY OUTCOMES TO SHOW MARKET POWER

Market power is not defined in terms of strategies and proof of its existence should not be attempted by analyzing strategies directly. Rather their consequences for market outcomes and consequences should be analyzed. This section will ignore the consequence of profitability (which will be assumed) to focus on the role of the price-quantity outcome on determining market power.

Generally, if any of the four price-quantity differences ($\Delta P_M$, $\Delta Q_W$, $\Delta P_{distort}$, and $\Delta Q_{distort}$) are positive, market power has been exercised and if market power has been exercised, all are positive. There are two exceptions. If the supply curve is vertical, market power can be exercised without causing quantity distortion, and if the demand curve is horizontal, a positive markup does not indicate that market power has been exercised. These relationships are summarized as follows:

**Table 4-1.1** Price-Quantity Outcomes and the Exercise of Market Power*

| | | | | |
|---|---|---|---|---|
| $\Delta Q_W > 0$ $\Longleftrightarrow$ market power | | | $\Delta P_{distort} > 0$ $\Longleftrightarrow$ market power | |
| $\Delta P_M > 0$ $\Longleftarrow$ market power | | | $\Delta Q_{distort} > 0$ $\Longrightarrow$ market power | |

\* Profitability is assumed.

The two most fundamental connections with market power involve *price distortion*, which is part of the definition of market power, and *quantity withheld*. In a power market, there is no chance of the demand curve being horizontal, so a positive markup or Lerner index proves market power, provided the profitability condition is met. Quantity distortion is most weakly linked to market power. When it does not occur there will be no dead-weight loss, but market power can still exist and transfer wealth.

**Constant-Output Withholding**

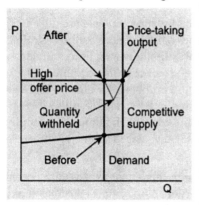

As the figure at the left shows, market power can be exercised even though the *quantity distortion* is zero. In this case there is still a positive *quantity withheld* because some supplier is not acting as a competitive supplier but is producing less than the competitive output at the market price. The amount less is the quantity withheld.

The quantity withheld is also a crucial indicator of the importance of an exercise of market power. If a supplier raises its offer on the last 1% of output by $1,000/MWh, and this sets the price, it will look like an enormous exercise of market power to the regulator that ignores the quantity withheld. But when that quantity difference is examined and found to be 1%. The astute regulator will realize that it cannot estimate generating capacity that accurately and has no case. The marginal cost of emergency output really could be $1,000/MWh and the regulator cannot know the location of the emergency operating range within 1%. If however the gap were found to be 500 MW, there would be no mistaking the exercise of market power.

In power markets, supply curves are often vertical or nearly vertical. In this case marginal cost can be very difficult to pin down, so markup is hard to measure. This is what lies behind the case of 1% withholding. Marginal cost actually change extremely rapidly and is very difficult to estimate for the last 2% or 3% of output, so in this region, the quantity withheld, $Q^*(P^e) - Q^e$, can be more accurately measured than marginal cost. (In PJM, generators report 3% of total capacity as being in their **emergency operating range**.) Also, with vertical supply curves, the market price is often depressed below the competitive price by auction rules, and this causes confusion when computing the markup. Again the quantity withheld is a more reliable indicator of market power.

**Why Monopoly Power Must Involve Quantity Withheld.**   Competition is defined by price-taking behavior which means producing as if the market price could not be affected. Market power cannot be exercised while acting as a price taker, that is, being a perfect competitor. To exercise market power, a supplier must produce less than the competitive (price taking) output level (producing more would lower the price and reduce profit). The gap between this lower amount and the price-taking output at the market price is the quantity withheld. Thus, if monopoly power has been exercised, $\Delta Q_W$ must be positive. This confirms the result reported in Table 4-1.1.

| | | |
|---|---|---|
| **Result** | **4-1.1** | **Monopoly Power Always Causes the Quantity Withheld to be Positive** |

**Monopoly Power Always Causes the Quantity Withheld to be Positive**
The exercise of monopoly power causes actual output to be less than the competitive output at the market price. Similarly market price will be higher than the competitive price.

**The Competitive Equilibrium.**   The definition of market power refers to altering prices *away* from the competitive price level, which is the price determined by a competitive equilibrium. (See Section 3-8.2.) "Altering prices away" is used instead of "raising price" to include the exercise of market power by the demand side, monopsony power, which means reducing price below the competitive level.

**Definition**           **Competitive Equilibrium**
A market condition in which supply equals demand and traders are price takers.

If price had been reduced by monopsony power, and the supply side brought price back up toward the optimal (competitive) level, that would not cause a problem. Certainly it would take some "muscle" to raise the price in opposition to the demand-side's exercise of market power. Economists have found it more useful to define market power, not just in terms of the ability to change the market price, but to change it in a problematic way and benefit from the change. It is this set of circumstances that causes market problems that may require attention. Other definitions are possible, but technical terms are most useful when they have a clear meaning, so there is a great advantage in using those generally agreed upon.

If market price has been reduced below the competitive level by actions of the system operator, this must be accounted for when analyzing market power. But the distortion of the market price by demand-side flaws can be ignored.

**Result**       **4-1.2**      **When Assessing Monopoly Power, Ignore Demand-Side Flaws**
Demand-side flaws prevent the market price from equaling the price that would result from a fully competitive market. This will only tend to raise the market price except when a price cap is in effect. Consequently any increase in price will alter price away from the competitive equilibrium and the demand-side flaws can be ignored in the analysis of monopoly power.

## 4-1.5    MONOPOLY POWER IN A POWER AUCTION

### Negative Market Power

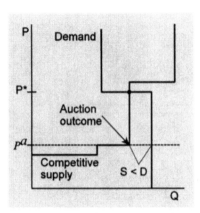

Many power markets are run as auctions, and auctions set the price with formal rules which may or may not specify a market clearing price. A common example is the auction rule that sets the market price at the offer price of the last accepted supply bid. Sometimes this clears the market and sometimes it does not. When it does not, it is because demand is greater than supply at this price (shown in the figure at the left). At the market price determined by the auction, $P^a$, demand is greater than supply, so the market price is not a market-clearing price. It *is* the price paid by all load and paid to all suppliers. Because it is not market clearing it cannot be the competitive price. That price is $P^*$ and is considerably higher.

Note that, although this auction does not use a competitive market price, this does no harm in the short run. Exactly the same trade is made at either price. The difference is only a matter of money. Of course this may raise fairness issues and there may be a consequence for investment. Most likely such rules do very little harm.

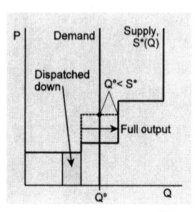

If the supplier whose bid set the auction price were to raise its bid, that would raise the market price, but it would raise the price toward the competitive price, not away from it. This is the opposite of exercising market power. Raising the bid price above $P^*$ would raise the market price above the competitive level. This would constitute an exercise of market power if it were profitable.[4]

In one instance, this effect can be truly dramatic. If an auction has a price cap of $10,000/MWh, and the highest bid of any generator is $150, and demand exceeds supply, then the supply and demand curves intersect at $10,000/MWh, yet the market price is set to $150. In this case, a supplier can raise the price all the way to $10,000/MWh and all it has done is brought it up to the competitive price level. It has exercised no market power.[5]

| Result | 4-1.3 | **Profitably Raising the Market Price May Not Be Market Power** |
|---|---|---|

If the auction sets market price equal to the price of the highest accepted bid, the competitive price may be much higher than the market price. In this case, bidding higher and raising the market price to a level below the competitive price is not an exercise of market power.

---

4. If it were not profitable, it would simply be a mistake. While mistakes usually do harm, they are automatically punished by a reduction in profit.

5. This example is similar to PJM in the summer of 1999, except that PJM had a price cap of only $1,000/MWh.

**Making Someone Else Withhold.**  In a market with a clearing price, the supplier that exercises market power has a positive markup and also violates the price-taking assumption—it withholds output. In some auctions, this same supplier can produce its price-taking level of output while exercising market power. Consider the case of a vertical demand curve (though this assumption is not required). Let this intersect the middle of a 200-MW supply bid with an offer price of $30/MWh. If the bid is raised to $50/MWh, and there is no other unaccepted bid this low or lower, the market price will increase from $30 to $50/MWh and the supplier will profit. This is market power, but will this supplier withhold? If the supplier were a price taker, at a market price of $50MWh he would certainly supply another 100 MW. But power markets must balance supply and demand.

What actually happens depends on the details of the market. Most likely the system operator will be forced to accept a "decremental" bid. If the generator in question does not offer such a bid, and the next cheaper generator offers to decrement 100 MW for $28/MWh, the system operator will accept this bid and the cheaper generator will back down. This allows the generator that is exercising market power to produce at full output, which is price-taking behavior. But according to the definition, there is still a positive quantity withheld. At the market price, the competitive supply curve is clearly to the right of the actual level of output. In this case, the system operator has been induced to pay a cheaper generator to withhold by the auction rule, the need for balancing, and the behavior of the monopolistic generator.

As a further complication, it is entirely possible that the system operator will set the market price to $28 because this is system marginal cost. In this case the $30 generator may be paid $50/MWh (perhaps in the form of a side payment), while the market price decreases from $30/MWh to $28/MWh. The generator has raised its price but lowered the market price, so it has technically not exercised market power. In this case an exception to the standard definition may be needed as it was not intended to cover markets as peculiar as power pools.

## 4-1.6  MARKET POWER ON THE DEMAND SIDE

Monopsony power, as opposed to monopoly power, is market power exercised on behalf of, or by customers. It is difficult to exercise in a power market because it requires withholding of demand or something equivalent, and demand is generally uncontrollable.

Like monopoly power, monopsony power can be exercised by physical withholding or by bidding a price lower than the power's marginal value. But for monopsony, there is a third alternative. A monopsonist can generate power at a cost greater than the market price in order to depress the market price. The right diagram in Figure 4-1.2 shows this possibility. The load-serving entity owns an expensive generator, which it bids in at a very low price or simply turns it on. The effect is to shift the supply curve to the left and lower the price dramatically. If it is buying more power than it is generating, this will be profitable. Exercising monopsony power moves the market price lower and away from the competitive level.

**Figure 4-1.2**

Exercising monopsony power by withholding demand and by generating at a cost above the market price.

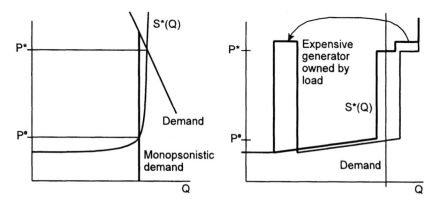

The left diagram in Figure 4-1.2 shows the more standard approach of withholding demand. This can be very effective when the supply curve is steep, but in a power market there are few ways to accomplish it.

The most straightforward approach to monopsony power is for the ISO to withhold **interruptible load**. There are various types of interruptible load contracts and all have some price, either implicit or explicit. If that price is above the competitive price, the load should not be interrupted, but the system operator could interrupt it anyway, pay the cost, and lower the market price, thereby saving much more for customers than the cost of interrupting the load. This is an exercise of monopsony power.

Finally, the ISO may exercise monopsony power by controlling imports and exports. If the internal market price is $150 and the external price is $200/MWh, the ISO can curtail exports. More power will be available internally, reducing the price for the internal market. If this saves more than it costs, monopsony power has been exercised.

So far this discussion has ignored the possible existence of monopoly power. If this has raised the market price above the competitive level, and the techniques just described are utilized to bring the market price down toward but not past the competitive level, then they are not exercises of market power. In this case they can be beneficial. If they are used to reduce the market price below the competitive level, they are an exercise of market power, and they are harmful. Although monopsony power that counteracts monopoly power will save consumers money, it will still reduce the total surplus.

# Exercising Market Power

*People of the same trade seldom meet together, even for a merriment or diversion, but the conversation ends in a conspiracy against the public, or in some contrivance to raise prices.*

Adam Smith
*The Wealth of Nations*
1776

M ONOPOLY POWER ALWAYS RESULTS IN COMPETITIVE SUPPLY BEING GREATER THAN DEMAND AT THE MARKET PRICE—A POSITIVE QUANTITY WITHHELD. This signature of market power must be checked to determine the nature of any observed price increase and to determine its significance. A market that cannot tolerate a few hundred megawatts of withholding cannot withstand a hot afternoon or a generator outage. Such a market should be shut down and redesigned.

**Chapter Summary 4-2:** Market power should be looked for only in the real-time (RT) markets. It should be looked for among inframarginal as well as marginal generators. The amount of withholding should always be examined along with the price increase.

**Section 1: Market Power and Forward Markets.** If prices in a forward market are too high, customers can wait for the next forward market or for real time. In the RT market they can wait no longer. Consequently, market power cannot be exercised in forward markets; this includes day-ahead (DA) markets. Market power exercised in real time is reflected into forward-market prices through arbitrage.

**Section 2: Long-Run Reactions to Market Power.** The threat of new supply entering can discipline the exercise of market power because those exercising it fear high prices will attract too many future competitors. Suppliers may also choose to exercise less market power for fear of retaliation by regulators.

**Section 3: Marginal and Inframarginal Market Power.** Frequently the generator that sets the market price is not the one exercising market power. Often it is some inframarginal generator, possibly one that would be marginal had it not priced itself out of the market.

**Section 4: The Two Effects of Market Power.**   Market power raises price and thereby transfers wealth from customers to all suppliers in the market, not just to suppliers who exercise power. A secondary effect is inefficiency. A high price causes consumers to buy less power than they would if they had to pay only the marginal cost of production; however, because consumers of power are relatively insensitive to price, this effect is quite small.

**Section 5: Long-Run and Short-Run Market Power.**   Withholding raises the market price and attracts investment in new power plants which lowers the market price. The result is too much installed capacity which must be paid for by customers. Barriers to the entry of new power plants cause a lower level of equilibrium capacity and higher profits which compensate for the cost of the barriers.

**Section 6: Is a 1000% Markup too Much?**   Markups by themselves tell little; the percent increase in retail cost caused by the average markup is informative. If the amount of withholding is less than typical weather-induced load fluctuations, and this is still sufficient to transfer significant wealth, the market's profit function needs to be redesigned.

---

## 4-2.1   MARKET POWER AND FORWARD MARKETS

Unless a regulator deliberately stifles arbitrage, it is essentially impossible for suppliers to exercise market power in a forward market.[1] Physical withholding is both normal and expected in these markets. In many forward markets there is as much "withholding" as there is participation—many buyers and sellers just don't use them. In other forward markets there may be more traded because of speculation than will be delivered. PJM's DA market sells more power than is produced in real time. If a farmer does not sell his wheat in a forward market, it is not an attempt to exercise market power. It just means he intends to sell later.

The price in forward markets is derived from the RT price, just as the forward contract itself is a financial product derived from the physical product. Forward markets are derivative markets. Their price is set relative to the expected RT price and according to the riskiness of these markets.

If half of the load does not purchase in a forward market, this will not drive the price to near zero in all hours; it will simply cause an equal number of suppliers to wait for the RT market. They may do this by bidding their marginal cost, knowing their bid will not be accepted, or they may choose not to bid. Similarly if half the suppliers do not make offers in a forward market, half the load will wait for the RT market.

If a supplier, because of foolish bids by customers, were able to raise the price in the forward market, load would quickly see the price difference between forward and real time and migrate to the RT market. This would solve the problem. Before

---

1. Cornering the market is a possibility in forward markets, but is unlikely and will be ignored.

real time, the opposing side of the market always has an alternative—it can wait. In the RT market, it cannot.

---

**Result    4-2.1    Market Power Cannot Be Exercised in Day-Ahead Power Markets**
Provided the market rules do not inhibit arbitrage between the DA market and the RT market, market power cannot be exercised in the DA market.

---

Day-ahead markets will suffer from market power exercised in the RT market. Any increase in the RT price will be translated into the forward markets. Consequently, markups that look identical to those seen when market power is exercised will be seen in the DA market. These are only reflections of problems in the RT market. In forward markets, suppliers always have the excuse of opportunity costs in the RT market, and the excuse is usually valid. Real problems occur only in real time.

This conclusion does not mean that forward-market trading can be ignored when assessing market power. Since raising the price in the RT market increases the price in forward markets, any trader that expects to be selling in forward markets has a greater incentive to raise the RT price. When calculating the profitability of an action affecting the RT price, expected forward trading must be taken into account. (See "Why Shorter Term Forwards Matter Less" in Section 4-3.3.)

## 4-2.2  LONG-RUN REACTIONS TO MARKET POWER

Suppliers consider the impact on forward prices when exercising market power in the spot market, but this is not the only future consequence they consider. High spot prices also cause demand-side, supply-side, and regulatory responses that reduce future spot prices. A supplier that intends to continue exercising market power will consider these reactions.

Suppliers will realize that if prices are persistently high, this will act as an inducement for competitors to enter and for regulators to take restrictive actions. Both of these effects are well recognized. The first is referred to as the "threat of entry," and it is the central insight of the theory of contestable markets. This threat effect is different from the effect of actual entry. The threat of entry causes incumbent suppliers to exercise less market power than they otherwise would. This holds down prices and in the process slows actual entry. Nonetheless, the net effect is beneficial. Because the threat of entry helps reduce current spot-market prices and current prices of long-term forward contracts, making entry easier can provide relief before the entry actually occurs.

The effect of the regulatory threat has not been formalized but has been statistically detected in the British Pool by Wolfram (1999b). In extreme circumstances long-run considerations can also lead to perverse short-run behavior. If generators conclude that re-regulation or stricter regulation is inevitable, they will act to maximize the short-run exercise of market power with no thought for future consequences. In other words, if generators are sure there will be, or sure there

**Figure 4-2.1**
An inframarginal
exercise of market
power.

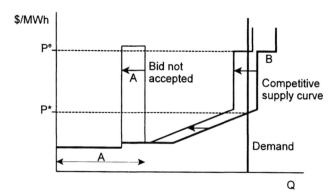

will not be, future consequences, they will behave myopically. They will take account of the future only if they believe it is conditional on their present behavior.

Demand also responds to spot prices that it perceives to be persistently high—it takes long-run conservation measures. This response was surprisingly strong in California. The high prices of 2000 caused a demand reduction of nearly 10% in 2001. Estimates of the one-year household price elasticities have ranged between almost 0.15 and 0.35, while, as expected, longer-term elasticities have been estimated to be higher. (See, for example, Reiss and White, 2001.) If suppliers are rational they will take long-run elasticities into account when exercising market power. In a new and turbulent market, this is unlikely, but in a more stable environment, it may help to reduce market power.

## 4-2.3   MARGINAL AND NONMARGINAL GENERATORS

Most DA and RT power auctions set a clearing price that equals the bid of the "marginal" generator. Usually a single generator sets the market price. Since market power requires the "ability to alter prices," it seems natural to conclude that only the price-setting marginal generator can exercise market power. Consequently, many attempts to measure or study market power focus on the marginal generator.

**Result    4-2.2**    **The Bid that Raises the Price May Not Set the Price**
Bidding high may result in a bid not being accepted. This has the same effect as physical withholding. The result is a high price set by some other generator. A supplier with several generators in the market may find this profitable.

Figure 4-2.1 shows typical supply and demand curves with the bids of two of the many suppliers labeled A and B. Supplier A owns two relatively cheap plants, and generator B owns one expensive plant. Only A exercises market power, and it does so by raising the offer price, and thereby withholding the output from its higher-marginal-cost plant which earns less short-run profit. This withdrawal shifts the more expensive part of the supply curve left by the amount of capacity withdrawn. The vertical demand curve then intersects the shifted supply curve in the output region of supplier B which has a marginal cost of $P^e$, and this sets the market

**Figure 4-2.2**

Wealth transfer and
dead-weight loss caused
by the exercise of
monopoly power.

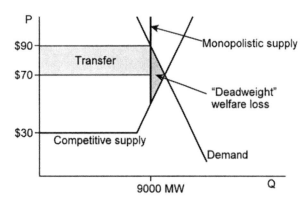

price. Generator A's withdrawal of capacity has caused the market price to increase from the competitive level $P^*$ to the much higher level $P^e$. This is profitable, and A has exercised market power.

In this example the marginal generator has bid its true marginal cost and has withheld no output. If only marginal units were examined for market power, none would be found. Supplier A has "altered" (raised) the market price without being the generator that sets the market price. Supplier A could withhold power either physically or financially; it would make no difference.

## 4-2.4   THE TWO EFFECTS OF MARKET POWER

Figure 4-2.2 shows the result of withholding supply by a generator and the consequent increase in price. Although there are various methods of withholding supply, including bidding too high, every exercise of market power is, in essence, the result of withholding output.[2]

**The Transfer of Wealth.**   When monopoly power is exercised, the most obvious effect is the transfer of wealth from consumers to suppliers. Economics often ignores the problem of wealth transfer, not because it believes it is unimportant, but because it has no way to evaluate it. However, this does not preclude the policy maker from taking it into account.

**The Deadweight Welfare Loss of Monopoly.**   Economics is more concerned with efficiency and in this case, is concerned that high monopoly prices will lead to inefficiently low levels of consumption. The price increase reduces benefit to consumers more than it increases profits. This is called a deadweight loss to society, and it is what is meant by "inefficient."

Figure 4-2.2 shows the deadweight loss from an exercise of market power. Note that the main effect of the price increase is to transfer $180,000/h of revenue from consumers to the supplier. The consumers' willingness to pay is represented by the demand function which varies in height from $70 to $90/MWh over the interval

---

2. Even strategies such as quality reduction can be interpreted as withholding of the high-quality product. See Chapter 4-1 for the equivalence of high offer prices and withholding. Strategies such as raising the cost of competitors generally falls under the classification of anticompetitive behavior and not the exercise of market power because they tend to raise the short-run competitive price.

**Figure 4-2.3**

Equilibrium installed capacity with market power.

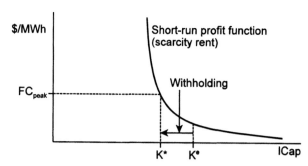

of output reduction. The lost benefit to consumers is the area under the demand curve from Q = 9000 to Q = 10,000 MW, but the reduction in benefit is accompanied by a reduction in the cost of production. This cost savings is the area under the supply curve in the same output interval. The deadweight loss is the lost benefit minus the cost savings, and this is exactly the area of the shaded triangle in Figure 4-2.2. This area is half of 1000 MW times $40/MWh, or $20,000/h.

Deadweight loss is the result of production that could have been carried out at a cost less than its value but was not. When demand elasticity is low, as it is in power markets, the deadweight loss is typically low compared with the transfer of wealth.

There is a second source of dead-weight loss in an oligopoly market. A monopolist always withdraws the most expensive production units as shown in Figure 4-2.2, but when there are several suppliers, the one exercising market power may not own these units. In this case units remaining in the market after the exercise of market power will not be the cheapest and this will cause productive inefficiency. Given the inelasticity of demand, this may be a larger source of welfare loss than is represented by the standard dead-weight loss triangle.

## 4-2.5   LONG-RUN AND SHORT-RUN MARKET POWER

Exercising market power by financial or physical withholding is a short-run strategy but one that may be repeated over the long run. If it is repeated, there will be investment consequences that have not yet been considered.

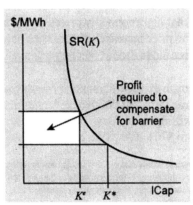

When one supplier raises the market price, all suppliers benefit. When this is repeated over a long period of time, it makes the market more attractive to all potential entrants and thus increases entry—investment in new generating facilities or expansion of old facilities. The result is a higher equilibrium level of installed capacity, ICap. The new equilibrium is illustrated in Figure 4-2.3. (See Chapter 2-7 for a discussion of the short-run profit function.)

If the equilibrium capacity, $K^e$, were equal to the optimal competitive value $K^*$, then withholding would reduce the level of available capacity to below $K^*$, and this would raise profits above normal levels. This would attract entry. The process continues until equilibrium capacity, $K^e$, is enough above the optimal level, $K^*$, that profits return to normal. This is the point at which $K^e$ minus withholding equals $K^*$.

**Barriers to Entry.** If there is a large cost to generation entering the market, this has an effect similar to market power. It can reduce capacity and cause the market price to increase. If the barrier is created by one of the suppliers in order to raise the market price, and the strategy is profitable, it might be considered an exercise of long-run market power. Long-run market power might be defined as the ability to raise the market price above the long-run competitive price. The discussion of Chapter 4-1 does not apply to this definition, but a definition of long-run market power may deserve exploration. More typically, barriers to entry are not created by market participants and would not fit under any definition of market power. One such barrier is the cost of obtaining permission to build generation.

For simplicity, assume that a barrier exists and is costly enough to prevent investment unless the investor expects a profit of $6/MWh in excess of normal profits. The long-run effect of this is to create a shortage of installed capacity that is just sufficient to produce the profit required to compensate for the cost of the barrier. For installed capacity the result of a barrier to entry is opposite to the result of withholding. The equilibrium level of capacity is too low. This will not result in dramatic reliability problems as it is exactly what would happen if the fixed cost of generator were a bit higher than it is, as it was historically. Consumers will pay to much, but the market should operator reasonably unless the barriers are dramatic.

**Summary.** Withholding causes inefficiency by causing too much generation to be built. This generation will be paid for by customers because profits remain normal. The investment that is attracted by the high prices caused by market power will be investment by small competitive suppliers, not by the large suppliers exercising market power. This will reduce the market share of the larger suppliers. Thus new entry has two effects: it decreases the competitive price by increasing supply and it reduces market power. Of course there may be offsetting effects such as mergers.

The cost of barriers is passed through to customers, but there is an additional cost due to the reduced level of ICap. Depending on the steepness of the profit function this may be very small. More likely the profit function will simply be adjusted so that $K^e$ is correct for reliability purposes and $K^*$ is too high relative to an optimal market. With this adjustment, the only loss is the cost of the barriers themselves.

If both barriers and withholding occur together, the profit function can still be adjusted so that $K^e$ minus withholding is optimal for reliability. (See Chapters 2-7 and 2-8 for methods of adjusting the profit function.) In this case, actual capacity, $K^e$, will be too high, so this cost plus the cost of the barriers will be paid for by customers. In the long run, barriers to entry cause higher prices, and withholding causes wasteful excess capacity. The cost of excess capacity is a second-order effect. A marginal increase in capacity above the optimal level produces just as much benefit from extra reliability as it adds to cost.[3] But nonmarginal increases are costly, and each time the excess capacity doubles, the inefficiency quadruples.

---

3. This assumes that all capacity contributes to reliability, because there is no withholding after price reaches the price cap.

## 4-2.6    IS A 1000% MARKUP TOO MUCH?

The Department of Justice discusses a hypothetical monopolist that charges 5% above marginal cost. In power markets, charging $660/MWh instead of $60/MWh is not uncommon—that's 1000% higher than marginal cost. Sometimes this is just scarcity rent, but sometimes a gas turbine, known to have a marginal cost of $60/MWh, bids $660/MWh and refuses to start when the price hits $400/MWh. Then the ISO sets the price to $660/Mh, and it starts up within five minutes.

When this evidence is combined with the fact that the supplier is long in the spot market, there can be little doubt that the market power has been exercised, and that it raised prices 1000%. Action is rarely, if ever, taken in these cases. If 5% is worth worrying about, isn't 200 times that much a disaster? There seems to be no regulatory standard, and with this level of ambiguity, it is no wonder policy has been erratic.

Power markets are prone to very short but high price spikes, which are partly caused by market and partly by scarcity. As a consequence, markups by themselves are meaningless. There are three approaches to evaluating the seriousness of a market power problem: (1) average wealth transfer, (2) average deadweight loss, and (3) effective level of withholding. Wealth transfer can be stated as a percentage of the retail price in order to put it in perspective. A 5% average increase in the retail price is a meaningful number. Exactly what should be tolerated depends on the cost of reducing it. Deadweight loss will be of interest mainly to economists.

The effective level of withholding is useful for putting the problem in perspective. The profit function indicates how much profit will increase if capacity is decreased by any given percent. Consider a 2% decrease. That could come about by load growing 2% more than expected, or by ICap decreasing 2% when a nuclear plant is retired, or by hot weather that pushed peak loads up by perhaps 2.5% while leaving average load the same, or by *withholding* 2% during peak hours. All of these disturbances would have roughly the same impact on profits and thus on the cost of power.

Suppose the market is analyzed, and it is found that a 2% decrease in ICap would cause a $10/MWh average increase in prices. This is about a 12% increase in the retail cost of power. Such an increase might be considered intolerable, but in that case, normal fluctuations in weather, load, generation retirements, and outages must also be considered intolerable. It may be argued that these are transitory while market power is not. There is some truth to this, but load growth and plant retirements are not transitory by nature and their effect is transitory only because the market responds. The previous section demonstrated that the same adjustment mechanism that makes ICap track load also makes it track and remove market power problems.

The conclusion should be that levels of withholding that are smaller than the normal fluctuations in weather, load and outages may as well be tolerated. The market will remove them soon enough, and they are no more problematic than several other problems that must be accepted. If very small amounts of withholding produce problematic cost increases, the correct conclusion is that the profit function needs to be re-designed. (See Part 2.)

*Chapter* *4-3*
# Modeling Market Power

*Merchants are occupied solely with crushing each other:*
*such is the effect of free competition.*

Charles Fourier
(1772-1837)

*Like many businessmen of genius he learned that free competition was wasteful,*
*monopoly efficient.*

Mario Puzo
*The Godfather*
1969

MODELING MARKET POWER HELPS EXPLAIN THE FACTORS THAT
CONTROL THE EXERCISE OF MARKET POWER. The models, however, are
not accurate predictors of market power. Except in the case of pure monopoly,
market power is not well understood because it involves strategic behavior by
several competitors—an **oligopoly**. Game theory is subtle, and the game of
oligopolistic competition has complex rules. By abstracting from much of its
complexity, models of market power explain its main features.

Elasticity of demand is the most important factor in present power markets. The
distribution of the size of competitors is also a key factor. Even this is extremely
hard to compute in a power market because it depends on barriers to trade which
are very complex and vary continuously, often dramatically. The style of competition is also crucial. Economics can model competition based on price or quantity,
but in power markets suppliers compete using "supply curves" that combine the
two. The theory of supply-curve competition (discussed in Chapter 4-4), while
promising, is not yet well developed.

**Chapter Summary 4-3:** The market power of a monopolist is limited by
demand elasticity. An oligopoly's market power is also limited by the number and
relative size of competitors. Market share is a supplier's sales divided by total trade
in the market. The sum of the squares of market shares is called the
Herfindahl–Hirschman Index (HHI) and is one of three factors determining market
power in the Cournot model.

**Section 1: Monopoly and the Lerner Index.** A monopolist will raise price
to maximize profit but could raise it even higher. The more customers respond to
high prices by curtailing demand, the less a monopolist will raise the price. The

Lerner index, $L_X$, measures the markup of price above marginal cost. Under monopoly, $L_X = 1/e$, where $e$ is the demand "elasticity" (responsiveness) of consumers.

**Section 2: The Cournot Model.** The Cournot model of oligopolistic competition predicts that the markup of a supplier depends not only on demand elasticity but also on the supplier's share of market output. The larger the share, the more market power a supplier has and the greater his markup.

**Section 3: Unilateral Action and the HHI.** The HHI is computed as the sum of the squares of market shares of suppliers. FERC and DOJ not withstanding, market shares are always defined as the fraction of market output produced by a particular supplier. The HHI ranges from zero to one, but in legal documents it is multiplied by 10,000 (because it is computed by squaring percentages instead of fractional shares). Unilateral action involves no collusion, and the Cournot model is one model of such behavior. The HHI is related to unilateral action by the fact that when divided by demand elasticity it predicts the average relative markup ($L_X$ in a Cournot market.

---

## 4-3.1 MONOPOLY AND THE LERNER INDEX

The simplest case of market power, and one of the few that is well understood, is that of a single supplier, a monopolist, who supplies the entire market. Even this case is only understood when buyers are competitive and cannot exercise market power of their own. Like a competitive supplier, a monopolist maximizes profit, but unlike a competitive supplier it can affect the price and takes this into account. As a consequence, a monopolist with constant marginal cost has this short-run profit function: $SR_\pi(q) = P(q) \times q - MC \times q$, where $q$ is its output, $MC$ its marginal cost, and $P(q)$, the inverse demand function, describes how market price depends on the quantity supplied. The demand curve, described by $P(q)$, tells the monopolist how much a reduction in output will raise the market price.

Figure 4-3.1 shows how such a monopolist would maximize profits with a linear demand curve. Note that the monopolist sets price well above marginal cost but not nearly as high as it could. Monopoly pricing is restrained by the response of customers. The more they respond to a price increase by cutting back their demand, the less profitable it is for a monopolist to raise price and the lower the monopoly price will be. This is the most crucial lesson for power market design. Because power markets have almost no price responsiveness on the demand side, suppliers can have enormous market power.

### What Determines the Price-Cost Markup?

**The Lerner Index for a Monopoly.** The exercise of market power results in a market price, $P$, that is greater than both the competitive price and the marginal cost, $MC$, of production. The most common measure of this difference is the price-cost margin, also called the Lerner index, which is defined as follows.

**Figure 4-3.1**

Profit maximization by a monopolist.

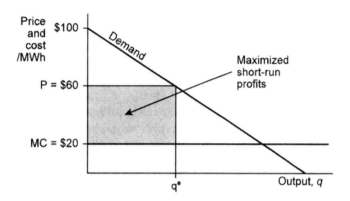

| Definition | **Lerner Index or Price-Cost Margin ($L_X$)** |
|---|---|

If $MC$ is a supplier's marginal cost calculated at its actual level of output and P is the market price, then the Lerner index is defined by $L_X = (P - MC)/P$.

The Lerner index of the monopolist in Figure 4-3.1 is $(60 - 20)/60 = 0.66$, while the largest possible Lerner index is 1.

The price responsiveness of demand is called the "elasticity" of demand, and it plays a crucial rule in the theory of market power. It is defined as the percentage by which demand decreases when price increases 1%. If such a price increase causes a half percent decrease in demand, then the elasticity of demand is 0.5 (not ½ of 1%). This is defined more formally as follows.

| Definition | **Elasticity of Demand (e)** |
|---|---|

The elasticity of demand is given by

$$e = -\Delta Q\% / \Delta P\%$$

$$e = -(\Delta Q/Q)/(\Delta P/P)$$

or $\qquad\qquad e = -(dQ/dP)(P/Q)$

where $\Delta Q$ is the change in demand caused by a very small change in price, $\Delta P$. (In economic texts the minus sign is often omitted, which makes all demand elasticities negative.).

Having defined the **Lerner index**, $L_X$, and the elasticity of demand, $e$, the basic law of monopoly pricing can be expressed as follows.

$$L_X = \frac{P - MC}{P} = \frac{1}{e} \qquad\qquad (4\text{-}3.1)$$

The monopolist withholds output causing a price-cost markup (relative markup) equal to the reciprocal of the demand elasticity. This is simply the result of profit maximization. Any other attainable price-quantity combination is less profitable.

A higher price could be obtained but only by giving up profit on the attendant reduction in sales. The monopolist of Figure 4-3.1 chooses an output for which the demand elasticity is 1.5. That means a 1% change in price would cause a 1.5% change in quantity. Note that elasticity is not constant along a linear demand curve.

Consider the implications of a monopolist facing a demand curve with constant elasticity. If $e = 2$ then $L_X = \frac{1}{2}$, which, according to Equation 4-3.1, means $P = 2 \times MC$, so instead of price equaling marginal cost as it does under competition, price is twice as high as marginal cost. If $e = \frac{1}{2}$, then $L = 2$, which, by Equation 4-3.1, implies $P = -MC$, which is impossible. To understand this result, Equation 4-3.1 can be rearranged to give $P = [e/(1 - e)] MC$. As the elasticity of demand decreases toward 1, the denominator $(1 - e)$ approaches zero and price approaches infinity. For any values of elasticity less than 1, price would be infinite, and the negative value for price ($-MC$) warns of this result. This is important for power markets because the short-run demand elasticity in power markets is often quite close to zero (far below 1) and even the one-year elasticity is well below 1. A monopolist in a power market could earn nearly unlimited profits for more than a year.

| $e$ | $P$ |
|------|-----------|
| 2. | $2 \times MC$ |
| 1.5 | $3 \times MC$ |
| 1.2 | $6 \times MC$ |
| 1.1 | $11 \times MC$ |
| 1.01 | $101 \times MC$ |
| 1.00 | infinity |

In reality, various regulatory and/or legal limits would keep the price finite. But it is instructive to understand how a market would avoid infinite prices if it were allowed to function without any price restriction, and demand was extremely inelastic but able to respond to price. Consider the consequence of the monopolist cutting supply by 10% when demand elasticity is 1. The result is a 10% increase in price, so revenue remains essentially constant. Because production cost decreases, profit increases. As long as elasticity of demand remains at 1, the monopolist can keep raising price and increasing profit. Eventually, when price becomes high enough, demand will react strongly, and at this point, demand elasticity will become greater than 1. As price rises, eventually a low elasticity market turns into a high elasticity market ($e > 1$) and brings an end to monopolistic price increases. So every monopolist operates in the elastic region ($e > 1$) of the market's demand curve. A power market does not become elastic until price exceeds \$1,000/MWh, perhaps by a great deal.

## 4-3.2   THE COURNOT MODEL

In 1838, Cournot published a description of the exercise of market power by a group of uncoordinated suppliers, a noncollusive **oligopoly**. Although it has many shortcomings, Cournot's model takes a necessary step beyond the monopoly model and explains much about the role of market share in the determination of market power. According to the Cournot model, all suppliers choose a level of output and then market price is determined by the interaction of aggregate supply and the market demand curve.

Modern economics describes the Cournot model as a game played in such a way that the outcome is the kind known as a Nash equilibrium. Within the game-theory framework, the players of the Cournot game are the suppliers in the market. Each player is allowed only one move which is to choose an output quantity, $q$.

Each supplier's goal is to maximize profit from selling its output at the market clearing price and paying its costs of production.

---

**Definition**  **Cournot Competition**

Suppliers choose their quantity outputs. Price is determined by total supply and the consumers' demand curve. Suppliers maximize profits under the assumption that all other suppliers will keep their outputs fixed.

---

This defines the game's mechanics, but there is another crucial element, the *information set*. Each player has its own information set, but a Cournot model assumes that all players have the same information including knowledge of the market demand function and of the production cost functions of all other players (suppliers). This is an heroic assumption, but it represents the fact that suppliers are long-term players in the market and learn a great deal about their competitors.

The markup of a monopoly is determined simply by demand elasticity, but in an oligopoly, the size of the oligopolistic competitor matters. The most useful measure of size is **market share**, which is defined as the output of a supplier divided by the total market output. The larger the share of a competitor, the greater its markup over marginal cost.

---

**Definition**  **Market Share**

The output of a supplier divided by the total market output.

---

**The Lerner Index for a Cournot Oligopoly.**  The above monopoly result, Equation 4-3.1, can be generalized to a Cournot oligopoly if the supply of an individual supplier, $q$, is distinguished from the market supply, $Q$. For a Cournot oligopolist

$$L_X = s/e, \quad \text{where} \quad s = q/Q \qquad (4\text{-}3.2)$$

Notice that the formula for the Lerner index for a single Cournot supplier is the same as for a monopolist except that it is multiplied by $s$, the supplier's share of total output. For a monopolist $s = 1$, and the Cournot formula reduces to the monopoly formula. For an oligopoly consisting of five equal suppliers, each will have a relative markup, $L_X$, that is $\frac{1}{5}$ of the monopolist's relative markup. As before, market elasticity, $e$, plays a crucial role. In the case of five identical firms, if the elasticity were held down to 0.2 or less, the price would rise without limit. Or put more realistically, the price would rise until the elasticity of demand increased to something more than 0.2. Again note that with a short-run elasticity near zero (the case in many power markets), the Cournot model predicts that even five or ten noncollusive suppliers would push the price level extremely high.

The above result applies to a single supplier, but it is also useful to consider the average markup of all suppliers. This depends on all of their market shares. In the case of a Cournot oligopoly, a set of market shares can be summarized in a single index that measures industry "concentration" and allows a simple calculation of the average Lerner index.

## 4-3.3    UNILATERAL ACTION AND THE HHI

The Herfindahl–Hirschman index (HHI) is the most popular indicator of market power and is used by DOJ and even more heavily by FERC in evaluating the effect of mergers on market power.

Originally designed simply as a measure of concentration, and used as an indicator of the potential for collusion, it now has been adopted as an indicator of market power exercised through *unilateral action*, a regulatory term indicating that no collusion, tacit or explicit, is involved. There is one justification for the use of HHI as such an indicator: its connection with the Cournot model which is a model of unilateral action. The **HHI** is defined simply as the sum of the squares of the market shares of all suppliers.

---

**Definition**            **Herfindahl–Hirschman Index (HHI)**

If $s_i$ is the market share of the $i^{th}$ supplier, then HHI $= \sum (s_i)^2$ over all suppliers. HHI can take any value between 0 and 1, but in legal documents it is multiplied by 10,000.

---

The relationship between HHI and the Cournot model is derived in the Technical Supplement, Section 4-3.4. That derivation relies only on the Cournot markup equation, $L_X = s/e$, which holds for each supplier.

## The HHI Result

---

**Result     4-3.1**       **The Average Lerner Index Equals HHI over Demand Elasticity**
$$\text{Average } L_X = \text{HHI}/e \qquad (4\text{-}3.3)$$
The share-weighted average Learner index in a Cournot oligopoly is given by HHI/$e$.

---

The result shown above means that the average price-cost markup in a Cournot oligopoly (weighted by market share) is given by the HHI divided by the demand elasticity at the equilibrium price and output level. This provides some justification for using HHI as an indicator of unilateral market power because the price-cost markup is the immediate consequence of market power. Note, however, that HHI by itself does not determine average markup. Demand elasticity and the assumption of Cournot competition play equally important roles. Two markets that have the same HHI may have different levels of market power if they differ either in their demand elasticity or in their mode of oligopolistic competition. Economists tend not to rely on the HHI as an indicator of market power; however, its relative ease of computation and its appearance of definiteness make it popular with regulators.

Regulators often redefine "share" in order to change the value of the HHI.[1] The definition of a market share, as given in every text on industrial organization, is the output of a supplier divided by the total market output. This was the definition used by Herfindahl, and it is the definition for which HHI has at least some theoretical connection (Equation 4-3.3) to markup. Definitions of HHI that use a different definition of "share" are not the HHI as defined by Herfindahl, and they are not the HHI that is related to average markup. There is nothing wrong with defining new indices, but they should not be called HHI—that name is taken. Before new definitions are used, some study should be made of the relationship of the defined statistic to market power. In most cases the relationship is essentially nonexistent.

## 4-3.4  TECHNICAL SUPPLEMENT: MARKUP DETERMINATION

This supplement derives the three numbered equations, 4-3.1, 4-3.2, and 4-3.3.

**Markup for a Monopolist.**   A monopolist differs from a competitive firm in that it can affect market price. Thus it views price as a function of its output. Consequently its short-run profit function is written as $SR_\pi(Q) = P(Q) \times Q - c(Q)$. Note that $Q$ represents both the monopolist's output and the market supply, which equals market demand in equilibrium. The Lerner index, $L_X = (P - MC)/P$, that results from profit maximization, is found as follows.

$$dSR_\pi(Q)/dQ = 0$$
$$(dP/dQ)\,Q + P - dc/dQ = 0$$
$$(Q/P)\,(dP/dQ) + 1 = MC/P$$
$$- 1/e = MC/P - 1$$
$$L_X = 1/e$$

The fourth step uses the fact that by definition, $e = (dQ/dP)(P/Q)$, so $(Q/P)(dP/dQ) = 1/e$. The fifth step uses the fact that $MC/P - 1 = -(P - MC)/P$.

**Markup for a Cournot Supplier.**   To generalize the above result to a Cournot oligopoly, it is only necessary to distinguish between market supply, $Q$, and the supply of an individual supplier, $q$.

$$dSR_\pi(Q)/dQ = 0$$
$$(dP/dq)\,q + P - dc/dq = 0$$
$$(q/P)\,(dP/d\,q) + 1 = MC/P$$
$$(q/Q)\,(Q/P)\,(dP/d\,q) = MC/P - 1$$
$$(q/Q)\,(- 1/e) = MC/P - 1$$

---

1. FERC's Order 592 loosely defines four "capacity measures" and then states "concentration statistics should be calculated using the capacity measures discussed above .... Both HHIs and single firm market share statistics should be presented." FERC is in effect creating four new definitions of both the term, "market share" and the HHI. (FERC 1996b, 67, 70.)

$$- s/e = MC/P - 1$$

$$L_X = s/e, \quad \text{where} \quad s = q/Q$$

The ratio q/Q is a supplier's market share.

**Average Markup for a Cournot Oligopoly.**   In a Cournot market, each supplier's Lerner index is given by $L_X = s/e$. Multiplying both sides by $s$ and summing over all suppliers gives

$$\sum s_i \, L_X(\text{I}) = \sum (s_i)^2 / e$$

$$\sum s_i \, L_X(\text{I}) = \text{HHI}/e$$

$$\text{average}(L_X) = \text{HHI}/e$$

The index, i, runs over all suppliers. Note that this relationship holds only if $s$ is defined to be market share and that market share has a specific meaning, $q/Q$. If HHI is misdefined as being computed from other ratios, such as the ratio of a supplier's capacity to the total market capacity, this relationship will not hold for the resulting imposter "HHI."

*Chapter* *4-4*
# Designing to Reduce Market Power

*Monopoly, in all its forms, is the taxation of the industrious for the support of indolence, if not of plunder.*

John Stuart Mill
*Principles of Political Economy*
1848

MARKET POWER CAN BE CONTROLLED BY COMPETITION OR BY MONITORING AND ENFORCEMENT. Competition is preferable but does not automatically reach satisfactory levels. The four key determinants of the competitiveness of a power market are: (1) demand elasticity, (2) supplier concentration, (3) the extent of long-term contracting, and (4) the extent of supply-curve bidding.

All four of these are partially susceptible to regulatory policy and/or market design. None are particularly easy to affect, with the exception of the system operator's demand elasticity for operating reserves which translates directly into elasticity of demand for power. Since lack of demand elasticity is the primary source of market power in the industry, this simple change is long overdue. Fortunately, demand elasticity of large customers is also relatively easy to increase, once policy makers understand its importance and how to minimize the risks of real-time (RT) pricing.

Supplier concentration is quite difficult to decrease, but merger policy should prevent its increase at least until the demand elasticity problem has been fixed. There is little theory on how to increase long-term contracting except at the time of divestiture when vesting contracts can be required. Supply-curve bidding is naturally encouraged by uncertainty in the demand level. A bid that must span the entire day effectively increases this uncertainty. Supply-curve bids tend to be very elastic at lower output levels and very inelastic at high output levels, which explains why power markets have relatively little trouble with market power off peak. Research is needed to find a policy to harness this effect.

**Chapter Summary 4-4:** An ounce of prevention is worth a pound of cure and strengthening the normal forms of market-power mitigation may be the best form of prevention. Demand elasticity tops the list. Forward contracting should be encouraged rather than inhibited. Supply-curve bidding should be encouraged by slowing the re-bidding process to more closely reflect the speed of cost changes instead of the speed with which suppliers would like to change market-power strategies. Until more research is undertaken into how to encourage supply-side bidding and long-term contracting, careful policy recommendations will remain vague.

**Section 1: Demand Elasticity and Supplier Concentration.** The most important design consideration for increasing competition and curbing market power is structural, not architectural. Demand elasticity should be increased. This is easily done by making the system operator's demand for operating reserves elastic, but the effect is modest. The next step is to replace time-of-use rates, which are completely price-inelastic, with true RT prices for large customers.

**Section 2: What Keeps Prices Down?** Except when demand is very high, current power markets behave much better than predicted by a simple Cournot model. At least two market features mitigate market power in electricity markets and show some potential for enhancement by appropriate design: (1) forward contracts of suppliers, and (2) uncertainty of demand coupled with supply-curve bidding.

**Section 3: How Forward Contracts Reduce Market Power.** The most effective form of forward contracting is long-term forward contracting. The regulatory requirement to serve native load, as well as some long-term contracts, effectively reduces market share. Medium-term contracts, on the order of a year, work only to the extent that suppliers do not believe forward contract prices equal the average level of recent spot prices.

**Section 4: Demand Uncertainty and Supply-Curve Bidding.** Uncertainty of demand causes suppliers to enter elastic supply-curve bids. These increase the elasticity of residual demand as seen by other suppliers and decreases their market power.

**Section 5: Technical Supplement.** This solves the example of Section 3.

## 4-4.1   DEMAND ELASTICITY AND SUPPLIER CONCENTRATION

The two key requirements for competition are high demand elasticity and low supplier concentration. Compared with other industries, the power industry has reasonably low supplier concentration in most regions, but it is desperately short of demand elasticity. The first step is easy. System operators should stop pretending their demand for operating reserves is completely inelastic. This is wrong; it is never economical to buy a service based on a demand curve that mis-values it. While learning to purchase reserves more wisely, system operators can impart some elasticity to 5 or 10 percent of the load.

The second step is to recognize that time-of-use pricing, although it raises price at approximately the right time, does nothing for demand elasticity. If a supplier raises price through an exercise of market power, time-of-use prices do not respond. Real-time prices do. Time-of-use prices should be replaced with RT prices. Retail customers can be faced with RT rates but still be guaranteed they will pay no more than under the time-of-use rate provided their load-profile is at least as good as average. This removes the risk of large month-to-month fluctuations, but leaves them with the right price incentives. In order to get most of the benefit, such rates need only be applied to the largest customers who account for 25% of consumption.

Although supplier concentration is reasonable, market power could be reduced still further by reducing concentration, and in many cases this would harm efficiency very little. While it is extremely difficult to divest large suppliers of generation, mergers should be carefully monitored.

## 4-4.2   WHAT KEEPS PRICES DOWN?

Power markets lack demand elasticity but have other features that work naturally to limit market power. Understanding these provides guidance in designing markets so they enhance rather than suppress competitive forces.

Why are power-market prices so low? Most have an HHI no lower than 0.1 (1000), and their short-run price elasticity is generally near zero. As demonstrated in the previous chapter, under Cournot competition, price would increase without limit. According to this model, price should increase until the short-run elasticity increases to something over 10% in the case of an HHI equal to 0.1. Such a high elasticity would occur only at prices many times higher than the competitive price. Occasionally, power markets do produce such price spikes but usually only when demand is extremely high. A naive application of Cournot theory predicts that market price should constantly be many times the competitive price. What has been omitted? At least three considerations mitigate market power in power markets:

1. Forward contracts and obligations of suppliers.
2. Uncertainty of demand which causes supply-curve bidding.
3. Long-run consequences.

The first is most important and sometimes can be controlled by policy. The next three sections analyze forward contracts, explain supply-curve bidding, and discuss long-run reactions that inhibit the exercise of short-run market power.

## 4-4.3   FORWARD CONTRACTS AND OBLIGATIONS

The best example of the benefits of forward contracts is provided by the residual obligations of regulated utilities to serve their "native load." In newly restructured power markets, it is not uncommon for utilities to be required (or to choose) to divest generation. This typically leaves them in the position of being net buyers.

It is as if they had sold more than 100% of their potential output in long-term forward contracts.

This is an extreme case of forward contracting, but not an uncommon one. Suppose the utility had a 2000-MW load obligation and 1000 MW of generating capacity, as might be the case for a utility that had been required to divest half of its generation. With such a large excess of load it would probably sell none of its power and would be forced to purchase up to 1000 MW in the market. Some of this would be purchased in the spot market. Because it is only a buyer it will always want a low price and will have no incentive to exercise monopoly power. Being more than 100% forward contracted takes away all monopoly power and gives the supplier an incentive to depress the price by exercising monopsony power. This effect has been a strong stabilizing force in the PJM and CA-ISO markets.

When a supplier has no retail customers it will usually sell much of its power forward. As a second example, consider two suppliers, both with 1000 MW of capacity but the first with a forward contract for 900 MW at a fixed price. The second supplier has sold no power forward and sells all its power in the spot market. Say that by withholding 50 MW from the market, either supplier can drive the market price up from $100 to $120/MWh. The supplier without a forward contract would profit by $20/MWh on its remaining 950 MW, or $19,000/h. But if the supplier with the forward contract withholds 50 MW, it will make no extra profit on the 900 MW it sold forward and will profit by $20/MWh only on its remaining 50 MW of net supply, or $1000/h. Both will lose perhaps $700/MWh on the 50 MW withheld (assuming a marginal cost of $30/MWh), or $3,500/h. This amounts to a net loss of $2,500/h. The supplier with the forward contract has no reason to withhold output, while other suppliers will find its exercise of market power quite profitable.

## Market Power Reduced On Peak

The effect just described can help explain why market power is exercised primarily during times of peak demand. Consider a 50-GW market in which four utilities have each sold 20% of their capacity to a fifth supplier with the result that each has 10 GW of generating capacity. The four utilities each retain a peak load obligation of 12.5 GW. All suppliers have marginal costs of $20/MWh up to their capacity limits.

On the demand side, assume that a $1,000 price increase will reduce demand by 10% and the demand function is linear. This is more elasticity than is present in most new power markets, but this extra demand elasticity can be thought of as a proxy for some supply-side elasticity such as the elasticity of imports, or supply-curve bidding.

Consider the hour in which load reaches its peak level of 50 GW. At this time, the four utilities, with a capacity of 10 GW and load obligation of 12.5 GW, will want to obtain the lowest possible price on the 2.5 GW that they must purchase from the fifth generator. Consequently they will produce at maximum capacity for a total of 40 GW. This leaves the fifth generator facing a residual demand of 10 GW and acting as a monopolist with respect to this residual demand. By reducing

**Figure 4-4.1**

Long-term obligations:
one reason market power
is more of a problem
during peak hours.

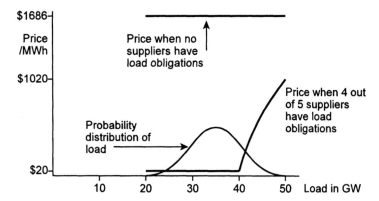

its output to 5 GW, it will force prices to $1,020/MWh and earn a profit of
$5,000,000 per hour. Any higher price would reduce its profit because of decreased
sales.

   At lower levels of demand, the fifth supplier cannot profitably raise the price
as high. For example, when demand is at 44 GW, the fifth generator would be
supplying only 4 GW and would be unable to reduce its output by the 4.5 GW
required to raise the price to $1,020/MWh. Some lesser price will prove most
profitable. Figure 4-4.1 shows this steep decline in market power as demand
decreases. Consequently, there is only a problem with market power for unusually
high levels of demand.

   If all generation were divested to five suppliers <u>without</u> load obligations and
without forward contracts, all five would want a higher market price at all load
levels. The upper price line in Figure 4-4.1 shows the markup under this scenario.
Market power is much more problematic, even though the HHI is the same as
before. Load obligations and long-term contracts are a crucial part of the market-
power puzzle.

## Why Shorter-Term Forwards Matter Less

The above analysis took no account of the length of the obligation but considered
only a single hour at a time. Such a myopic analysis is appropriate if the obligation
to load cannot change. It may also be appropriate for a new power market in which
suppliers are still unsophisticated.

   One-year forward contracts require a more dynamic analysis. Consider a supplier
with 11,000 MW of capacity that sells 10,000 MW of power forward as baseload
power (during all hours) at a fixed price. From the above argument, one might
assume this supplier would have little interest in raising the spot-market price
because most of its output has already been sold. But the supplier will think of the
fixed price contract it intends to sell for next year's power. If this year's spot prices
are high, buyers will anticipate high prices next year and will be willing to pay more
for a fixed-price forward contract for next year's power.

   But how much is it worth to the supplier to raise the price by $10/MWh for one
hour this year? Taking into account the number of hours in a year, the effect of
this would be to raise this year's average price by $(10/8760)/MWh, or about a

tenth of a cent per megawatt hour. This seems too small to matter, but it will raise the expectation of next year's price by the same amount, which would on average allow the supplier to sell its 10,000-MW one-year contract for an extra

$$10,000 \text{ MW} \times 8760 \text{ hours} \times \$(10/8760)/\text{MWh}.$$

This equals $100,000. That is exactly the same as the profit that would have been earned on $10,000 MW in one hour if it were sold in the spot market and the spot price were raised by $10/MWh. In other words, this simplified calculation indicates that a supplier that is 90% covered by a one-year contract has exactly the same motivation to raise prices as one with no contract cover. This is a discouraging result.

The full picture is not quite so bleak. When power is sold a year ahead, the supplier does not receive payment for a year, so the payment is discounted. More importantly, when customers evaluate future prices, they will base their estimate partly on this year's prices and partly on other information. Perhaps only half of this year's price increase translates into higher expectations of next year's prices. In this case, selling most of its power forward in one-year contracts could cut a supplier's market power in half. Little is known about the impact of current prices on future expectations, but the evidence from California is that it can be substantial. After a year of high prices, California's governor signed contracts for another 10 years of power at extraordinarily high prices only to see prices drop to rock-bottom levels a few months later. Of course, this behavior may not be indicative of how expectations will be formed in a mature market or by astute traders.

## Forward Contracts and Market Design

While forward contracts are highly desirable because of their effect on market power, it is difficult for the market designer to influence the extent of forward contracting. The exception is that when regulated utilities divest their generating assets, the purchasers can be required to sign "vesting contracts" with the utilities to sell back most of their output at a fixed or indexed price. These contracts can be quite long term. Such arrangements are typical and can be beneficial for reducing market power when a market is first started. However, when these contracts begin to expire they may not be replaced fully and this can lead to an increase in market power.

Vesting contracts help solve a second problem, the riskiness of fixed-price retail contracts. Utilities are in a risky position when they divest generation and are still required to serve their load at a regulated price. In California the utilities were initially discouraged from hedging these risks but later when they were encouraged, they demurred for reasons that are not well understood. While it is possible to fully hedge such a position by buying forwards for a couple of years, forward purchases become increasingly difficult beyond that time period. The simplest solution is to hedge their position with required vesting contracts. If these are indexed to fuel costs, the owners of the divested generation will be exposed to relatively little risk. The arrangement provides a large amount of forward contracting at the time of divestiture.

Two central questions of power markets remain open: How much forward contracting will occur naturally and what term structure will it have? If the answer is that 95% of total energy produced will be sold forward with a term of more than two years, power markets may have little trouble with market power. If the answer is that only 80% will be sold forward with a term of a year or more, then power markets may face serious problems. A related question asks if anything can be done to increase forward contracting. The most popular answer seems to be to increase the riskiness of the market. This is potentially quite costly, and it would be one of the few cases where deliberately causing problems proves beneficial. All of these questions deserve far more attention, and until they are answered, power markets will need close monitoring.

## 4-4.4  DEMAND UNCERTAINTY AND SUPPLY-CURVE BIDDING

A second factor reducing market power below the level predicted by the Cournot model is that suppliers do not bid in the Cournot style. They do not simply bid quantities but instead bid supply curves. "Supply-curve bidding," however, refers to more than just a formal bidding procedure in which a price is specified at a number of output levels.

**What Is Supply-Curve Bidding?**  Supply-curve bids come in two types: the Cournot type, and the true supply-curve type. A Cournot bid curve will set price to marginal cost up to some Cournot quantity limit and then set price at the price cap beyond that.[1] True supply-curve bidding differs from this prescription by setting price above marginal cost but below the price cap for some quantities. This is a qualitatively different strategy and is referred to as "supply-curve bidding." It is important because the use of this bidding strategy reduces market power relative to the Cournot strategy.

Like Cournot competition, supply-curve competition is defined in terms of a game played between suppliers.[2] In a supply-curve equilibrium, just as in a Cournot equilibrium, each supplier plays a strategy that is optimal given the other player's strategy. This is the standard Nash equilibrium concept.

**Why Suppliers Bid Supply Curves.**  The first step in analyzing a supply-curve equilibrium is to understand why suppliers would use a supply-curve strategy instead of a Cournot strategy. This is puzzling because it leads to a less-profitable outcome for all suppliers than if all use Cournot bids. Supply-curve bids are used to exercise market power in the face of uncertain demand. The supplier's problem is that different demand levels require different Cournot bid quantities, but the level of demand is unknown at the time bids are submitted. Generally if demand is low, the Cournot quantity is low, and if it is high, the Cournot quantity is high. By bidding a supply curve, it is possible to exercise some market power in either case

---

1. These Cournot bid curves result in the same equilibria as do standard Cournot bids, except when the bid cap is encountered.

2. The original paper on supply-curve bidding is by Klemperer and Meyer (1989). For subsequent developments in supply-curve analysis see Green (1992; 1996) and Baldick and Hogan (2001).

**Figure 4-4.2**

Finding residual demand for suppliers other than S1.

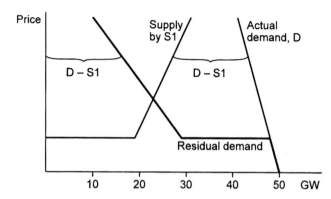

and, thereby, increase profit above the level achieved by any one Cournot strategy, which will necessarily prove wrong for most realized levels of demand.

This suggests that one supplier will find a supply-curve bid to be more profitable than a Cournot bid, and that is correct. But, if all suppliers use supply-curve bidding, all of their profits are reduced. Suppliers will fall into this trap as follows: Say a Cournot strategy would produce a profit of 10 for every supplier. If one defects and uses a supply-curve bid, that supplier will increase its profit, perhaps to say 12, but this will reduce the profits of others to, say 8. This logic causes all to defect, and the result is increased competition and lower profits for all. This is a variation on the "prisoner's dilemma" in which all suppliers are prisoners. Supply-curve bidding just increases competition.

**Why Supply-Curve Bids Increase Competition.**   When one supplier bids a supply curve instead of a fixed quantity, it reduces the market power of other suppliers. This is best understood by using the concept of a "residual demand curve." Figure 4-4.2 shows a very steep demand curve and a supply-curve bid by one particular supplier, S1. If the market has two suppliers, the other can analyze the market by taking both the demand curve and the supply of S1 into account. When price increases, demand decreases, and supply by S1 increases. Subtracting the supply of S1 from demand gives the residual demand that can be served by all other suppliers. This residual demand curve is also shown in Figure 4-4.2.

The reduced slope of the residual demand curve relative to the original demand curve indicates that it is much more price sensitive (elastic) than the true demand curve. When price increases, residual demand decreases for two reasons: (1) True demand decreases, and (2) the supply from other suppliers increases. Cournot competition is equivalent to the limiting case of supply-curve bidding where, because S1 is vertical, the residual demand curve is no more price sensitive than the true demand curve. But with supply-curve competition, each supplier makes the other supplier's residual demand curve more sensitive to price. With several suppliers, each must subtract the supply curves of all other suppliers from the demand curve to find its own residual demand curve. Each finds that increasing its price reduces its sales more with supply-curve competition than with Cournot competition. This discourages the exercise of market power.

**How to Encourage Competitive Bidding.**  The more uncertain suppliers are about the level of demand, the less their supply-curve bids resemble a Cournot strategy and the more they resemble competitive marginal cost curve bids. Fortunately, there is a second phenomenon that has the same effect as demand uncertainty. Demand fluctuates significantly over the course of a day, so if a supplier is forced to submit a single bid schedule for the entire day, it must accommodate a wide range of demand levels. Ideally the supplier would like to bid more aggressively on peak and less aggressively off peak, but the restriction to one bid schedule makes this impossible. The effect is the same as increasing the bidder's uncertainty about the level of demand.

It may be possible for a market designer to take advantage of supply-curve bidding by requiring bidders to submit a single price schedule for the entire day. This schedule should be used for the day-ahead market, the hourly market, and the RT market. The principle argument against such a rule is that variable costs may change during the day, and the primary example is the variation of heat rate with temperature. This phenomenon is well understood and can easily be handled with a nonmanipulable correction factor or by allowing modest last-minute changes in the bids.

Unfortunately, even if bids must be submitted for an entire day, they can be effectively manipulated by physical withholding. This can turn them back into RT Cournot bids. The net result is not obvious.

The NYISO carries this approach one step further. Although it probably is not consciously basing its approach on the theory of supply-curve bidding, that is what underlies its effectiveness. NYISO establishes reference levels by looking at a supplier's past bids, and uses those reference levels to reset bids if necessary to conform to the past bid levels. This is a highly complex and interventionist approach, but it is based on the same concept. If suppliers are forced to live with their bids over a long period of time during which demand fluctuates, they will not be able to exercise as much market power as if they could tune their bids to the particular market circumstances.

As with the one-day-long bids, NYISO's approach deserves more research. Again, there is a question of physical withholding. There are also questions of whether more conventional and less interventionist rules could not serve as well. This is not intended to suggest a conclusion. The market-power problems of NYISO may warrant this approach, or such an approach could even be optimal. Policies to induce more elastic supply-curve bids deserve more investigation.

## 4-4.5   TECHNICAL SUPPLEMENT: CALCULATIONS FOR SECTION 4-4.3

This supplement derives the results presented in the example of Section 4-4.3, "Market Power Reduced On Peak." Four incumbents utilities each have a peak load obligation of 12.5 GW. Each of these utilities and a fifth supplier without any obligation to serve load or any forward contracts have a capacity of 10,000 MW and a constant marginal cost of $20/MWh up to its capacity limit. Prices are all measured in dollars per megawatt-hour and quantities in megawatts.

The demand curve is linear, characterized by the shift parameter $Q_0$, and defined by two facts: $P(Q_0) = 20$ and $P(0.9 \times Q_0) = 1020$. This implies the following:

Inverse demand function:        $P(Q) = b \times (Q_0 - Q) + 20$

where                           $b = 10{,}000/Q_0$

$Q_0$ is the demand when price is equal to marginal cost. The demand function is found by solving for $Q(P)$ from the inverse demand function.

Demand function:                $Q(P) = Q_0 - (P - 20)/b$

Assuming that load is shared equally by the four utilities at all load levels, as long as it is below 40,000 MW, no utility will pay more than $20/MWh because it can self-supply for that cost. When demand is higher, they will produce at full output, and supplier 5 will face a residual demand curve of

Residual demand function:       $q(P) = Q_0 - (P - 20)/b - 40{,}000$

Inverse residual demand:        $P(q) = b \times (Q_0 - 40{,}000 - q) + 20$

where $q$ is the output of the fifth supplier.

The fifth supplier uses the inverse residual demand function to determine its profits from its output. Its short-run profits as a function of output are

Short-run profit:               $SR_\pi(q) = P(q) \times q - MC \times q$

Profits are maximized by optimizing $q$, which requires differentiation with respect to $q$ and setting the result equal to zero. This gives

Profit maximizing condition:    $0 = 20 + b \times (Q_0 - 40{,}000) - 2bq - 20$

or,                             $q = (Q_0 - 40{,}000)/2$

Substituting this value into $P(q)$ gives the market price as a result of the fifth supplier's exercise of market power

Monopoly equilibrium price:     $P = b \times (Q_0 - 40{,}000)/2 + 20$

Using the fact that $b = 10{,}000/Q_0$, $P$ can be plotted as a function of $Q_0$ as shown in Figure 4-4.1.

For the sake of comparison, the market price for a five-supplier oligopoly without forward contracts or other obligations to loads can be computed as follows. This case can be solved from the Cournot markup condition, Equation 4-4.2.

Cournot condition:   $L_X = (P - MC)/P = s/e$

$s$ is known to be 0.2 because there are five identical suppliers, but elasticity $e$ must be determined from the inverse demand function.

Elasticity definition:          $e = -(P/Q)/(dP/Q) = P/bQ$

Substituting the values of $s$ and $e$ into the Cournot equilibrium condition gives

Cournot condition:        $P(Q) - MC = 0.2\,bQ$

The market must still be at a point on the demand curve so the inverse demand function holds and can be used to replace $P(Q)$ in the Cournot condition giving:

$$b \times (Q_0 - Q) = 0.2\, b\, Q$$

This can be solved to find $Q = Q_0/1.2$. Substituting the value of $Q$ and the formula for $b$ into the inverse demand function gives $P = \$1{,}686/\text{MWh}$. This is a constant and does not depend on the level of demand as parameterized by $Q_0$. This constant value is plotted in Figure 4-4.1 for the levels of demand experienced by this market.

*Chapter  4-5*
# Predicting Market Power

*A friend in the market is better than money in the chest.*

Proverb
collected in Thomas Fuller, *Gnomologia*
1732

$S$TANDARD "WISDOM" HOLDS THAT HHIS BELOW **1000** ARE CER-
TAINLY SAFE—THEY ARE NOT. The HHI accounts for only one factor, concentration, out of five key economic factors that determine the extent of market power.
The other four, demand elasticity, style of competition, forward contracting, and
geographical extent of the market, can each affect market power by an order of
magnitude.

**Chapter Summary 4-5 :**   The HHI misses most of the action in power markets.
Cournot models can capture much more but still miss the mark widely. Results
are often reported in terms of the Lerner index, which frequently reports a decline
in marginal cost as if it were an increase in price. A combination of estimating
absolute market power and predicting relative market power may answer some
questions a little more accurately. These include the impact of mergers, transmission
upgrades and transmission pricing.

**Section 1: Four Factors which HHI Ignores.**   The Herfindahl-Hirschman
index is computed from the market shares of suppliers. It takes no account of
demand inelasticity which, other things being equal, makes market power at least
10 times worse in power markets than in most other markets. It predicts only the
Lerner index, which is loosely connected to market power. It takes no account of
the style of competition, the extent of forward contracting, or the geographical
extent of the market.

**Section 2: Difficulties Interpreting the Lerner Index.**   While the definition
of market power compares market price with the competitive price, the Lerner index
compares market price with marginal cost in the uncompetitive market. The Lerner
index combines the effects of reduced marginal cost and increased price. In a power

market, marginal-cost reduction can dominate this measure of market power, yet reducing marginal cost causes no harm.

**Section 3: Estimating Market Power.** The best current approach, Cournot modeling, is far superior to the HHI. But competition in power markets does not follow the Cournot model, and the demand elasticity is essentially unknown. This leaves Cournot predictions uncertain by an order of magnitude. Questions that ask about relative market power, for example before and after a merger, can take advantage of the fact that the merger leaves unchanged the two most problematic factors: style of competition and demand elasticity.

**Section 4: Technical Supplement.** This derives a formula for the Lerner index of a supplier with forward contracts in a Cournot market.

## 4-5.1   FOUR FACTORS THAT HHI IGNORES

The HHI is easily computed and, in a Cournot market, it is directly related to the Lerner index, a common measure of market power. In spite of its popularity, it provides almost no guidance when used in a power market. HHI only takes account of supplier concentration; this section discusses the four factors it fails to consider.

**Demand Elasticity.** Because of the extreme inelasticity of short-run demand in the power industry, HHI is particularly misleading. An HHI of 1000 indicates a Lerner index of 10% (price is 11% higher than marginal cost) if demand elasticity is 1. For this reason, an HHI of 1000 is said to assure a "workably-competitive" market. But with short-run demand elasticity near zero, as is the case for most power markets, Cournot competitors would set price at least 100 times higher than marginal cost. This happens from time to time in some power markets and tenfold markups (1000%) happen in all of them, but most of the time the markup is very low. HHI alone explains none of this, but demand inelasticity provides a large part of the answer.

**The Style of Competition.** Cournot competition is only one possible style of competition. Cournot suppliers compete by choosing a level of output, and they base that choice on the assumption that the choices of other suppliers will not be affected by their decision. For example, they assume that if they reduce their output, their competitor will not choose to produce more as a result of their reduction. Power-market competition is a more complex process.

In power auctions, suppliers often bid upward-sloping supply curves, so if one supplier cuts back, other suppliers will produce more. The price will rise but not nearly as much as if the other suppliers had held to a Cournot strategy. This is supply-curve bidding and, as explained in Section 4-4.4, it can greatly reduce the exercise of market power relative to Cournot competition.

Other styles of competition range from outright collusion to Bertrand competition which causes an oligopoly of only two suppliers to act as perfect competitors. The HHI fails to account for the style of competition.

**Forward Contracting.**   Section 4-4.3 demonstrated the importance of forward contracting to power markets. Given their low elasticities, all power markets would "melt down" were it not for the forward contracts and obligations of suppliers. As a first approximation, the market power problem in electricity markets should be thought of as follows: supplier concentration, measured by the HHI, is roughly normal, but demand elasticity is roughly nonexistent and forward contracts play a large role. Relative to a typical market, demand elasticity increases market power by a factor of 10 to 100, while forward contracts decrease it by roughly a factor of 10. To depend on the HHI is to guess that the two larger factors, which it ignores, will cancel each other out. HHI plays a smaller role than either demand elasticity or forward contracting.

**The Geographical Extent of the Market.**   To compute the HHI, a boundary must be drawn around the market, but the HHI provides no guidance on this problem. Calculating an HHI within the geographical boundary of a privatized utility may give an HHI of nearly 1 (10,000), while calculating it within the entire interconnection may give one of 0.01 (100). If it were not for transmission constraints the latter would be correct, but such constraints are numerous and complex.

This problem cannot be ignored because those who compute the sum of squared market shares must decide which shares to sum, so it has received regulatory attention from both DOJ and FERC. DOJ invented the "hypothetical monopolist" test for drawing market boundaries, while FERC invented the now defunct Hub-and-Spoke procedure. While DOJ's procedure has the appearance of economic logic, neither has been proven to have any particular economic property, and neither has been demonstrated to work for any example problems. An HHI computed within the hypothetical monopolist boundary still fails to account for three key factors, and at best, accounts for the fourth only crudely.

## 4-5.2   WHY THE LERNER INDEX IS UNRELIABLE

The HHI is often treated as a measure of market power, and in one circumstance it is. When (1) demand elasticity is one, (2) the style of competition is Cournot, (3) there are no forward contracts, and (4) the market is effectively located at a single point, the HHI is equal to the average Lerner index. This a measure of market power—of how far price is above marginal cost.

**Measuring Decreased Marginal Cost.**   How well does $L_X$, or its share-weighted average, measure market power? It compares price to marginal cost but measures both after the exercise of market power. In most markets, marginal cost remains relatively constant when market power is exercised, and $L_X$ increases almost entirely because the price increases. But what if market power predominately caused a decrease in marginal cost?

Consider a power market with a little demand elasticity and a supply curve that is horizontal at $30 out to 40 GW and then becomes vertical as shown in Figure 4-5.1. In a tight market the demand curve might intersect the supply curve at $1,000/MWh, and that would be the competitive price. Say a supplier withheld

**Figure 4-5.1**

A high markup and high $L_X$ can result from a price increase or a reduction in marginal cost.

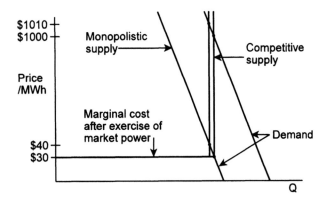

100 MW of supply and raised the market price to $1,010/MWh, a 5% increase. The result would be $L_X = (1010 - 30)/1010$, which is about 97%, very near the maximum possible value of 100% which indicates infinite market power. For comparison, say the price was raised from $30/MWh to $40/MWh, again a $10/MWh increase. The resulting $L_X$ would be $(40 - 30)/40$, or 25%.

In the above example the Lerner index indicates that raising the price from $1,000 to $1,010/MWh is a huge exercise of market power while raising it from $30 to $40/MWh is a moderate increase of market power. The first increase is flagged as extreme not because of the price increase but because of the marginal-cost decrease; yet the decrease harms no one.

**The Extent of the Problem.** The effect of changes in marginal cost on the Lerner index depends on the relative steepness of the supply and demand curves at the market-power equilibrium. If they are equally steep, then withholding will lower marginal cost as much as it raises price. Half of the value of $L_X$ will be erroneous. If the supply curve is only half as steep as the demand curve, only $\frac{1}{3}$ of the value of $L_X$ will be in error.

Unfortunately, most market power is exercised where the supply curve is steepest. Moreover, it is not the demand curve but the residual demand curve that matters. The residual demand curve includes the responses of all other suppliers when one exercises market power. It determines how much price will increase when the supplier in question withholds output. As explained in Section 4-4.4, residual demand curves are much more elastic and less steep than true demand curves. Consequently, when market power is exercised it is quite likely that the supply curve will be steeper than the residual demand curve. In this case, the Lerner index will indicate mainly the reduction in marginal cost.

**A Lerner Index with the Wrong Sign.** Consider a market with seven suppliers. Two are large utilities that have divested some of their generation to the other five with the result that the two now have market shares of 30% each. The other five each have 8% of the market. The two utilities have retained all of their obligations to serve native load, and the five independent power producers (IPPs) have sold no long-term contracts. This example is similar to the California market in the summer of 2000.

The HHI of this market is $2(30)^2 + 5(8)^2$, or 2120. Standard wisdom suggests this is a bit too high as 1800 is taken to be the safe cutoff.[1] But the two utilities are short of generation and must serve their customers at a regulated price. Consequently they have an incentive to lower price, not to raise it. In spite of this, they contribute 1800 out of the 2120 points to the HHI. Perversely, the high value of HHI is almost entirely due to the presence of suppliers who are doing their best to lower the market price.

In a Cournot market without forward contracts, the HHI is equal to the average Lerner index, but when forward contracts are present, they are not equal. The average Lerner index for this market can be calculated from the Lerner indexes of individual Cournot competitors some of which have forward contracts (see derivation in Technical Supplement). Assuming the forward contracts are long term, an individual supplier's Lerner index is given by,

$$L_X = ss/e, \quad \text{where } ss = (q - q_F)/Q \qquad (4\text{-}5.1)$$

$Q$ is the total quantity produced, or the sum of all $q$'s, and $q_F$ is the quantity sold forward by the supplier in question. This is the same formula for $L_X$ as given in Result 4-3.1 except that share, $s$, has been replaced by spot-market share, $ss$. When nothing is sold forward, the two agree.

The IPPs have no forward sales, so their spot shares are just their market shares, or 8% each. The utilities have obligations to their loads that are equivalent to forward contracts in the amount of 50% of total supply, so their spot share is 30% – 50% or –20%. The supplier's shares of total market supply are still 30% for utilities and 8% for IPPs, so

$$\text{Average } (e \times L_X) = 2 \times .30 \times (-.20) + 5 \times .8 \times .8$$
$$= -.1200 + .0320 = -.0880$$

and            $$\text{Average}(L_X) = -.088/e$$

The share-weighted average Lerner index is *negative* 0.088 divided by the elasticity of demand. The negative contribution comes entirely from the utilities and is due to their acting as oligopsonists and producing power at a loss in order to reduce the market price for all the power they buy from the IPPs. Producing extra raises their marginal cost above the market price and gives them a negative Lerner index. This leaves unanswered the question of whether the market price, with the utilities trying to lower it and the IPPs trying to raise it, is higher or lower than the competitive price. The Lerner index counts the monopsony power as negative and the monopoly power as positive, but does it get the right sign for the net effect?

The direction of the price change is determined by the net withholding. Do the IPPs withhold more than the extra power produced by the utilities? If the supply curves are flat to full capacity and then become vertical, the utilities can produce no extra power and the IPPs will clearly control the price. More realistically, the supply curves will display a very rapidly increasing marginal cost near full capacity. Generally, an increase in output will increase marginal cost more than a decrease

---

1. An 1800 cutoff is a compromise between those who thought five firms provided sufficient competition and others who thought six firms were needed. This logic did not consider power markets or their unusual structure.

in output will decrease it. So, although the utilities can increase production a little, their change in output will be more limited. Thus the market power of the IPPs will dominate, and the market price will increase above the competitive level.

After divestiture, market price will increase due to the exercise of market power by the IPPs. The increase in market power is accompanied by a decline in the Lerner index from zero to some negative value. The Lerner index has the wrong sign.

Another problem of the Lerner index occurs when price goes negative as it sometimes does in power markets. When price goes negative, the denominator of $L_X$ passes through zero, making $L_X$ first approach negative infinity and then positive infinity. When averaging the Lerner index over a year, these meaningless extreme values can dominate the result.

**What to Use Instead of the Lerner Index.**  Although the Lerner index is a reasonable indicator of market power in most markets, its sensitivity to changes in marginal cost makes it unsuitable for use in power markets. There is an appropriate replacement. Use the price increase, that is, the difference between the actual price, $P^e$, and the competitive price, $P^*$. The actual price is the monopolistic, equilibrium, market price determined by the exercise of market power. This is the difference on which the definition of market power is based and, as explained in Chapter 4-1, it is called the monopoly price distortion.

When price distortion, $\Delta P_{\text{distort}}$, is averaged over time, it gives the average price increase without any of the difficulties experienced by $L_X$. The average price distortion can easily be compared with either the average retail price or the average wholesale price, and the ratio of price increases to one of these prices can be used if a percentage is desired. The price distortion does not fit conveniently with Cournot theory, but its interpretation is completely straightforward. Sometimes the easy answer is the best answer.

---

| Result | 4-5.1 | **Price Distortion Measures Market Power Better** |
|---|---|---|

In power markets where marginal cost changes quickly, price distortion, the actual market price minus the competitive price, is a better indicator of market power than the Lerner index. For comparisons between markets, the ratio of price distortion to the long-run average price should be used.

---

## 4-5.3   ESTIMATING MARKET POWER

**The Cournot Approach.**  In power markets, the HHI is nearly irrelevant as an indicator of market power. What should be used instead? No currently known method of calculation gives a reasonably good prediction of market power. The Cournot model is probably the best available model. It can account for transmission constraints that cause the market to have geographic extent. It can also account for forward contracts, as shown in the Technical Supplement, and demand elasticity. Its principal shortcoming, in theory, is that competition in a power market is some type of supply-curve competition and not Cournot competition. Unfortunately models of supply-curve competition are still ambiguous and intractable, though development of practical approaches to their use is underway.

An equally important practical shortcoming is that demand elasticity is unknown. Perhaps the most influential study to adopt this approach (Borenstein and Bushnell, 1998, 16) states "We have run simulations for elasticities 0.1, 0.4, and 1.0, a range covering most current estimates of short-run and long-run price elasticity." This is a broad range, but perhaps not broad enough. A supply elasticity of 0.1 means that raising the price by 50% would produce a 5% decline in demand. For example, if the price in PJM were raised from $200/MWh to $300/MWh, demand would decrease by 5% of say 40,000 MW or by 2000 MW. PJM's dispatchers might suggest a demand elasticity several times lower as more realistic. In fact many consider the one-day demand elasticity to be effectively zero.

The consequence of using this range of values is demonstrated in the study's estimates. Table 4 of their report shows the values for the Lerner index given in Table 4-5.1 under peak load conditions in September when the exercise of market power is most crucial.

**Table 4-5.1** Effect of Elasticity Uncertainty

| Elasticity | Lerner Index | Price (*MC* = $50/MWh*) | Price Distortion (increase) |
|---|---|---|---|
| 1.0 | 15% | $59/MWh | $9/MWh |
| 0.4 | 64% | $139/MWh | $89/MWh |
| 0.1 | 99% | $5,000/MWh | $4,950/MWh |

* The "price" and "price distortion" values are not reported by Borenstein and Bushnell (1998, 23). These are calculated from an arbitrarily assumed marginal cost of $50/MWh and included only to aid in the interpretation of the Lerner index, which is highly nonlinear.

Part of the problem here may be with the misleading nature of the Lerner index as explained above. Borenstein and Bushnell (1998) do not distinguish between reductions in marginal cost and increases in market price, so the 500-fold range of price increase overstates the ambiguity of the results by an unknown amount. At other times the variation in markups is much less dramatic, and in some circumstance useful conclusions can be drawn though they might be invalidated if the possibility of lower elasticities were recognized.

**Relative Market Power.**  The Cournot approach is, by definition, unable to account for the style of competition and, because of lack of data, unable to account for demand elasticity. Although these key factors are unknown, one fact is known; these factors are quite constant relative to many changes in market structure. For example, if two suppliers merge, this will not change the demand elasticity and will probably have little, if any, impact on the style of competition. If suppliers bid supply curves before, they will most likely continue to bid supply curves after the merger. The same observation applies to the change in structure caused by building a new transmission line. It will change market power, but it will not change demand elasticity or style of competition. Similarly, a change in forward contracting or in transmission pricing will change neither demand elasticity nor competition.

These observations open the door to investigating a particular type of market-power question without relying on either the assumption of Cournot competition or a guess at true demand elasticity. The suggested approach is to use a Cournot model in both cases with demand elasticity held constant. This differs from a strict Cournot approach by using an estimate of market power to select a demand *elasticity parameter* that would partially account for actual demand elasticity and partially correct for the fact that competition in power markets does not follow the Cournot model. This helps overcome both major deficiencies of the strict Cournot approach. Because the approach is used to compare two markets with the same style of competition and elasticity, the predicted change in market power will be relatively insensitive to errors in the estimated elasticity parameter.

Why might using the *elasticity parameter*, which is a deliberate mis-estimate of the true demand elasticity, improve market-power predictions? Cournot competition assumes that other suppliers do not change output in response to a supplier that exercises market power. Consequently the full response must come from load.[2] Supply-curve bidding suggests that other suppliers will respond according to their supply curves. This response is quite similar to demand elasticity. Using a value of demand elasticity that compensates for the lack of supply response in the Cournot model should improve the performance of the model.

How could this approach be carried out? First, one caveat must be reiterated. This is not a method of predicting market power but of predicting changes in market power when certain circumstances change while style of competition and demand elasticity remain constant. It asks a less-ambitious question than the strict Cournot approach attempts to answer.

Suppose the present state of the market is A and that B is the proposed state after some change, for example a merger. The first step would make at least a rough estimate of the market power in state A. This is not a prediction based on a Cournot model; it is an estimate based on actual data such as found in Borenstein, Bushnell, and Wolak (2000) and Joskow and Kahn (2001a; 2001b). The second step is to model the market in states A and B with a Cournot model as was done by Borenstein and Bushnell (1998). The third step is to select the demand *elasticity parameter* that makes the Cournot model of state A reproduce the market power found in state A by the step-1 estimate. The final step is to use that elasticity parameter in the Cournot model of state B to estimate the market power in that state and compare it with the Cournot predictions (using the same elasticity parameter) for state A.

## 4-5.4   TECHNICAL SUPPLEMENT: MARKET POWER AND FORWARD CONTRACTS

When a supplier has sold forward a quantity of its output, that sale will affect its future exercise of market power in the spot market. If the supplier does not anticipate that today's energy price will affect tomorrow's price of forward contracts, or if the forwards are all very long term so there will be no repeat sales for a long time, the profit function can be written as

---

2. This describes a pure Cournot model. Borenstein and Bushnell (1998) use a more sophisticated approach that includes an elastic "competitive fringe" of small price-taking suppliers.

$$SR_\pi(q) = P(Q)(q - q_F) + p_F q_F - c(q)$$

where $p_F$ is the forward price, $q$ is the supplier's output, and $q_F$ is the quantity of power sold forward which may be greater than its output. From here, proceed as with a normal Cournot supplier to find

$$\frac{d\,SR(q)}{dq} = P(Q) + \frac{d\,P(Q)}{dq}(q - q_F) - \frac{d\,c(q)}{dq} = 0$$

$$1 + \frac{(q - q_F)}{Q}\frac{Q}{P}\frac{dP}{dq} = \frac{MC(q)}{P}$$

$$ss\frac{1}{-e} = \frac{MC}{P} - 1$$

$$L_X = \frac{ss}{e}, \quad \text{where} \quad ss = (q - q_F)/Q$$

Note that this derivation again uses the facts that $dP/dq = dP/dQ$ and that demand elasticity is defined to be positive. Spot share, $ss$, plays the same role in the determination of $L_X$ as market share $s$ played previously. If all load is under forward contract, then the sum of $ss$ over all suppliers is zero instead of one. Also note that because $ss$ can be negative, the Lerner index of the corresponding Cournot competitor will be negative. This indicates that it will overproduce, causing its marginal cost to rise above the market price. This is an example of monopsony power as discussed in Section 4-1.3. These producers are net buyers, and they exercise market power in order to push the price down.

# Monitoring Market Power

*Why is there only one monopolies commission?*

Anonymous

$D$ISCOVERING MARKET POWER OFTEN REVEALS MARKET FLAWS. While fixing these flaws, temporary restraints may need to be placed on market participants, but the goal of fixing the market should be kept in focus. If the rules are flawed, repairs can be made quickly, but when the architecture or structure is flawed, the required changes can take years. Then market monitors are forced to spend too much effort controlling an unruly market.

**Chapter Summary 4-6:** The trick to market monitoring is to ignore vague definitions and rigorously apply the economic definition of market power. If the market price is above the competitive level, then necessarily, some supplier is not acting as a price taker—somewhere there is a gap between the profit-maximizing supply of a price taker and the actual supply. That gap is termed the "quantity withheld," and it is proof of market power. Observing a high price is not proof, nor is observing a price higher than left-hand marginal cost. Observing the real-time (RT) price above right-hand marginal cost demonstrates short-run withholding. In the RT market, this proves market power, except for unusual cases of opportunity cost such as exhibited by hydrogenerators.

**Section 1: FERC's Ambiguous Standard.** A popular misinterpretation of the DOJ/FTC definition of market power, abetted by DOJ's Guidelines, requires the price increase to be significant in magnitude as well as duration. FERC's requirement that "no market power" be exercised makes sense only under this misinterpretation; otherwise, the requirement would be an impossibility. As a consequence, FERC's policies regarding market power are highly ambiguous, and its pronouncements on the need to eliminate all market power cause confusion.

**Section 2: Market Monitoring.** Power markets need monitoring because of two structural problems: (1) electric energy cannot be stored, and (2) RT demand is extremely inelastic. Market power is not necessary, not beneficial, and can be measured. It can be detected by finding instances when power could have been, but was not, sold at a profit, taking the market price as fixed. Opportunity costs cause little difficulty in the RT market, and average profits are irrelevant.

## 4-6.1    FERC's Ambiguous Standard

Market power is the ability to affect profitably the market price even a little, and even for a few minutes. This is the economic definition given in Chapter 4-1 and used throughout this book. Under this definition, it is unnecessary and impossible to eliminate all market power from electricity markets. But DOJ defines market power as being "significant," and FERC uses such a definition to call for its complete elimination. This would be an extreme and counterproductive position if FERC were relying on the economic definition.

### The DOJ/FTC Definition of Market Power

In their Horizontal Merger Guidelines, the U.S. Department of Justice (DOJ) and the Federal Trade Commission (FTC) have redefined the term "market power." Their definition follows.

| | |
|---|---|
| **Definition** <br> ·(regulatory) | **Market Power** <br> "Market power to a seller is the ability profitably to maintain prices above competitive levels for a significant period of time."[2] |

This definition rules out cases that economics considers to be market power but regulators consider to be insignificant. It narrows the definition and makes it vague. The Guidelines use the problematic phrase "significant period of time" only in the definition of market power and never discuss its meaning; however, the similar phrase, "small but significant and non-transitory," is used more than 20 times and set off in quotes. Because of the emphasis on this alternative phrase and the complete lack of attention to the defining phrase, "small but significant and non-transitory" has come to be regarded (incorrectly) as part of DOJ's market-power definition.

This key phrase describes the type of price increase that a hypothetical supplier must cause in order to meet the DOJ/FTC test for a "hypothetical monopolist." It is a small step to assume that what turns a hypothetical supplier into a hypothetical monopolist is market power. If this were so, "small but significant and non-transitory" would be part of DOJ's definition of market power. But this is not DOJ's intention.[2]

---

1. The Horizontal Merger Guidelines (DOJ 1997).

2. Dr. Gregory Werden of DOJ states that the Guideline's definition "includes a significance concept only for duration." Personal communication August 14, 2000.

Although the Guidelines give no interpretation of "significant period of time," they do interpret "small but significant and non-transitory."

> *In attempting to determine objectively the effect of a "small but significant and nontransitory" increase in price, the Agency, in most contexts, will use a price increase of five percent lasting for the foreseeable future.*

Despite much circumstantial evidence to the contrary, the DOJ/FTC definition of market power requires no test of significance for the price increase and its suggested 5% rule does not apply to its definition of market power, only to its rule for determining market boundaries. Moreover, the "significant period of time" phrase that is part of the DOJ/FTC definition may not have been intended as a modification of the standard economic definition. DOJ's Dr. Gregory Werden has explained that "While perhaps inappropriate, the duration significance test was, I think, meant only to distinguish purely transitory phenomena resulting from imperfect information or market inertia."[3] If the Guidelines intended only this distinction, then they did not intend to narrow the economic definition, only to clarify it.

## FERC's Use of the Term "Market Power"

FERC's use of the term "market power" makes sense only if the phrase, "small but significant and nontransitory increase in price," is included in the definition. Under either the economic definition or the literal DOJ/FTC definition, FERC's frequent requirements of "no market power" would be unreasonable. All current and foreseeable power markets include the opportunity for suppliers to exercise market power periodically. Only under the misinterpretation of DOJ's definition could FERC have reasonably approved any market-based rates, given its standard of "no market power."

---

| Definition (implied by FERC) | **Market Power (popular misinterpretation of DOJ/FTC):** Market power to a seller is the ability profitably to maintain a small but significant and nontransitory increase in price above competitive levels. |
|---|---|

---

If "significant and nontransitory" is interpreted to mean "so large and long lasting as to be unjust and unreasonable," then it makes perfect sense for FERC to require a standard of "no market power." This interpretation would make sense of the following passages from Order 2000 (emphasis added) (FERC 1999, 144, 464):

> *The Commission has a responsibility under FPA sections 205 and 206 to ensure that rates for wholesale power sales are just and reasonable, and has found that market-based rates can be just and reasonable where the seller has **no market power**. Under the FPA, the Commission has the primary responsibility to ensure that regional wholesale electricity markets served by RTOs **operate without market power**.*

---

3. Personal communication, August 14, 2000.

This pattern of misinterpretation dates back to 1993, the year after DOJ and FTC issued their Guidelines.[4]

The ambiguity in the misinterpretation facilitates writing regulations that have only the appearance of clarity. Each time FERC requires that "affiliates lack market power" or that a party "mitigate any market power that they may have possessed," it appears that they have issued a clear and precise directive when, in fact, interpretation must be based on an undefined phrase in the misinterpretation of DOJ's definition of market power.[5]

The second problem with using the popular misinterpretation is that it presents an easy target to attack. Those opposed to market monitoring typically insist that the standard of "no market power" is unrealistic and leads to harmful regulatory meddling. By either the economic definition or the literal DOJ definition, they are right, "no market power" is an unreasonable standard. This lends credence to their argument even though FERC was not using these standard definitions. As is explained in the next section, power markets need to be monitored, but the goal should never be the prevention of all market power. Market monitors, as well as FERC, need to make a hard decision: How much market power is too much?

## 4-6.2  MARKET MONITORING

Most markets are not overseen by market monitors. Generally, in the United States, this function is left to the legal system, the Department of Justice and the Federal Trade Commission. But there are major exceptions, the stock markets being the most prominent.

Power markets deserve special treatment, first because new power markets are often poorly designed. (Both the California and New England markets have experienced prices of $10,000 for a commodity that normally trades for under $10, and both concluded this was due to market power made possible by a flaw in the market's design.[6]) Second, two attributes of market structure are extremely conducive to market power. Electric energy is almost impossible to store, and the RT elasticity of demand is exceedingly low.

### Market Monitoring Tasks

Market monitoring comprises six distinct tasks:

1. Checking for the exercise of market power and for inefficiency.
2. Discovering inappropriate market rules.
3. Suggesting improvements to market rules.
4. Checking for and penalizing those who violate rules.

---

4. Louisville Gas & Electric Company, 62 FERC 61,016 at 61,146 (1993); Southwestern Public Service Company, 72 FERC 61,208 at 61,966-67 (1995), reh'g pending; Louisiana Energy and Power Authority v. FERC, 141 F.3d 364 (D.C. Cir. 1998).

5. Quoted phrases are from FERC's Order 888 (FERC 1996b, 27)

6. In California the commodity was the lowest grade of operating reserves, and in ISO-NE it was a right to recall installed capacity. See Cramton and Lien (2000) for a description of ISO-NE market flaws.

5. Re-pricing market transactions deemed not to be just and reasonable.

6. Penalizing those who exercise market power.

The first three of these are crucial for the development of efficient power markets. The fourth is crucial for efficient operation, but the task should be carried out by a team different from the one that carries out the first three. Separation is essential because those charged with market analysis and design need to have the trust of market participants, and those charged with rule enforcement are likely to be viewed with suspicion.

The fifth and sixth tasks are most controversial. Both ISO-NE and the NYISO have engaged in considerable after-the-fact re-pricing, and FERC has done it in California, but actual penalties have been rare. In the United States, tasks five and six are complicated by the fact that electricity pricing is governed by the Federal Power Act which requires rates to be "just and reasonable." In an unregulated market, FERC has interpreted this to mean that prices are competitive and "the seller has no market power."[7] The vagueness of the definitions make carrying out tasks five and six controversial. The rest of this section will concentrate on detecting and proving market power.

## Monitoring Fallacies

Data limitations can make even large exercises of market power impossible to detect, which lends credibility to those who claim the task is theoretically impossible. But an even more essential criticism of market monitoring must be addressed before considering these practicalities.

**Is Market Power Essential?**   The most basic claim of those who oppose market monitoring is that market power is actually beneficial and in fact necessary for the efficient operation of the market.[8] This is generally based on three fallacies discussed in Sections 2-2.1 and 2-2.2 and in Chapter 3-9. The first fallacy states that fixed costs cannot be covered by a competitive market, the second that competitive recovery of fixed costs will cause inefficiency, and the third that startup costs cannot be covered. Although the opponents of market monitoring have an abiding faith in *markets*, they are suspicious of *competition* where their own product or client is concerned. This is natural as competition holds down profits. The most fundamental result of economics states that competition makes a market efficient except in the case of natural monopoly.[9]

---

7. This is why the exercise of market power is often actionable in power markets although it is not in other markets. "While it is not illegal to have a monopoly position in a market, . . ." From "Maintaining or Creating a Monopoly," by the FTC, www.ftc.gov/bc/compguide/maintain.htm.

8. This is refuted by Borenstein, Bushnell, and Knittel (1999, 7).

9. There are other exceptions, but the only one market power might fix would be a problem recovering startup costs. This problem is discussed in Chapter 3-8, and market power is not required or helpful.

## The Market-Power Fallacy

---

*Fallacy*  **4-6.1**    **Some Market Power Is Needed and Beneficial**
Because competitive price cannot cover fixed costs, or perhaps because it cannot cover startup costs, market power is needed to keep a sufficient number of generators in business.

---

**Can Market Power Be Measured?** The second argument against market monitoring proceeds in two steps. First, it claims that because scarcity or opportunity costs can cause high prices, the existence of high prices cannot prove market power. The second claim is implicit: Because high prices cannot prove market power, market power is unprovable. The first claim is true; the second is false. Market power cannot be proven by observing high prices, but it can be proven by observing a quantity of output withheld. As explained in Chapter 4-1, withholding can result from either a high offer-price strategy or a physical withholding strategy. The consequent withholding is measured by the difference between what a price taker would produce and what actually is produced. Observing that a supplier is not acting as a price taker proves market power. The effect of the withholding on prices can be used to compute the resulting price increase.

---

*Fallacy*  **4-6.2**    **Market Power Cannot Be Proven**
Because high prices can be caused by scarcity or by opportunity costs, it is not possible to prove that they have in fact been caused by market power. In addition, no market-power determination can be made without waiting a year to see if profits are higher than normal.

---

### How to Detect and Prove Market Power

Consider a realistic example. A gas turbine generator (GT), known to have a cost of about $65/MWh, has been producing at that price for years. One day it bids $850/MWh into the RT market. The market price then goes to $400/MWh and the gas turbine does not start up for over an hour. After two more price increases, the ISO sets the RT price to $850 and in less than 10 minutes the GT is producing at full output. The owner of the GT was long in the RT market on that day, so raising the market price affected another 500 MW of its output. Later the price went higher and when it returned to $400/MWh the generator continued to produce.[10]

Once the GT was producing at full output and the system needed its output and more, the GT was no longer exercising market power even though the market price was $850 and higher. The high price is not evidence of market power, and it is not the setting of the market price by a high bid that constitutes the exercise of market power. Refusing to produce when the price is already well above marginal cost—the

---

10. Why it did not exercise market power as the price fell is puzzling. This might have been due to asymmetries concerning who can set the market price, the behavior of the dispatchers, or the psychology of the supplier.

withholding of output—constitutes an exercise of market power if the supplier is rational.

As explained in Section 4-1.2, depending on the auction rules, market power can still be in effect after the GT sets the price and while it is producing at full output. But to focus on that complexity is to miss the fundamental point. During the period after $P$ first exceeds the GT's $MC$ and before the GT starts up and sets the market price with its high bid, market power is being exercised by the GT (financially) withholding itself from the market. The dispatcher will probably wait to start it, because of its high price, until most of its capacity is needed. Soon, if not immediately, after setting the price, all of its output will be needed and it will no longer be withholding and no longer exercising market power. Once it starts up it is very difficult to prove market power because price does not prove it. Before it starts up and sets the price, it is relatively easy to prove.

---

**Result**   **4-6.1**   **To Detect Market Power, Look for "Missed" Opportunities**
If a supplier would profit (in expectation) from the sale of an additional unit, assuming the market price would not change and the supplier chooses not to sell, it has exercised market power.

---

This rule is not a definition of market power, and it might not provide the proof needed in court, but it misses very few cases and wrongly detects market power only when a supplier makes a costly error. If a supplier refuses to make a sale that is itself profitable, there is only one good reason: to raise the market price so that it will make more on other sales.

**Opportunity Costs.**  Opportunity costs are accounted for by the parenthetical "in expectation." It might be that a supplier is expecting another opportunity to sell at a better price, so a present sale at the market price would not be *expected* to increase its profits. In this case the supplier has not exercised market power. Sometimes evaluating the expected opportunity is difficult; sometimes it is easy. In forward markets it is always difficult, but as Section 4-2.1 explains, market power cannot be exercised there. In real time, the opportunity to shut down is usually not valuable. It can be for hydro resources that expect to sell at a higher price later, and it can be for generators with a fixed allotment of emission allowances. In unusual circumstances, staying off can be an opportunity to defer maintenance. But, except for hydro, almost all generators at almost all times prefer to run rather than not run if they are paid just a little more than their variable costs (including startup if they are off line). In real time, opportunity costs are usually minimal.

**Proving Market Power.**  To turn market power detection into a proof of market power it is necessary to show that the supplier profited from withholding. Without this step, the supplier could claim it withheld by mistake and lost money. There are three approaches to proving profitability: (1) the rationality assumption, (2) accounting, and (3) statistics.

The fundamental assumption of economics is that market participants are rational. For suppliers this means that they maximize profits. If a firm is observed passing up opportunities to sell power at a significant profit under the price-taking assumption, there is only one rational, profit-maximizing explanation. It must be profiting under a more realistic assumption; it must be profiting by raising the market price. Thus, when coupled with the rationality assumption, the repeated detection of market power using Result 4-6.1 is itself proof of market power.

A second approach is simply to learn the facts of a supplier's position in the market and its production costs and compute the profit. If the exercise of market power is minimally profitable, this is difficult; if it is hugely profitable it is usually easy.

The third approach is statistical. If a supplier's withholding behavior is predictable in a statistically significant way, and there is no explanation but market power, this is proof. Suppose the effect on market price of a certain generator's breaking down is calculated and then correlated with claimed breakdowns. If the probability of a breakdown proves to be 90% when the resulting increase in price is $100/MWh or more, and 10% when the resulting increase would be $10/MWh or less, that is proof. There is simply no physical reason that could explain this pattern of breakdowns. It is statistically possible to compute a probability that the pattern would occur by chance. Since the windows of opportunity for large price increases are narrow, the possibility of a dozen chance occurrences could easily be less than one in a billion. Many more subtle applications of the statistical approach are possible.

**Profit Does Not Prove Market Power Nor Does Lack of it Prove Innocence.**   Many factors, such as scarcity, cost of inputs, and technical efficiency, affect profits. These can easily reverse the gains from market power or make a fully competitive firm exceedingly profitable. A supplier that makes double the normal rate of profit five years in a row may not be exercising any market power. A supplier that habitually exercises market power may lose money five years in a row. Moreover, a supplier who exercises market power raises the profits of every supplier in the market, and since there is a cost to exercising market power, those who profit from it *without* exercising it should be most profitable.

# *Part 5*
# Locational Pricing

# Power Transmission and Losses

*ELECTRICITY. n.s. [from electrick. See ELECTRE.] A property in some bodies, whereby, when rubbed so as to grow warm, they draw little bits of paper, or such like substances, to them.*

Samuel Johnson
*The Dictionary of the English Language,* 1755

*ELECTRICITY, n. The power that causes all natural phenomena not known to be caused by something else.*

Ambrose Bierce
*The Devil's Dictionary,* 1881–1906

**W**ATTS MEASURE POWER, BUT VOLTS AND AMPS ARE THE NUTS AND BOLTS OF ELECTRICITY. The economics of power flows can be understood without their help, yet they underlie every important physical phenomenon in the power marketplace. This chapter uses them to explain power flow, transmission losses, and the reason Westinghouse's AC networks triumphed over Edison's DC networks.

**Chapter Summary 5-1:** Voltage is pressure and electrical current is like a flow of water; with more pressure, more current flows. Power delivered is voltage times current (volts times amps) and is measured in watts. The power lost in a transmission line of a given voltage is proportional to the square of the power flow, but if the voltage is doubled the same power can be delivered with $\frac{1}{2}$ the current and $\frac{1}{4}$ the loss. Transformers make it easy to raise an AC voltage and this allows the transmission of power with very little loss.

**Section 1: DC Power lines.** Direct-current power lines, used in the past and again gaining importance, provide the simplest example of power transmission. Power transfer, $W$, equals $V \times I$, which is the voltage, $V$, of the transmission line times the current flow, $I$. Transmission losses are proportional to the square of the delivered power and inversely proportional to the square of the power-line voltage. Consequently, two equal loads cause four times the loss caused by one. This makes a meaningful assignment of losses to loads impossible.

**Section 2: AC Power lines.** In the United States, alternating current (AC) completes a *cycle*, two reversals of direction, 60 times per second. Transformers,

**Figure 5-1.1**

Transmission of power over a DC line.

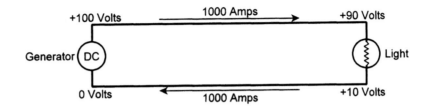

which only work with AC, are used to increase the voltage before it is transmitted, and decrease it once it arrives at the distribution area. The ability to transmit power great distances with little loss makes huge generators and long-distance trade economical.

## 5-1.1 DC POWER LINES

Thomas Edison's downfall as a power marketer was his commitment to direct current (DC). It is conceptually simple; it flows in only one direction and its voltage is quite constant. Alternating current (AC), the type available from electric outlets in every home, is more complex but can easily be "transformed" from one voltage to another. This provides enormous advantages for long-distance transmission. Ironically, DC may be the wave of the future. Solid-state technology now makes high-voltage DC power lines economical on occasion, and they are far more controllable.

### Kirchhoff's Laws

Imagine a DC generator sending power down a long transmission line to a small group of houses equipped with Edison's light bulbs. The voltage produced by the generator is low, perhaps 100 V, and the generator produces only 100 kW. All the lights together can be thought of as a single large light bulb as shown in Figure 5-1.1.

Electricity is like water, voltage is like water pressure, and a generator is like a water pump. This analogy, based on the physics of electricity is extremely useful. Electrons move as freely inside a copper wire as do water molecules inside a pipe, and one electron pushes against its neighbor in almost the same way that one water molecule pushes against the next. The generator creates a flow of electricity called an electrical current. This flow leaves the positive-pressure side of the generator, flows through one transmission wire, then through the light bulb and then back through the other wire and into the low-pressure side of the pump.[1]

This system is like a perfect water system as there can be no net loss of the electrical fluid. Wires cannot lose their electrons. If you push some in one end,

---

1. Benjamin Franklin, knowing nothing about electrons, defined electric current as flowing from plus to minus. Electrons proved to be negative, so they flow from minus to plus.

the same number will be pushed out the other end.[2] The current flow into the light exactly equals the current flow out of the light. Electrical power is used up in the light and lost in transmission, but current is not. (Water power is used up by a water wheel, but the water is not.) The current that flows out of any point (node) in a network must be replaced instantly by an equal amount of current flowing in. This is Kirchhoff's first law.

The light can be turned off by disconnecting one wire from the light with a switch. This is like turning off a faucet because this stops the flow of current. Electrons cannot normally jump through the air.[3] The switch will work just as well on either side of the light. Providing power is not a matter of electricity getting *to* the light but of flowing *through* the light, and to do this it must make a complete *circuit* and return to the generator. This is an *electrical circuit*. A *short circuit* occurs when the two transmission lines touch in the middle, and the current goes half way to the light then turns around and comes back the short way.

---

| Result       5-1.1 | **Kirchhoff's Laws** |
|---|---|
| First Law: | The current flow into any point (node) in a circuit equals the current flow out. |
| Second Law: | The voltage drops around any loop sum to zero. |

---

Kirchhoff's second law is a simple consequence of the fact that voltage is like pressure or elevation. It says that if you walk around any complete circuit, ending where you started, you will walk down (to lower voltage) exactly as much as you walk up (to higher voltage). In Figure 5-1.1, $(100 - 90) + (90 - 10) + (10 - 0) + (0 - 100) = 0$.

## The Power Law

The filament of a light bulb is very, very thin, and consequently electrons flowing through it must flow more rapidly, just as a river flows most rapidly where it is narrowest. Flowing faster, the electrons collide more forcefully with the tungsten atoms of the filament and make the filament white hot. This friction also causes a loss in electrical pressure between the inflow side and the outflow side of the light, from 90 V down to 10 V in Figure 5-1.1. Just as there is a loss in elevation from the inflow to the outflow of the narrows in a river, pushing electrons rapidly through a narrow wire, or water rapidly through a narrow channel, takes power. In the case of electricity, the power used is measured in watts and the formula is given in Result 5-1.1.

---

2. Small violations of this law are possible due to "capacitance" but not enough to be noticeable in a DC system.

3. They can if the voltage is greater than approximately 10,000 V/cm, so high voltage lines must be kept far enough apart to prevent arcing (sparks).

| Result | 5-1.2 | **Power Equals Voltage Times Current (Volts × Amps)** |
|---|---|---|

The electrical power, $W$, measured in watts, consumed by any element of an electrical circuit equals the voltage drop, $V$, across that element times the current, $I$, flowing through that element.

Power, which is measured in **watts**, is traditionally denoted by $P$, but because $P$ is so frequently used for price, $W$ will be used instead. Voltage is most traditionally denoted by $E$ (for electromotive force), but now $V$ is often used. Electrical current flow is measured in **amperes**, or amps, but the traditional symbol for current is $I$, while for amps, the units of current, the symbol is A. In Figure 5-1.1, the power used by the light is 80 V × 1000 A, or 80,000 watts, or 80 kW. Similarly the power output of the generator is 100 V × 1000 A, or 100 kW. The discrepancy is due to transmission losses which follow the same law as any other use of power. Each transmission line uses 10 V × 1000 A, or 10 kW.

Electrical friction makes it more difficult to push the electric current through the filament of the light bulb than through the much larger transmission lines. There must be a larger voltage (pressure) drop across the filament to achieve the same current as flows through the transmission lines. This relationship between voltage, current and friction is called Ohm's law. Electrical friction, or the difficulty of pushing current through a wire or appliance, is known as resistance, denoted by $R$, and measured in **ohms**.

| Result | 5-1.3 | **Ohm's Law Is Voltage Equals Current Times Resistance ($I × R$)** |
|---|---|---|

The electrical current, $I$, flowing through a conductor equals the voltage drop across the conductor, $V$, divided by the resistance, $R$, of the conductor: $V = I \times R$. (Voltage is measured in volts, current in amps and resistance in ohms.)

Ohm's law is useful because resistances tend to stay constant, especially for wires. Consequently, for the current through a wire to double, the voltage drop between its ends must also double.

## Transmission Losses: Why Power Lines Are High Voltage

Combining the power law with Ohm's law can explain why transmission lines use high voltage. As a first step, compute the resistance of the transmission lines of Figure 5-1.1. Applying Ohm's law ($V = I \times R$) to one wire of the transmission line shows that $10 = 1000 \times R$. This means that $R = 0.01$ ohms (the unit of resistance). Now imagine that the generator of Figure 5-1.1 is changed to produce 1000 V and only 100 A. Its power output would be unchanged at 100 kW. What would happen to line losses? According to Ohm's law, $V = 100 \times 0.01 = 1$ volt. So the voltage drop from end to end of one line is 10 times less than in Figure 5-1.1. But the current (100 A) is also 10 times less, so the power loss is only $L = 1\,V \times 100\,A = 100\,W$, or 100 times less. Raising the voltage by 10 times reduces transmission line losses by a factor of 100 with no reduction in power delivered. Of the 20 W formerly lost, 19.8 can now be delivered to the load.

| | | |
|---|---|---|
| **Result** | **5-1.4** | **Transmission Losses Are Proportional to Power$^2$/ Voltage$^2$** |

$$L = a W^2, \text{ where } a = R_T/V^2$$

Transmission losses, $L$, are proportional to the square of the power consumed by the load, $W^2$, and the line resistance, $R_T$, and inversely proportional to the square of the line voltage, $V$. (This is accurate for a DC line and a good approximation for AC).

---

When this relationship is formalized it not only explains how losses decrease with voltage, but also how power losses are determined. This is the key input to the pricing of losses discussed in Chapters 5-7 and 5-8.

To find this relationship, assume that the system operator keeps the voltage at the load constant as its power usage changes. The power law indicates that power consumed by the load will equal the voltage at the load times the current flow through the load, or $W = V \times I$. But the current flow through the transmission line is the same as the current through the load. So the line current, $I_T$, equals $I$, equals $W/V$. From Ohm's law, the voltage difference between the generator and load ends of the transmission line is given by

$$V_T = I \times R_T, \text{ where } R_T \text{ is the line's resistance.}$$

$$L = V_T \times I.$$

$$L = I^2 \times R_T = (W/V)^2 \times R_T.$$

Transmission losses are proportional to the square of the power transmitted and inversely proportional to the square of the transmission voltage. By raising the voltage of transmission lines from about 200 V DC to about 500 kV AC, transmission losses were reduced by a factor of about six million. This makes possible the long-distance transmission of large quantities of power.

Because losses are proportional to the square of the power used by the load, two equal loads will together create four times the losses of either load separately. This means that the loss is not entirely attributable to individual loads but is partly due to their interaction. Neither can losses be attributed to particular generators. Nonetheless, it is possible to develop efficient loss charges.

For the DC system of Figure 5-1.1, the power loss equation is extremely precise, the main discrepancy being caused by a slight change in $R_T$ as the wires warm from more power. AC power flow is much more complex due to reactive power flows, but the equation is still accurate enough for most economic purposes.

## 5-1.2   AC POWER LINES

In the United States, alternating current reverses directions 120 times every second and since it takes two reversals to complete a cycle, the frequency of AC is 60 cycles per second. In 1933, Heinrich Hertz was honored by having the unit of a

**Figure 5-1.2**

120 V AC with 1 A of
current flowing.

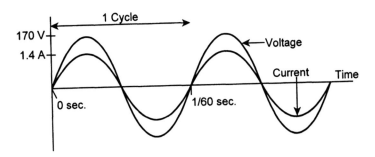

cycle per second named for him, so AC frequency in the United States is said to
be 60 Hz; in Europe it is 50 Hz.[4]

Alternating current changes direction gradually; the current slows down, stops,
and then the flow increases in the opposite direction. Graphing the current flow
results in the sine wave shown in Figure 5-1.2. The AC voltage also follows a sine
wave and, in the figure, is perfectly synchronized with the current. The next chapter
explains why they usually are not exactly in step.

Notice that although the figure is labeled "120 V AC," the peak voltage is
approximately 170 V because the "root-mean-square" (RMS) average voltage is
always reported instead of the peak voltage. Similarly the peak current flow is 1.4 A
while the RMS current flow is only 1 A. Using RMS average values allows AC
power to be computed by simply multiplying volts times amps, just as DC power
flows are computed. Of course, this power flow is itself an average power flow.
One hundred and twenty times per second the voltage and current are both zero,
and there is no power flow at all.

The simplest view of AC power flow is that it is just a sequence of DC power
flows that alternates directions. Reversing both the voltage and the current in Figure
5-1.1 has no effect on the power flow; it still flows from the generator to the light.
Most of the basic properties of AC power flows that are needed to design markets
can be understood in terms of this essentially DC model, but some important
phenomenon are purely AC in nature.[5]

## Transformers

Transformers make power markets possible by creating the high voltages needed
for long-distance transmission and then reducing them to safe levels for consumption. A remarkably simple device, a basic transformer consists of an iron ring with
two coils of wire wrapped around it. It has no moving parts and requires little
maintenance.[6] AC power flows into one coil, and the same amount of AC power
flows out of the other. Amazingly, the two coils are electrically disconnected so

---

4. In the late 1880s, the German physicist, Heinrich Hertz, demonstrated experimentally Maxwell's
prediction that electromagnetic (radio) waves are a form of light.

5. The AC power described here is called "two phase," while the type used for power transmission is
"three phase." This explains why power lines always come in groups of three. The three lines are each 120°
out of phase with the other two. The basic principles are the same for three-phase AC.

6. Large power transformers often have cooling systems that require pumps. Some have mechanisms for
changing how many turns are used on one side. This allows their output voltage to be adjusted. The
distribution transformers on local power polls are more like the illustration.

that no electric current can flow from one to the other. What is useful about this device is that it can change the voltage.

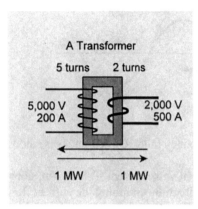

A Transformer

5 turns     2 turns

5,000 V         2,000 V
200 A            500 A

1 MW      1 MW

If power comes into a coil with 100 turns (loops) of wire and goes out of a coil with 2000 turns of wire, then the voltage of the outgoing power will be 20 times greater than the voltage of the incoming power. Obviously transformers do not produce free power, so when the voltage increases, the current must decrease. For the present example, the current would be 20 times less. Because power is voltage times current, the power remains unchanged, except for losses which are on the order of 1%.

While it is economical to transmit power at hundreds of thousands of volts, such high voltages are too dangerous for consumers. This problem is solved by using a "step-down" transformer that reduces the high voltage of transmission to the low voltage of residential and commercial uses. A step-down transformer is essentially the same as the "step-up" transformer described above except that power flows into the side with a high number of turns. The same amount of low voltage power is delivered from the side with fewer turns. Because generators do not produce extremely high voltages, step-up transformers are used between generators and transmission lines to raise the voltage.

If a very large increase or decrease in voltage is needed, two or more transformers can be used in sequence. For example, two transformers of the type just described could step 50,000 V down to 2500 V and then to 125 V. As power flows from a very high-voltage transmission line, say 500 kV, it is typically stepped down to something like 128 kV or 65 kV and then down to 15 kV, which is a typical voltage used along residential streets. Finally it is transformed down to 120 V for home use.

**How a Transformer Works.**   If DC power is put into a transformer, it will typically burn out the coil of wire it flows into, and even if it does not, no power will flow out of the output coil. DC fails because it is changing currents, not steady currents, that make the transformer work. The following nontechnical explanation of a transformer is intended to provide nonengineers with a feeling for the subtletly of an AC power system. Other such complex physical phenomenon can close down a power market in a second.

As the current begins to flow in the transformer's input coil, a magnetic field is created in the transformer's iron "core." This magnetic field increases in strength as the input current increases. Because this magnetic field is carried in the iron core, it also passes through the output coil. As the magnetic field increases in strength in the output coil, it creates a current in that coil. (A constant, in other words DC, field would create no current.) It is impossible to keep the input current increasing forever, so the only way to keep the magnetic field strength changing is to begin decreasing the input current. The magnetic field decreases and produces an output current that flows in the opposite direction. This only explains how it is possible for power to get from input to output without a direct electrical connection. (The power flows through the magnetic field.) It does not explain the increase or decrease in voltage, which is the real purpose of the transformer.

## Electrons Move Slowly
## Power Flows at the Speed of Light

No wire connects the input and the output coils of a standard transformer. Instead the power is carried from input to output by an electromagnetic field—not by electrons. Because the power flow on transmission lines is proportional to the flow of electrons in the wires, electrons are suspected of carrying the power. The power is thought to flow inside the wires. But the flows of electrons produce electric and magnetic fields, and these fields carry the power.

Transmission lines are surrounded by magnetic fields that are proportional in strength to the current flow. Between the lines, there is an electric field that is proportional in strength to the voltage. The interaction of these two fields carries the power in the space around the power lines. Very little power flows inside the wires. Instead the wires guide the surrounding power to the appropriate destination.

Light is a self-reproducing electromagnetic field, and power flow is essentially a form of light that is guided by power lines. Just as light travels more slowly in glass than in a vacuum, so electric power on power lines travels more slowly than the speed of light in a vacuum—about $^2/_3$ that speed, over 100,000 miles per second. By comparison, the electrons in the wire are standing still. Most of their motion is random and at speeds of only a few hundred miles per hour (not per second). With direct current, electrons make some progress, but with alternating current they reverse direction every $^1/_{120}$ of a second and go nowhere at all.

## Chapter 5-2
# Physical Transmission Limits

*We're set up for direct current in America.*
*People like it, and it's all I'll fool with.*

Thomas Edison
1884

**W**ITHOUT TRANSMISSION LIMITS, POWER MARKETS WOULD HAVE AMPLE COMPETITION AND NO NEED FOR CONGESTION PRICING. Designing power markets would be far easier. A competitive market will only account for the physical limits on all power lines and transformers if property rights are designed to represent these limitations properly.

Transmission limits are of two types: (1) physical limits, and (2) contingency limits. Physical limits are the basis of the contingency limits, but contingency limits are stricter and are the relevant limits for trading. A contingency limit ensures that a line's physical limit will not be violated if some other line or generator goes out of service unexpectedly. All limits, whether physical or based on contingencies, can be expressed at any point in time as a simple megawatt limit on power flow that is allowed over the power line or transformer in question. But contingency limits present a complex problem for trade because a trade can affect the contingency limit on lines it uses. (This problem is not discussed in Part 5.)

Two facts about limits are economically important, the megawatt limit itself, and how predictable it is. Unpredictability makes forward trading difficult, and this chapter explains some of the reasons that limits vary.

**Chapter Summary 5-2:** Power lines have physical limits that restrict the amount of power they are allowed to carry. These limits prevent overheating of the wires, instability of the power flow, and low-voltage conditions at the load end of the lines. Any of these limits can be thought of as a simple limit on real power flow, but because the basis of these limits is complex, they may vary according to how the system is operating at a particular time.

**Section 1: Thermal Limits on Power Lines.** Line losses can heat the power lines to the point where they sag or even melt and break. To avoid this, limits are imposed on the amount of real power the system operator can schedule on any line.

**Section 2: Reactive Power and Thermal Limits.** *Real* power, the type that's traded, flows only from generators to loads. Because voltage and current are not always synchronized, there are also *reactive* power flows that carry no net real power in either direction. These flows are not *called* "real" but they are real, and the currents they cause heat up transmission wires and generators, thereby causing loss of "real" power. The thermal limits on generators and power lines when expressed as real power limits depend on the amount of reactive flow and so are somewhat variable.

**Section 3: Stability Limits on Power Lines.** Power flows through AC power lines because the voltage at the generator end reaches its maximum slightly ahead of the voltage at the load end. This difference in timing means the voltages at the two ends are different, and this causes current to flow, and together these make power flow. The more the generator voltage leads the load voltage, the more power is transferred, up to the point where the generator voltage leads by $\frac{1}{4}$ cycle or 90°. If the phase difference ever becomes greater, the voltage collapses at the load end in about a second. To avoid such a disaster, stability limits on power lines specify a maximum phase difference and a corresponding maximum power flow. This limit tends to be binding on long power lines in the Western United States, while thermal limits tend to restrict flows on the shorter Eastern power lines.

## 5-2.1 THERMAL LIMITS ON POWER LINES

Power flow causes a loss of electrical power, and the "lost" power heats the power lines causing the copper to expand and the line to sag. At a higher temperature the sag becomes permanent and eventually the copper melts and the line breaks. During this process, the small increase in the line's resistance does not protect it from damage. Instead the system operator must know the line limits and make sure the system is run in such a way that they are not exceeded.

Electrical current determines line losses and thus the thermal limits on power lines. Power, $W = V \times I$, so current, $I = W/V$. Power lines are kept near their rated voltage, so the current can be limited by limiting the power flow, $W$. A line's thermal limit can be thought of as a limit on the power that can be allowed to flow over the line. The outputs of generators must be controlled (the system dispatched) to prevent this limit from being violated for any line in the system.

Because of reactive power flows (described in the next section), expressing the line's limit as a real-power-flow limit leaves the limit more variable than expressing it as a limit on current. But the market trades power, not current, so for

**Figure 5-2.1**

The real and reactive parts of total power flow.

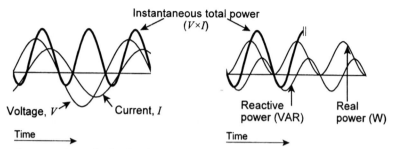

trade and economic analysis, the limit needs to be expressed as a limit on the power transfer capability.

The view of thermal limits as fixed limits on power flow is sufficient to understand all of the principles of congestion pricing, and the remaining chapters require no more. But some forms of transmission rights depend on transmission limits being constant and predictable. The remainder of the chapter explains the complexities that cause the physical transmission limits to vary.

The first cause of variation in the thermal limit is ambient temperature. Because heat is the problem, the higher the air temperature, the lower the thermal limit. Typically, reactive power flows cause larger and quicker changes in the real-power-flow limits, but thermal limits are fairly stable.

## 5-2.2   REACTIVE POWER AND THERMAL LIMITS

Generators produce and consumers use both *real* and *reactive* power. **Reactive power** is a necessary part of the transmission of real AC power and has no counterpart in DC power flows. **Real power** is simply the normal electrical power traded in power markets.

Reactive power is best understood by calculating the instantaneous total power flow that occurs when voltage and current are "out of phase," as shown in Figure 5-2.1. This means that peak voltage occurs before or after peak current flow, typically before. Because they are not synchronized, there will be times when one is positive and the other negative. Because power is voltage times current, this implies a *negative* power flow, power actually flows back into the generator. When voltage and current are only a little out of phase, as is normal, the negative power flow is brief and small.

Although the total instantaneous power flow shown in Figure 5-2.1 is completely "real," it is not what is meant by real power. The instantaneous flow fluctuates rapidly and changes sign 240 times per second. This is too much detail even for most engineering purposes, so the total flow is divided into two parts one of which is always positive (the real power flow) and one which averages out to zero (the reactive power flow). Both still fluctuate, but this is ignored in favor of the average of the real power flow and something a little less than the average absolute value of the reactive power flow. To summarize, real power only flows from generators to load, and it delivers the service of electric power. Reactive power flows back and forth in equal amounts and supplies no energy.

In spite of going nowhere, reactive power is useful (at least in helping to solve problems caused by reactive power). It helps keep the voltage at the load end of the line at its proper level. It also has a cost. Reactive power flows cause very real currents which heat wires and waste power—reactive power contributes to losses. Because it flows into generators, and through transformers as well as on power lines, it causes losses and produces heat in each of these.

## Thermal Limits

A limiting factor of most electrical components is the heat they can tolerate, and the primary source of heat is the friction of electric currents in the wires. The heat produced is determined by the current flow, $I$, which is proportional to the real power flow if there is no reactive flow. But when there is both real power, $W$, and reactive power, $Q$, current is proportional to $\sqrt{W^2 + Q^2}$, which is called *apparent power*. A thermal limit that involves both $W$ and $Q$ is not economically useful and should be reformulated as a real-power limit which depends on $Q$ as described by Result 5-2.1.

---

**Result    5-2.1**

**Thermal Limits Depend on Real and Reactive Power Flows**
If the maximum real power that a system element can handle, without any reactive power flow, is the thermal power limit, $TL_W$, then with reactive power flow $Q$, the thermal limit is

$$W < TL_W \times PF, \text{ where } PF = W / \sqrt{W^2 + Q^2}$$

$PF$ is the power factor and equals 1 when the power flow is purely real. An equivalent formulation is $W < TL_W \sqrt{1 - (Q/TL_W)^2}$ .

---

If $Q$ is 10% of $TL_W$, the real-power thermal limit is reduced by $\frac{1}{2}$ of 1%, while if $Q$ is 50% of $TL_W$, a large value for $Q$ under normal conditions, the limit is reduced by 13.4%. The fundamental point is that the thermal limit on real power flow depends on the amount of reactive power flowing at the same time, but the dependence is usually not dramatic.

## Use and Production of Reactive Power

Although reactive power is a symmetrical back-and-forth flow of real power, there are two types of reactive flows. One occurs when the voltage peak occurs just before the maximum current flow, as shown in Figure 5-2.1, and the other when the voltage lags the current. If a system element causes the current to lag the voltage, it is said to use reactive power, while if it causes the reverse, it is said to supply it. Loads and transmission tend to use more reactive power than they produce, so some system resources must be dedicated to producing it.

**Capacitors.** Capacitors attached to power lines *produce* reactive power and require no external energy input. This is possible because reactive power flows carry no energy in either direction.

**Synchronous Condensers.**  A more flexible source of reactive power is a synchronous condenser, so-called because it acts like a capacitor ("condenser" is the old term for capacitor), but like a generator, it must spin in synchrony with the AC power flow. Although the design is more subtle, it can be thought of as a generator run by an electric motor and adjusted to produce reactive power. It requires no external fuel source, but it uses some real power while it produces reactive power. Some generators are built to act as synchronous condensers when not being used as generators.

**Generators.**  Most generators cannot be run as synchronous condensers, but they can produce (or absorb) reactive power. Because reactive power flows reduce the generator's thermal limit, there is an opportunity cost to producing reactive power when the market price is above the generator's marginal cost. There is also some cost to building generators with the ability to produce reactive power.

**Motors and Transformers.**  Most motors consume significant amounts of reactive power. The magnetic ballasts of fluorescent lights behave similarly. Such appliances are said to have a poor "power factor," and they place a burden on power systems that goes beyond the real power they use.

### Transmission of Reactive Power

When real power is transmitted, some is lost. When reactive power is transmitted it increases the loss of real power, but in addition, reactive power is also lost. (This is the same as the use of reactive power mentioned above.) Because the losses of reactive power are relatively great, it cannot be transmitted over great distances and most of it must be produced locally. This limits competition in any market for reactive power.

## 5-2.3   STABILITY LIMITS ON POWER LINES

Like a thermal limit, a stability limit specifies a maximum flow of real power. Although the underlying problem is far more complex, the application of the limit is the same. Unlike a thermal limit which depends on the thickness of the wires, and not at all on their length, a stability limit depends on the length of a transmission line. The longer the line, the lower the limit. Consequently the relevant power-flow limits for short lines tend to be thermal limits and for long lines tend to be stability limits. In the United States, Eastern lines tend to have thermal limits while stability limits are more likely to be found on Western lines.

Stability limits are the result of the way power is pushed down an AC power line. Like DC power transmission, this is done with voltage differences, but unlike DC transmission, it does not require a lower voltage at the load end of the line. The relevant voltage difference is caused by the voltage at the load being out of phase with the voltage at the generator. The greater this phase difference, the greater the power flow. But voltage is sinusoidal and when two sign waves get out of phase by a full cycle, they are exactly in phase again. A full cycle is measured as 360°,

**Figure 5-2.2**

Voltages with a 36°
phase difference between
the generator and load
end of a transmission
line.

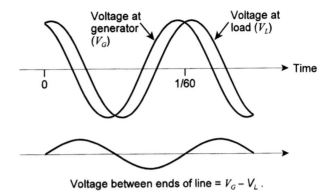

Voltage between ends of line = $V_G - V_L$ .

and power flow only increases until the phase difference between the two voltages reaches 90°. This puts a limit on how much power can be carried by a long power line.

Unfortunately, passing the 90° phase limit causes the power transfer to become unstable and the voltage to drop rapidly. Consequently, power lines are never deliberately run near this limit, instead a lower limit on the phase difference, of say 30° or 45°, is imposed as the line's stability limit. This corresponds to some real power flow, which becomes the stability limit on real power transfer.

## Why AC Power Flows on Transmission Lines

Reactive power flow is characterized by a phase difference between voltage and current at a given location; power flow is driven by a phase difference between two voltages at different locations. Figure 5-2.2 shows the voltage at the sending and receiving end of a power line. The sending voltage always leads the receiving voltage, and the difference is measured by the phase angle. As a consequence of the phase difference, even though both ends of the line experience an AC voltage peak of, for example, exactly 235, kV there is usually a voltage difference between the two ends of the line. That difference, shown at the bottom of the figure, is what makes the current flow in the line, and current flow along with voltage delivers the power.

The larger the phase difference, the greater the difference in voltage between the two ends of the line and the greater the current and power flow on the line. When the load uses power, the phase difference increases and more power is delivered.

**Table 5-2.1** Ohm's Law with Water and AC Power Analogies

| DC Current | Voltage ($V$) | Resistance ($R$) | Current ($I$) | $I = V/R$ |
|---|---|---|---|---|
| Water | Pressure ($p$) | Friction ($R$) | Flow ($F$) | $F = p/R$ |
| AC Power | Phase difference ($\varphi$) | Impedance ($X$) | Power ($W$) | $W = \varphi/X$ * |

* Power flow is proportional to $\varphi$ for small phase angles, relatively short lines and normal operating conditions. For larger angles replace $\varphi$ with $\sin(\varphi)$.

### Stability Limits and Voltage Collapse

When loads take more power, more power flows down the transmission line from the generator. This causes the phase difference between sending and receiving voltage to increase, and it tends to cause the voltage at the load to decrease just as happens with the DC transmission of power shown in Figure 5-1.1. The reduction of voltage is usually corrected by the injection of reactive power for voltage support. This allows the load to draw even more power and the phase difference increases more to transfer more power over the transmission line.

If the load continues to draw more power, one of two things will happen. It may become impossible to support the voltage at the load, in which case the voltage will sag and there will be a brown out. The system operator will not let this proceed too far and will shed load if necessary to maintain a safe voltage level.

The second possibility (assuming the thermal limit of the line is not reached) is that the voltage can be supported and the phase difference will continue to increase as more power is supplied to the load. At some point the system operator will stop this process because the danger of voltage collapse is too great. This is the stability limit. If the phase difference is allowed to get too high and a contingency occurs the phase difference will be pushed past 90°. This results in voltage collapse at the load end. It happens because, past 90°, power delivery is reduced, which causes a further increase in the phase difference, which causes a further reduction in delivered power, and so on.

Voltage collapse takes only a second, but it is one of the worst calamities that can befall a power system. Before generators had automatic protection from such disasters, the unstable power flow during a voltage collapse was known to actually break generator shafts—power flowing into a generator can apply force to a generator just as it does to an electric motor. The entire AC interconnection is one large machine, and electrical linkages are as powerful as mechanical linkages.

# Congestion Pricing Fundamentals

*The popular belief is that radium constantly produces heat and light without any appreciable loss in its weight. . . . there exists a form of energy of which we have as yet no knowledge, but which may yet become available to us as a result of further discoveries.*

George Westinghouse
1911

**P**HYSICAL IMPEDIMENTS TO TRADE CAUSE COMPETITIVE PRICES TO DIFFER; THE DIFFERENCE IS THE PRICE OF CONGESTION. Power lines, because of their limited capacity, often cause energy prices to differ between locations. Congestion prices were not invented for electricity grids, need not be centrally calculated, and occur on their own in competitive markets.

**Chapter Summary 5-3:** If there are binding physical transmission limits between different locations, a competitive bilateral market with physical transmission rights will trade power at different prices in different locations. These competitive locational prices of power (CLPs) are unique, are the same as nodal prices (LMPs or LBMPs), and are the only efficient prices.

**Section 1: Congestion Pricing is Competitive Pricing.** If the transmission line between two locations is inadequate to handle the desired trade between those two locations, the downstream location will be forced to buy power from more expensive local generators. This will raise the local price of power relative to the remote price, which is a standard competitive result and has nothing to do with centralized computation.

**Section 2: Benefits of Competitive Locational Pricing.** Like all competitive prices, CLPs minimize the cost of production and reveal to consumers the true cost of their consumption. CLPs at each location equal the system's marginal cost of providing power at that location.

## 5-3.1    CONGESTION PRICING IS COMPETITIVE PRICING

Transmission lines have capacity limits which must be enforced in order to protect the lines and the stability of the system. When these limits are binding, that is, when traders would like to use more capacity than is safely available, transmission is a scarce resource. Economics recommends that, whenever practical, a market be used to allocate what is scarce. One approach, though impractical in a large network, is particularly suited to displaying the fundamental nature of both congestion prices and locational prices. Tradeable physical transmission rights can be used by a classic, decentralized market to solve the congestion problem.

This section analyses a single transmission line that is sometimes a scarce resource and that is allocated by a competitive market in transmission rights. The objective will be to find what prices would emerge from such an ideal and fully decentralized competitive market. These prices provide a useful benchmark against which the price outcomes of proposed market architectures can be compared.

The market architecture considered here is often referred to as "bilateral trading" and is often considered the antithesis of the more centralized "nodal pricing." In spite of this, bilateral-trading prices are exactly those that nodal pricing is designed to produce. Both systems aim for the same set of prices, and the differences lie elsewhere.

### The Competitive Market Structure

Transmission rights can take many forms; some are more complex to administer, some more difficult to trade, and some allow more market power. In a perfectly competitive market with only a single transmission line, little can go wrong, and in this setting, the simplest right is a physical right to transmit power. The present example consists of two trading locations connected by a single 500-MW transmission line. To protect this line and the integrity of the system, 500 MW of physical transmission rights have been issued. These confer on their owner the right to transmit from the remote location, Bus 1, to the city, Bus 2.

---

**Definition**      **Buses and Nodes**

Transmission networks consist of high-voltage power lines that connect different locations. Where several lines meet or where a line terminates at a generator or load, there is a "bus." This is a piece of electrical equipment (a bus bar) that is used to make connections. A "node" is a more general mathematical term applied to the intersection of connecting paths in any type of network. The two terms are often used interchangeably in power system economics.

---

It does not matter who owns the transmission rights initially. They could be owned and auctioned off by the ISO, owned by third parties with no other role in the market, owned by power marketers, or owned by consumers or suppliers. All that matters is that the rights be tradeable and the owners not exercise market power.

**Figure 5-3.1**
A power market with two locations.

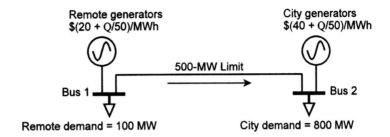

Remote generators
$(20 + Q/50)/MWh

City generators
$(40 + Q/50)/MWh

500-MW Limit

Bus 1

Bus 2

Remote demand = 100 MW

City demand = 800 MW

## Example

Bus 2, the load center, is a city that sometimes needs as much as 1 GW of power but at the time of the example, needs only 800 MW. To meet its load, the city has 500 MW of generation and a 500-MW power line connecting it to remote Bus 1 where there is more than 500 MW of available generation.

A large number of small generators with differing variable costs are located at each bus. The cost of generation in the city starts at $40/MWh and increases at the rate of $2/MWh for each additional 100 MW of output. The situation at the remote location is similar except that costs start at $20/MWh instead of $40/MWh. These two supply curves are summarized in Table 5-3.1.

**Table 5-3.1** Generation Supply Curves

| Location | Marginal Cost ($/MWh) |
|---|---|
| Remote (Bus 1): | $20 + Q_{\mathrm{L}}/50$ |
| City (Bus 2): | $40 + Q_{\mathrm{L}}/50$ |

## The Price of Transmission Rights

Transmission rights (TRs) are not bundled with energy but are traded separately. To sell and transmit power from Bus 1 to Bus 2, either the buyer or seller must own a transmission right in the amount of the power sold.

In a competitive market, city customers can always buy what they need at the city price, and remote generators can sell what they want at the remote price. If power is cheaper by $D/MWh at the remote bus, city customers will pay *at most* $D/MWh for a TR to gain access to remote power. Generators at Bus 1 will pay at most $D to gain access to the city customers. If the TR price were less than D, then remote generators would buy rights so they could sell to the city. This would drive up the price of transmission rights. Competitive pressures in the bilateral market will drive the price of transmission rights to the energy price difference between the remote bus and the city bus.

| Result   5-3.1 | **Transmission Price A→B Is the Power Price Difference, $P_B - P_A$**<br>If the market for transmission rights is competitive, the equilibrium price of a transmission right from A to B, $P_{AB}$ is equal to the price of power at B minus the price of power at A.<br><br>$$P_{AB} = P_B - P_A$$ |
|---|---|

## The Price of Power

The price of TRs in this example cannot be negative because supply and demand conditions assure that the price at Bus 2 cannot be less than the price at Bus 1, but is it positive or is it zero? This is the same as asking if the line is congested. The answer determines how to find the locational prices of power. If the line is congested, the line limit will play a role, and there will be two different prices. If not, the line limit is irrelevant and there is a single price of power.[1]

| Definition | **Congestion**<br>In a competitive market, the path from A to B is congested if the price of transmission rights from A to B is positive. This is equivalent to the price of power at B being greater than the price of power at A. If a line would be overused if its limit were not enforced, it is congested. |
|---|---|

To find if the line is congested, assume it is not and determine how much it would be used. If it is overused when the limit is ignored, then it is congested. Both remote and city customers would buy all their power at Bus 1 which would drive the price there to $(20 + 900/50)$/MWh. This is $38/MWh, which is less than the price at Bus 2 and confirms the fact that all power will be purchased at Bus 1. But city customers would want 800 MW of TRs to import the power they purchased and only 500 MW of TRs are available. This will force city customers to buy 300 MW of power from city generators and will drive the city price above the remote price. The line will be congested and there will be two prices.

With 300 MW purchased from city generators, the price there will be $(40 + 300/50)$/MWh, or $46/MWh. With only 600 MW purchased at the remote bus, the price there will be $(20 + 600/50)$/MWh, or $32/MWh. Applying Result 5-3.1 shows the prices of transmission rights is $(46 - $32)/MWh, or $14/MWh. Note that all of these prices have been determined by a perfectly competitive, fully decentralized, bilateral market and are not the result of a centralized optimization process.

---

1. In a network with loops, the "path" from A to B must be defined to include all wires over which power would flow from A to B.

**Table 5-3.2** Competitive Locational Prices

| Commodity | Symbol | Price |
|---|---|---|
| Power at the city bus (Bus 2) | $P_2$ | $46/MWh |
| Power at remote bus (Bus 1) | $P_1$ | $32/MWh |
| Transmission rights from Bus 1 to Bus 2 | $P_{12}$ | $14/MWh |

### Competitive Locational Prices (CLPs) in Context

Although there are a number of ways to structure a competitive market in the example network, all would lead to equivalent results. In all examples that do not contain an unusual coincidence, there is a unique set of competitive prices. These are determined by the normal laws of supply and demand and by the constraints that are reflected in the rules governing transmission rights which are a perfectly normal form of property right. The prices for the three commodities shown in Table 5-3.2 are the outcome of a normal competitive market.

Some descriptions of these prices emphasize the fact that they could be computed as the solution to a large system of equations derived from supply functions, demand functions, and network constraints. While it is interesting and useful to compute these prices, the emphasis on computation has obscured the fact that these prices would be the outcome of any well-functioning competitive market.

In PJM, competitive locational prices (CLPs) are called locational marginal prices (LMPs), while in New York they are called locational-based marginal prices (LBMPs). These names derive from the fact that CLPs equal marginal costs at the relevant location (marginal price is not a standard economic term).[2] The emphasis on "marginal" serves as a reminder of the computation process used by PJM and NYISO, but it also corresponds to a fundamental property of competitive prices.

## 5-3.2   BENEFITS OF COMPETITIVE LOCATIONAL PRICES

Because CLPs are normal competitive prices they have the standard properties promised by economics for all competitive prices. They cause suppliers to minimize the total cost of production, and are the only free-market prices capable of doing this. To be consistent, those who oppose "marginal-cost pricing" of power (another term for competitive locational pricing) must reject either cost minimization or free markets.[3]

The second benefit of CLPs is that they send the right signals to consumers. If an additional megawatt-hour would cost $X to produce and deliver to a particular consumer at a particular location, that consumer sees a price P of $X/MWh. As

---

2. The prices also equal marginal values to consumers. If marginal cost is ambiguous, price is determined by marginal value and vice versa.

3. Under very unusual circumstances, the marginal cost at a location can be ambiguous and then there is more than one price that will induce efficient production.

a consequence, consumers use only an additional megawatt of power when they value it more than its cost of production.

## Production Cost Minimization

To check that CLPs minimize the production cost in the example, ignore the question of price and look for the cheapest way to supply the load at Bus 1. The cheapest generators are the remote generators, so they should be used first. When these have been used to the greatest extent possible given the transmission limit, their marginal cost is still only $32, so all of these should be used. The cheapest 300 MW of city generation should be used to finish the job.

Will CLPs induce exactly these generators to run? Because the city CLP is $46/MWh only those generators with lower costs choose to run, and this is exactly the 300 MW of generation that should run. All other generators at this location would lose money if they produced. The situation is the same at Bus 1. Consequently, if one wants to induce the least-cost dispatch in the example network using a free market, one must use the CLPs. No other prices will do the job.

### Locational-Pricing Result #1

| Result | 5-3.2 | **Only Competitive Locational Prices Minimize Total Production Cost** |
|--------|-------|----------------------------------------------------------------------|
|        |       | If generators choose their production levels freely, based on market prices, only competitive locational prices will minimize total production costs by inducing the right set of generators to produce. (Very rarely, a CLP is ambiguous due to a marginal cost ambiguity.) |

## Demand Side Efficiency

Consumption is efficient if the CLPs reflect the cost of producing the next megawatt of power, but the cost of the next megawatt seems to depend on which bus it comes from. Choosing to consume another megawatt, a costumer at Bus 2 might purchase it from either bus. If purchased from Bus 2, the cost would be $46/h. If purchased from Bus 1, it would cost only $32/MWh to produce.

How can one CLP at Bus 2 signal both of these marginal costs? The $32/MWh cost of power from Bus 1 is an illusion because it ignores real opportunity costs. (The purchase at Bus 1 forces some other customer to give up a $14/MWh opportunity.) Purchasing a megawatt from Bus 1 requires buying a TR which means some other customer will sell the TR and be forced to switch 1 MW of consumption from a Bus 2 generator to a Bus 1 generator. The cost of serving this consumer will increase by $14. The total cost of buying an additional megawatt from Bus 1 is $32 for that megawatt plus $14 for the consumer that had to switch its purchase to a more costly generator. That comes to $46/MWh which is the CLP at Bus 2.

This is a general property of competitive locational prices. The CLP at location X equals the system marginal cost of supplying an additional megawatt at X. Consequently, CLPs are the only prices that send the right signals to consumers.

# Congestion Pricing Methods

*Investigations . . . indicated such great simplification in wireless telephone apparatus that we may, within the quite near future, have placed at our disposal a simple portable apparatus which will permit wireless conversation to be carried on over a considerable area. This will prove of great value in sparsely settled districts.*

George Westinghouse
1911

T HE POINT OF CENTRAL CALCULATION IS TO FIND THE PERFECTLY COMPETITIVE, BILATERAL-MARKET PRICES. If competition is strong in a centralized market, bids will be honest and the data used in the central computation accurate. Though bilateralists often object to "nodal" or "marginal-cost" locational prices, bilateral theorists know that theoretical nodal prices are exactly the prices a bilateral market would produce if it worked perfectly. The real debate is not over the prices but over which system will do a better job of finding them.

**Chapter Summary 5-4:** Central computation finds the optimal dispatch and then computes prices from the marginal benefits of a free megawatt at each location. Transmission constraints make power more valuable in some locations than others. Bilateral traders never consider the optimal dispatch but look only for profitable trades. Arbitrage produces a single price at each location, but transmission constraints can prevent it from leveling prices between locations. These two different processes lead to the same quantities being traded and to the same prices because perfectly competitive bilateral trade is efficient.

**Section 1: Centralized Computation of CLPs.** Central computation of CLPs makes the ISO the trading partner of every buyer and seller and does not price congestion separately. Power flows in a looped network are governed by the impedances (resistances) of the lines. This is illustrated with a three-line looped network with one constrained line. If the constraint is binding, all three buses will have different prices, and these will cause generators to minimize the total cost of production.

**Section 2: Comparing Bilateral and Central Congestion Pricing.** Competitive bilateral trading produces the same locational prices in the example network as central computation. The bilateral system collects congestion rent by selling transmission rights while the centralized market collects the same rent by selling power at a higher price than it pays for it. Central pricing is more complex for the system operator but places less burden on buyers and sellers. Consequently it allows fewer arbitrage opportunities from which marketers can profit.

## 5-4.1   CENTRALIZED COMPUTATION OF CLPs

Both PJM and the NYISO compute their CLPs centrally in the day-ahead market instead of letting bilateral traders find them. The central calculation is no different in principle from the calculation performed to solve the example of Chapter 5-3. Provided that generators and their customers tell the ISO their true supply and demand curves, the ISO will find the same CLPs that a bilateral market would find because it will calculate what the bilateral market would do if it existed and were competitive.

Actual CLP calculations in PJM and NYISO are complicated by solving the unit commitment problem as discussed in Part 3. This is not a necessary part of finding locational prices, and Part 5 will ignore those complexities and assume the generators simply bid a variable cost curve. If they did, the central calculations would still compute a complete and accurate set of CLPs.

### How Central Computation Works

With central calculation, there is no need to issue transmission rights; there is only an energy market. This makes trading simpler than in a bilateral market. Another simplification for traders is that they need not look for trading partners or engage in comparison shopping in the energy market; everyone trades with the ISO. Consequently, every load customer automatically gets the benefit of every supply bid and every supplier benefits from every demand bid.

Each generator and every load customer submits a bid to the ISO which specifies the location at which they will take or provide power. If, in the day-ahead market, they bid to supply power at A but actually supply it at B, they are required to compensate by buying power at B and selling it at A in the real-time market. For present purposes assume that all traders submit bids for their physical location. Then the distinction between the day-ahead and the real-time market can be ignored.

A generator's bid is typically a supply curve that offers to sell $q_1$ MW at a price of $p_1$, $q_2$ additional megawatts at a slightly higher price, and so on. Loads are generally allowed to bid similar demand curves, but sometimes in immature markets they are restricted to a certain quantity or not allowed to bid at all. When load does not bid, the ISO simply bids for the load based on its best expectation of real-time demand.

Once the ISO has collected the supply and demand bids, it *accepts* bids as described in Section 3-3.1. It maximizes total surplus and sets price equal to marginal surplus at every location. With supply- and demand-curve bids of the type assumed here, this clears the market at every location. But what does this mean when supply is located one place and demand another? When finding a supply-demand balance at Bus A, the supply at A does not refer only to generators located at A. The supply at Bus A allows power to be sent to Bus A from every generator in the system provided the combined flows violate no transmission constraint. If the market has cleared at A at a price of $30/MWh, it is impossible to supply another megawatt to A, in the sense just defined, for less than $30.

## The Three-Line Example

The network of Figure 5-4.1 is the simplest network that permits **loop flow**. It can occur only in a network that contains at least one loop, a term which has the same meaning for networks as for hiking trails. A network that is not looped is called **radial**, because it resembles a set of lines radiating from a hub (though these lines can branch). When power flows from point A to point B in a looped network, it can, in some cases, take more than one path, and if it can, it does. These flows are said to be parallel or to take parallel paths, even though the paths are not geometrically parallel. The way in which lines connect—three at one node, four at another, with three loops, or no loops—is the *topology* of the network. The length and direction of the lines are irrelevant to the network's topology.

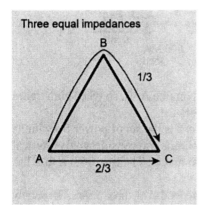

Three equal impedances

Besides topology, the impedance of the lines is important. **Impedance** is to AC lines as resistance is to DC lines. Impedance is a generalization of resistance that includes resistance, inductance, and capacitance (in power lines, mostly inductance). All that matters economically is that impedance describes how difficult it is for power to flow over a certain path. If there are two paths from A to C, and the impedance of the line directly from A to C is $Z$, and the impedance from A to B to C is $2Z$, then twice as much power will take the direct path as takes the indirect path from A to B to C. This is the case in the present example because each of the three power lines, A–B, B–C, and C–A, has the same impedance.

---

**Result      5-4.1**      **Power Flows Are Approximately Additive**
If a balanced trade causes power flows $(F_1, \ldots, F_N)$ on lines 1 through N, and another balanced trade causes flows $(G_1, \ldots, G_N)$, when the two trades take place simultaneously, the flows on the lines will be $(F_1 + G_1, \ldots, F_N + G_N)$.

---

The *principle of superposition* must be used along with the impedance rule when determining more complex power flows. The principle says that if trades are grouped so that each group is balanced (inflow equals outflow) the flow from all of the groups trading at once is the sum of the flows the balanced groups would produce individually. For example, if generators at A supply 300 MW and B uses 100 and C uses 200, this can be viewed as two simple trades: 100 from A to B and 200 from A to C. Two-thirds of the first trade (66.6 MW) flows on line A–B and

**Figure 5-4.1**

Competitive locational prices with loop flow and one transmission constraint.

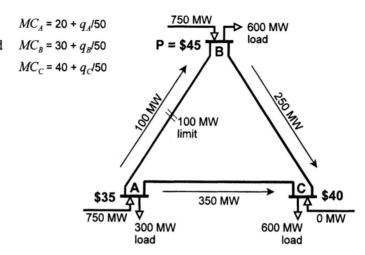

$MC_A = 20 + q_A/50$

$MC_B = 30 + q_B/50$

$MC_C = 40 + q_C/50$

one third of the second trade (66.6 MW) flows on line A–B, so when both trades are made at once, 133 MW flows over line A–B.[1]

The third relevant characteristic of a network is the set of power flow limits on the lines. This example ignores contingency limits and considers only physical line limits, only one of which is low enough to matter. The line from A to B has a power flow limit of 100 MW.

In the present example there is generation and load at all three buses. The supply curves have the same slope of $2/MWh for each additional 100 MW produced but with the initial megawatt costing $20 at Bus A, $30 at Bus B, and $40 at Bus C. There is 300 MW of load at Bus A and 600 MW at the other two buses.

Centralized computation finds the production levels at the three locations that minimize the total cost of production to meet the given load while respecting the 100 MW flow limit on line A–B. The unique solution to the problem is to produce 750 MW at both A and B and none at Bus C. After finding the least-cost dispatch, the central computation finds the three locational prices by computing the marginal surplus of an extra megawatt of generation at each of the three Buses: $P_A = \$35/MWh$, $P_B = \$45/MWh$, and $P_C = \$40/MWh$.

## Checking the Computed Prices

Depending on the algorithm, central computation usually computes prices and the optimal dispatch simultaneously. There is no choice as to the definition of the least cost dispatch (except for rare ambiguities). It is best to think of the central computation as first optimizing the dispatch and then finding the prices.

**Checking the Optimization.**  Could load be supplied more cheaply? If this were possible, it would be possible to back down an expensive generator and produce

---

1. Losses and large phase-angle differences between voltages reduce the accuracy of this approximation. At low power-flow levels it is very accurate and is reasonable for most realistic power flows. Because marginal changes in trade obey the principle of superposition exactly and prices are based on marginal changes, this principle is an accurate guide to many principles of locational pricing.

more with a cheaper generator. Bus C generators produce nothing, so they cannot be backed down. Bus A generators are cheapest and should not back down. Backing down a megawatt at B and producing another at C would save $5/h, but from a power-flow perspective, this is equivalent to injecting a megawatt at C and withdrawing it at B. This increases the flow on the congested line which is prohibited. Backing down B and increasing A saves $10/h, but this increases the flow on the congested line and is also prohibited. Consequently, no cheaper way can be found to produce the required power, and the dispatch is optimal.

**Checking the Prices.**   Price at each location is the value of an additional free megawatt at that location. Because demand is inelastic, accepting this megawatt requires an equal reduction in some generator's output. If the free megawatt is added at A, it cannot be removed at B or C without violating the line constraint, so the generators at A must be backed off. Their marginal cost is $20 + 750/50 = $35/MWh. That is the amount saved and the centrally computed price at A. If a megawatt is added at B, it could be removed at A or C, but it is most valuable if it replaces a megawatt of generation at B, which has a marginal cost of $45/MWh. That is the computed price at B.

   If a free megawatt is added at C, backing down generation at B would increase the flow on A–B, while backing down generation at A would decrease that flow an equal amount. Since backing down B generators is preferred, the best that can be done is to back down A and B generators equally. This results in no net change of flow on A–B and saves an average of $(35 + 45)/2 per megawatt-hour, so that is the computed price at C. This completes the check of the results presented in Figure 5-4.1 of centralized computation of CLPs.

## 5-4.2   BILATERAL PRICING COMPARED TO CENTRALIZED PRICING

The central computer worries about minimizing total cost of production; bilateral traders do not. They seek to maximize their own profits by continuing to trade until there are no profitable trades to be made. A bilateral equilibrium can be identified by the lack of such potential trades. Trading establishes market prices, and if all profitable trades have been made the prices are said to clear the market. The centralized results of the example are now checked for profitable trading opportunities. If none can be found the computed outcome is also the bilateral outcome.

### Are the Centralized Trades the Best Bilateral Trades?

Because one line is constrained, trading conditions are different at every location. To check all the trading opportunities systematically, each bus is examined in turn. If a more profitable trade can be made it must provide power more cheaply to one of the three buses. To ask if a better trade is possible, the current bilateral trades must be specified, something the central computation does not do. Many sets of bilateral trades are equivalent to the centralized trades, so the most convenient set, shown in Table 5-4.1, has been picked as the reference set.

The prices paid for energy depend on who owns the transmission rights. For convenience, assume all 100 MW of rights on A–B have been purchased by the generators at A who make Trade #1. The 100 MW of rights allows them to sell 150 MW of power because only 2/3 of it flows on line A–B. Other generators sell at marginal cost to load at their own buses, or to traders buying power for Bus C. The traders, having no transmission rights, buy equal amounts of power from A and B so the two flows to Bus C cancel out on line A–B. Each trade should be thought of as a large group of identical small trades made by independent generators or traders so the market is perfectly competitive.

**Table 5-4.1** Bilateral Trades Compatible with Figure 5-4.1

| Trade # | MW | From | To | Sale Price | TRs |
|---|---|---|---|---|---|
| 1 | 150 | A | B | $45 | 100 MW |
| 2 | 300 | A | A | $35 | 0 |
| 3 | 450 | B | B | $45 | 0 |
| 4 | 600 | A & B | C | $40 | 0 |

Trade #4 purchases equal amounts at A and B, so it causes no flow on line A–B when sending power to C.

**Bus-A Opportunities.**   Could load at Bus A purchase power more cheaply than $35/MWh? All idle generation has a marginal cost of $35/MWh or more, and no generator is selling its output for less. There are no profitable opportunities to sell power to the load at A, and supply equals demand at the local price.

**Bus-B Opportunities.**   Could load at Bus B purchase power more cheaply than $45/MWh? The marginal cost at Bus B is $45, so no local generator could profitably sell more at that price. Some generators at Bus A can produce for $35/MWh, $10/MWh below the price being paid by load at B. This is a potentially profitable trade opportunity, but such a trade would require a transmission right and those rights are worth $15/MWh to the generators using them because each megawatt of TR allows them to sell 1.5 MW of power at a profit of $10/MWh. The cost of TRs would exactly cancel the profit that could be made buying power at A for sale at B. A similar calculation shows that the TRs needed to import power from C would reduce the profits on such trades to zero. There are no profitable opportunities at B, and supply equals demand at the local price.

**Bus-C Opportunities.**   Could load at Bus C purchase power more cheaply than $40/MWh? The only cheaper generation is at Bus A. A $5/MWh saving or profit could be realized if power could be purchased from A without a TR. But one third of the power would flow over line A–B, so for every megawatt imported from A, a $\frac{1}{3}$-MW TR would be required. This would cost $\frac{1}{3}$ of $15/MWh or $5/MWh, which would again eat up all the profits. There are no profitable opportunities at C, and supply equals demand at the local price.

**Conclusion.**   There are many sets of trades that would correspond to the prices and quantities shown in Figure 5-4.1. No matter which set actually occurred, there would be no profitable trades remaining. These prices and quantities represent a bilateral equilibrium, that is, the final outcome of a perfectly competitive bilateral market. A perfect central computation finds exactly the same prices as a perfect bilateral market.

This is not surprising. The single most studied topic in economics is the behavior of perfectly competitive bilateral markets. As a result, economics has learned how supply and demand interact to determine competitive prices. The "locational" part of this market is nothing new and is easily handled by standard economic theory. Centralized pricing simply takes the supply and demand curves and computes what would happen were there a perfectly competitive bilateral market. The two sets of prices are the same because the point of central computation is to make them the same.

The central computation does not proceed by mimicking a large group of bilateral traders because economics has found short cuts. In particular, Adam Smith guessed, and modern economics has proven, that competitive bilateral trading results in efficient production. This is why it is desirable. Computation takes advantage of this fact and searches for the efficient production levels, knowing they will be the bilateral outcome. When the efficient dispatch is found, the prices are easily computed. They are the only prices that will induce generators to produce the optimal quantities voluntarily.

## Locational-Pricing Result #2

| | | |
|---|---|---|
| **Result** | **5-4.2** | **Competitive Bilateral Prices Equal Centralized Locational Prices**<br>Market prices in a perfectly competitive bilateral market would equal, at every location, perfectly competitive centralized nodal prices. This common set of prices is called the competitive locational prices, CLPs. |

## The Bilateral-Nodal Debate

There has been a long, hot debate over whether it is better to use a bilateral market or a centralized market. Many have argued against the centralized approach because of its prices. (Accusations against "marginal-cost prices" are discussed in Chapter 5-5.) Oddly, those most in favor of competitive markets have been most critical of competitive prices, often failing to realize that central computation simply computes the prices that would be produced by the ideal competitive market. They feel that competitive prices, because they are good prices, must equal the prices they feel are fair. Typically they want prices to be cost-based, in an average-cost sort of way. Since this is how regulators set prices, the complaint against centrally computed prices is essentially that competitive prices, CLPs, do not look like regulated prices. Of course those who complain do not understand that this is the nature of their complaint.

While most of the debate has taken place in this looking-glass world, the theorists on both sides have been well aware that their two systems were intended to produce exactly the same prices. They debated different questions: Which system would arrive at the prices more accurately? Which system was more susceptible to market power or bureaucratic rigidity? Which system would give rise to more innovations? These questions are important but difficult, and little progress has been made.

In real time, the centralized approach is the clear winner, while for markets taking place more than a day ahead, the bilateral approach seems well accepted. Section 3-5.2 suggests that the centralized approach is probably also better for a day-ahead market. That discussion will not be continued here, as the focus will remain on the properties of the prices themselves.

## More Comparisons

**How Many Prices?**   Centralized computation is frequently accused of calculating too many prices. Although it may be possible to get by with fewer, bilateral trading and centralized pricing produce the same prices and the same number of prices when they work perfectly.

In an imperfect world, bilateral trading could produce either more or fewer depending on the microstructure of the trading institutions. Probably, they would produce more. In a bilateral market if there are 20 trades at the same location for the same period of time, there may be 20 different prices. Centralized pricing produces 1 price per location at a given time.

Either a bilateral or a centralized market can be set up to find fewer but less accurate prices, either by using zones or some other approximation technique. The drawback would be less accuracy and, if badly designed, serious gaming opportunities. The advantage might be more liquid forward markets. This trade-off deserves investigation.

The total number of prices means little to market participants who are only interested in the number of prices they need to track. For them, centralized pricing is an enormous simplification. A generator or load is typically concerned only with a single price; they trade only with the ISO and only at one location. By contrast, in a bilateral market, every generator must know the price of TRs from its location to the location of every load with which it might trade. It also must negotiate a different price for energy at every location.

**Are the Prices Public?**   It is difficult to discover prices in a bilateral system where traders need to know hundreds of prices. Bilateral prices are often considered trade secrets, but centrally computed prices are published immediately to the Internet (for example, see www.pjm.com).

**Who Benefits from Complexity?**   Complexity provides the arbitrage opportunities on which marketers thrive. In an *efficient* market with many locations, the price at A will nearly equal the price at B plus the price of shipping from B to A (the price of a transmission right from B to A). This provides little opportunity for arbitrage. By contrast, an *inefficient* market will typically have many pricing discrepancies and will provide many profitable opportunities. Marketers thrive on market inefficiency and are paid for their role in making the prices converge, that is, in making the market more efficient. If the market is inherently efficient, little money will be made as central computation of prices reduces arbitrage opportunities.

**Would Zonal Prices Work?**   Zones reduce the number of locational prices below what would be produced by a perfectly competitive market taking account of all constraints. The hope is that zones can be structured so that zonal prices will almost reproduce CLPs. California broke itself into two internal competitive zones, two regulated zones, and many external pricing points that were essentially one-bus zones. By early 2001 it proposed to FERC the creation of 11 internal zones. While there were clear problems with the initial approximation, no general analysis of zonal approximation has yet been made. Those opposed to zones point out that they are not perfect and claim it is cheap to get exactly the right answer.

**Can an Auction Be a Competitive Market?**   Although the central computation of prices looks very different from a spontaneously organized market, it can exhibit all the properties of a competitive market. Everything depends on market power. Given small enough competitors, both approaches will be competitive. Whether a bilateral or a centralized market is more susceptible to market power remains an open question.

**Collecting Congestion Rent.**   In the present example, under centralized pricing, the ISO would buy power for $750 \times \$35/h + 750 \times \$45/h$ or $\$60,000/h$. It would sell power for $300 \times \$35/h + 600 \times \$45/h + 600 \times \$45/h$ or $\$61,500/h$. The difference of $\$1,500/h$ is called congestion rent.

In the bilateral market, there must be transmission rights and bilateral trading will bid their value up to $\$1,500/h$. With the rights and trades of Table 5-4.1, there would be 100 MW of TRs worth $\$15/MWh$. The initial owners of these rights would collect the congestion rent. If the ISO sold the rights initially, it would collect the same amount under either system. If it gave them away—grandfathered them—someone else would collect the congestion rent.

If the rights had been financial rights instead of physical rights, there would have been 150 MW of rights from A to B and they would have been worth $\$10/MWh$. The rights to use the lines have the same value in any competitive market no matter how they are defined.

*Chapter 5-5*
# Congestion Pricing Fallacies

*Furthermore, the use of electricity will conserve the coal deposits of the world.*

George Westinghouse
1911

NOT REALIZING THAT CONGESTION PRICING *IS* COMPETITIVE PRICING, ADVOCATES OF COMPETITION CRITICIZE IT AS UNFAIR. Competitive locational prices, like all competitive prices, contain scarcity rents that cover fixed costs of generators as well as congestion rents that cover the fixed costs of the grid. None of this revenue is wasted, and occasional high prices caused by congestion send the right signals to investors to build new generators, to customers to use less power, and to the ISO to build needed lines.

**Section 1: Are Competitive Locational Prices Too High?** Congestion is managed by redispatching expensive local generation in place of cheaper remote generation. Usually the additional production cost is much less than the cost to consumers, and this is often cited as a failure of competitive locational pricing. The high prices are necessary, however, to cover the fixed costs of an efficient set of generators and to cover fixed transmission costs.

**Section 2: Congestion Taxing.** Critics of congestion pricing usually suggest congestion taxing, instead of congestion pricing, though not by that name. This three-step plan first pretends congestion does not exist and finds a single price to clear the entire market. Then the ISO redispatches around the congestion, at a cost, and finally taxes consumers to pay the redispatch cost. If generators would bid honestly, the tax would be small, but the system induces dishonest bidding that results in a higher average price for power than does standard congestion pricing.

## 5-5.1    ARE COMPETITIVE LOCATIONAL PRICES TOO HIGH?

Fallacies of congestion pricing tend to focus on the occasional high prices caused by local scarcity when import lines are congested. Perhaps the most articulate exposition of these misconceptions is found in Rosenberg (2000).

The Rosenberg example below is similar to the example of Chapter 5-3 but with only two types of generators at the city bus (Bus 2). When congestion occurs, it causes just a little of the expensive generation to be used which makes the price increase dramatically while increasing production cost very little.

## Example: High Congestion Prices

Figure 5-5.1 shows the network of the Rosenberg example, except for Generator B which plays no role. Two cases are considered, the first without a power-flow limit and the second with a 50-MW limit on the line from Bus 1 to the city bus, Bus 2. In the first case the two cheap generators easily supply all the power, and the $25 generator is marginal and sets the price. In the second case, the line limit causes the city to use all of its cheap generation, and although it uses no power from expensive generator C, Rosenberg has that generator set the city price. As he points out, if city demand had been slightly greater, generator C would have set the price and complicated the arithmetic. So the reader is asked to imagine that some power is used from Generator C but not enough to show up in the numbers, only enough to set the price. This does no harm and is indicated by "0+" for Generator C in Table 5-5.1, which shows Rosenberg's results. They are correct.

**Table 5-5.1**  Producer and Consumer Cost with and without Congestion

| | Output Levels (MW) | | | | Price (/MWh) | | |
| Case | A ($24) | D ($25) | C ($50) | Producer cost | Bus 1 (100 MW) | Bus 2 (200 MW) | Consumer cost |
|---|---|---|---|---|---|---|---|
| No limit | 200 | 100 | 0 | $7300/h | $25 | $25 | $7500/h |
| 50-MW limit | 150 | 150 | 0+ | $7350/h | $24 | $50 | $12400/h |

## Congestion Prices Are Competitive Bilateral Prices

What if this market was a perfectly competitive bilateral market of the type discussed in Chapter 5-4? The bilateral prices would be exactly the same. Clearly this is right for the no-limit case, but the congested case should be checked. To discuss this as a competitive example, each of the three generators must be imagined to be a group of identical competing generators. This would also make the original example more convincing by explaining why the generators bid their marginal costs for the centralized computation. Often Generator A or D will be referred to as type-A generators or type-D generators as a reminder of this assumption of competition.

With congestion, no type-D generator will sell power for less than $50/MWh because customers have nowhere to turn but to an idle type-C generator. A type-C generator will sell for no less because that would be below cost, and it would not charge $51/MWh because there are many idle type-C generators ready to step in and take the sale away by offering to sell (at a profit) for $50.50. This competitive dynamic holds the price down to $50/MWh. City load cannot buy cheaper power from type-A generators at Bus 1 because all transmission rights are profitably in use. They are worth $26/MWh to the traders using them, and would not be sold

**Figure 5-5.1**

A power market with two locations.

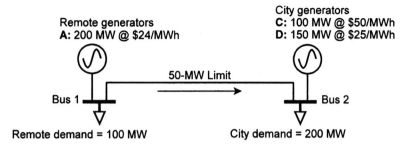

Remote generators
**A: 200 MW @ $24/MWh**

City generators
**C: 100 MW @ $50/MWh**
**D: 150 MW @ $25/MWh**

50-MW Limit

Bus 1

Bus 2

Remote demand = 100 MW

City demand = 200 MW

for less. Thus to buy power at Bus 1, load would need to buy a transmission right for $26 plus power for $24, for a total cost of $50/MWh. Once again a bilateral market would establish the same competitive locational prices as a competitive central-computation market.

## The Locational-Pricing Fallacy

*Fallacy*  **5-5.1**     **Congestion Rent > Redispatch Cost Is Unfair to Consumers**
Competitive locational prices charge loads more than they pay generators. This is congestion rent and it exceeds the cost of redispatch caused by the congestion. Surely this makes no sense and is unfair to consumers.

Rosenberg (2000, 38) refers to these prices as the "results of the Hogan LMP method." This is correct, but they are also the result of the Adam Smith competition method, and the method of competition prescribed by every economics text. These are the prices that cause supply to equal demand, minimize the cost of production and maximize the total surplus. These standard results can be checked once again.

Inspection of Figure 5-5.1 shows there is no cheaper way to supply the required power. Since a bit of generator C is in use, the marginal cost of power really is $50/MWh and consumers are being sent the right price signal. If their demand is elastic, they will make the right choice and this will maximize the total surplus. What then is the complaint against these particular competitive prices? Many have objected to them, and this objection is best articulated by Rosenberg (2000, 36) as follows:

> *The author finds it fascinating that while the actual redispatch cost to alleviate the congestion was only $50, the cost to the consumers on the downstream side of the congested interface was $5000, or 100 times that.*

This is the standard complaint. It focuses on the discrepancy between the high cost to consumers of congestion prices ($5,000) and the low cost ($50) of "fixing the problem" (alleviating the congestion). At first, it appears unfair to raise the price on all power at a location just because a little power costs more.[1] Wouldn't lower prices be better?

---

1. Most trades are hedged by forward purchases and are not directly affected by the real-time or day-ahead congestion price. But the cost of forward purchases is affected by the average of real-time congestion prices, so it is correct to think of these prices as affecting the price of all trades.

The first step toward understanding the high prices is to trace the flow of the extra $5,000/h of consumer cost at the city bus. Under nodal pricing, it is impossible to tell who is buying power from whom, but an arbitrary assignment will cause no trouble in the present example. Consider the financial and economic changes between the uncongested and congested dispatches. Assuming that city load bought 100 MW from the remote bus before the line was limited and 50 MW after, the changes in costs and profits are shown in Table 5-5.2.

**Table 5-5.2** Distribution of Additional Consumer Cost

| Production cost | Profit Increase | | Congestion | |
|-----------------|-----------------|---|------------|-------|
| Increase | A | D | Rent | Total |
| $50/h | − $100/h | + $3,750/h | $1,300/h | $5,000/h |

The extra 50 MW purchased from Generator D instead of from A raises production costs on those megawatts by $1/MWh, so total production cost increases by $50/h. Generator A was making $1/MWh selling 100 MW to the city, but after the line is limited its price falls to $24/MWh and it makes no profit. Generator D made nothing before the limit but makes $25/MWh on 150 MW of sales when the line is congested, while $26/MWh of rent is collected on the 50 MW flow on the congested line.

Rosenberg does not say what prices he would prefer, only that "Congestion cost must be limited to the actual cost consequences to the ISO of measures necessary to relieve the congestion." This suggests that approximately the entire $25 price increase is unwarranted, and the price at the city bus should be about $25/MWh as it was before the line became congested. Certainly this would be good for consumers in the short run, but shouldn't some consideration be given to generators? Perhaps they have some costs they need to cover.

First consider the congestion rent component of this transfer. If the congestion rent were not returned to customers, it would be used by the ISO for some purpose that benefits customers. Most likely it would help pay the cost of transmission lines and thereby reduce the charges for these lines that appear on customers' bills.

The key question remains: Do type-D generators need all the money they make from competitive pricing, or would they continue to prosper without it? Rosenberg observes that, with the line congested, a type-D generator "is now receiving $50 for output of 150 MWh, while its variable costs are $25/MWh." This is correct, and if the price were reduced to $25/MWh, these generators would cover their variable costs and no more. What about their fixed costs? When should these be recovered? Nowhere does Rosenberg mention fixed costs of generators, their recovery, or the consequences for future investment in generation if fixed costs are not recovered. His title refers to "monopoly pricing," but there is no exercise of monopoly power in this competitive market. The high prices are scarcity rents. Usually competitive scarcity rents cover fixed costs.

## A Broader View of the Congestion-Pricing "Critique"

The standard critique of congestion pricing described above has very little to do with congestion. Though it is not intended as such, it is a general attack on competitive pricing. The complaint is that consumers are overcharged by competitive prices. But congestion rents (as Rosenberg acknowledges in end note 8) "should be used to reduce the embedded revenue requirement." Thus congestion charges are returned to consumers. It is the high price paid to generators that is a net cost to the consumer. To understand the nature of high prices paid to generators, consider a small modification of the example.

Suppose the load at the city bus were reduced to 149 MW, and instead of a line limit of 50 MW, suppose there never was a line to the city. Now, on some hot afternoon, the load increases to 151 MW. The example's results change very little. Type-D generators will provide 150 MW at a variable cost of $25/MWh, while type-C generators will provide 1 MW at a marginal cost of $50/MWh. The "Hogan LMP method" will set the price for the whole city at $50/MWh.

Once again, the cost of providing the last megawatt is only $50/h, but the immediate cost to the city of LMP pricing is 150 MW × $25/MWh or $3,750/h. This price has nothing to do with pricing transmission; there are no transmission lines. Apparently marginal-cost pricing is the real problem, with or without congestion. Why should load pay an extra $3,750/h for 150 MW of cheap generation just because it needs 1 MW of $50 generation?

Once again the price in question is the standard competitive price. Draw the supply and demand curves; they intersect at $50/MWh. That price clears the market. That is the competitive price. "Supply equals demand" is not some perverse scheme invented recently by a Harvard professor.

Chapter 2-2 explains in detail why *prices must periodically rise above average variable cost in order to cover fixed costs*. If they are too high on average, more generation is built and the price is reduced; if they are too low, investment is delayed until increasing demand restores a normal profit level.

Possibly, some inappropriate restriction has been placed on building new generators in the city, so the type-D generators are inappropriately scarce and are being paid too much by competitive pricing. The $50/MWh price could occur too frequently and for too long. If so, the problem lies not with competitive pricing but with the restrictions on investment.

## 5-5.2   CONGESTION TAXING

Among those who dislike competitive locational prices, there is a standard approach to avoiding them. A three-step process, it amounts to congestion taxing:

1. Price power under the fiction that there is no congestion.
2. Redispatch using incremental and decremental generation bids.
3. Tax someone to pay for the redispatch costs.

Consider how this would work in the above example. Again, assume the load at the city bus is just over the stated 200 MW, so when the line is congested, the CLP will be $50/MWh.

The first step is to pretend there was no line limit as in the first case of the example in Section 5-5.1. The market-clearing price is again $25/MWh, and 100 MW flows over the line to the city bus. This is shown in the row labeled "Step 1" under "Congestion Taxing: Day 1" (in Table 5-5.3).

In Step 2, the ISO recognizes the line limit and redispatches to meet it. The negative (decremental) bid shown in "Step 2" in the table indicates that Generator A is willing to pay $24/MWh to be allowed to reduce its output below its Step 1 contract level while still being paid the full Step 1 contract price. The ISO accepts a 50-MW reduction and purchases a 50-MW increment from Generator D to compensate. The net cost is only $50/h. In Step 3 (not shown), the ISO taxes someone, probably city load or all load, to collect the $50/h. This is usually called uplift, but it is economically like a tax.

With honest bidding as assumed for Day 1, and with the ISO paying Step 2 bidders the price they bid and not a market-clearing price, this scheme works well to hold down prices.[2] If the ISO were successful, this scheme would reduce consumer costs substantially in the short run but would fail to cover the fixed costs of cheap type-D generators. No more of these would be built, and eventually the existing ones would be retired.

## Gaming the Congestion-Tax Scheme

Type-D generators soon notice that if they bid $49.90, they still sell their power. When the first one bids $49.90, it loses in Step 1 but is selected in Step 2 for the redispatch. Once all type-D generators follow suit, the market-clearing price in Step 1 becomes $49.90 (or $49.99 if bidding is to the nearest penny).

At the remote bus, generators know they must beat the type-D price because they lose in Step 1 and do not get a second chance in Step 2. Type-A generators bid, for example, $49.80 which insures their acceptance. The remote, type-A generators continue to submit **decremental bids** of $24. If a 1-MW remote generator is not decremented, it produces 1 MW at a cost of $24 and earns a profit of $24.80/h. If it is decremented, then it is paid $49.80 to produce and then must pay back $24 when it is decremented (curtailed). So it earns $24.80/h for doing nothing. This is equally profitable.

Consumers at both buses pay $49.90/MWh for their power. Local consumers are thus better off by $0.10/MWh than under CLP pricing, and remote consumers are worse off by $25.90/MWh, but this ignores the tax. The ISO must increment 50 MW of generation at the local bus at a cost of $49.90/MWh and must decrement 50 MW at the remote bus at a savings of $24/MWh. Consequently the ISO is out-of-pocket 50 × $25.90, or $1295/h. This will be collected via the congestion tax.

---

2. Following Rosenberg (2000, footnote 14), the true model consists of just over 200 MW of load at Bus 2, too little to effect calculation but enough to force marginal cost up to $50/MWh when the line is congested. This requires the ISO to accept a small incremental bid from the type-C generator and this would set the price for all incremental bids if a market-clearing price were used.

**Table 5-5.3** Congestion Taxing

| | Generator A | | Generator D | | Price | | Consumer cost |
|---|---|---|---|---|---|---|---|
| | Bid | Q taken | Bid | Q taken | Bus 1 | Bus 2 | |
| Competitive locational pricing | | | | | | | |
| | $24 | 150 MW | $25 | 150 MW | $24 | $50 | $12,400/h |
| Congestion taxing: day 1 | | | | | | | |
| Step 1 | $24 | 200 MW | $25 | 100 MW | $25 | $25 | $7500/h |
| Step 2 | – $24 | – 50 MW | $25 | 50 MW | | | $50/h |
| Congestion taxing: after gaming develops | | | | | | | |
| Step 1 | | 200 MW | $49.90 | 100 MW | $49.90 | $49.90 | $15,020 |
| | $49.80 | | | | | | |
| Step 2 | – $24 | – 50 MW | $49.90 | 50 MW | | | $1295 |

Load is 100 MW at Bus 1, and 200 MW at Bus 2. When generator D runs out of capacity, marginal cost is $50/MWh at Bus 2.

Consumers pay considerably more under this congestion-tax scheme although it is designed to hold prices down to variable cost. Unless a market is more tightly regulated, real scarcity undermines most price-cutting schemes.

## More Gaming

Unfortunately not even this dismal outcome is the end of the story. Investors will notice a peculiar opportunity for profit. They can build an ultra-poor-quality generator at the remote bus, one that is very cheap to build but expensive to run; they might pretend to rehabilitate a retired generator. They bid it in at $49.80, just as other remote generators are bid but will enter a decremental bid of $25 instead of $24. They will pay $25/MWh not to run. With this bid they are sure to be decremented, and their profit will be $49.80 minus $25, or $24.80/h—not a bad profit for a cheap generator that never runs.

*Chapter* 5-6
# Refunds and Taxes

*When asked by Gladstone if electricity would have any practical value, Faraday replied,*
*"Why sir, there is every possibility that you will soon be able to tax it."*

Michael Faraday
(1791-1867)

**H**OW SHOULD THE ISO REFUND CONGESTION RENT AND COLLECT
REVENUE SHORTFALLS? The answer is surprisingly simple. Over- and under-
collections of revenue should be added together and collected as a flat energy charge
or tax. Whether to collect it from generators or loads should be decided as matter
of convenience as the tax incidence of both choices is the same.

This answer depends on defining "prices" as charges intended to allocate
resources efficiently and "taxes" as charges intended to raise revenue. The answer
is most convincing if prices have been designed as efficiently as possible. Some
will argue that because prices are not yet right, taxes should be designed to compensate for their shortcomings. But if taxes begin to take on the role of prices, improved
prices will then be disputed on the basis that these would interfere with the incentives of the tax structure. The best designs will result from separating the two tasks:
taxes should only raise revenue to pay for fixed-costs of the system, while prices
should be designed with only efficiency in mind.

**Chapter Summary 5-6:** Pricing should be used to the greatest practical extent
for two reasons: (1) to maximize efficiency, and (2) to minimize taxes. Taxes should
be designed to minimize deadweight loss. This is accomplished easily with a flat
charge on energy.

**Section 1: Pricing versus Taxing.** System prices are defined as charges
intended to increase efficiency, while system taxes are defined as charges intended
to raise revenue. As many services and externalities as possible should be priced
efficiently, both to raise revenue and to increase efficiency. The inevitable revenue
shortfall must be recovered through taxes which should be designed to reduce
efficiency as little as possible. A flat energy tax does this so well there is little
chance for improvement.

**Section 2: Energy Taxes.** An energy-based (per megawatt-hour) charge, or flat energy tax, whether placed on load, generators, or some combination of the two, will have exactly the same effects, both financial and physical, as if placed fully on load.

# 5-6.1   PRICING VERSUS TAXING

An ISO has many charges for various services and uses of the system; setting them is always a contentious matter. The charges have two fundamentally different purposes, and setting them requires two different economic approaches. Some are intended to encourage the efficient use of system facilities, transmission and generation in particular, while others are intended to raise funds to pay for those facilities and various services.

To distinguish these two purposes, charges designed to promote efficient use will be called *prices*, and charges designed to raise revenue will be called *taxes*. While this is in general agreement with common usage, "tax" is more often defined as a charge levied by a government, but such charges are primarily intended to raise revenue and not to promote efficiency. Taxes on *externalities* are an exception to the present classification system. They are intended to equal the marginal value of the **externality** to its producer and the marginal cost imposed on those affected by it. This is exactly the role played by congestion prices. To avoid confusion, the present discussion will refer to efficiency-inducing charges on externalities as *prices, not taxes.*

| | |
|---|---|
| **Definitions** | **System Price**<br>A charge placed on market participants by the system operator for the purpose of promoting the efficient use of public or private system resources.<br>**System Tax**<br>A charge placed on market participants by the system operator for the purpose of collecting revenue. |

**Efficient Prices.** Efficient prices maximize social welfare by equating the marginal costs and values of producers and consumers. The value of the last megawatt consumed should equal the cost of the last megawatt produced. This equality is difficult to achieve without a competitive market.

The first step toward efficient charging is to design efficient pricing or efficient markets for as many services and externalities as possible. Use of the wires should be efficiently priced by market-determined congestion pricing. Losses should be efficiently priced. Use of reactive power and generator ramping might be priced efficiently in a more mature market. In each case, prices should be determined either by competition, or if that is impossible, by the system operator setting price as near marginal cost as feasible. *No attempt should be made to modify these prices to increase the revenue raised.*[1]

Although revenue should never be the purpose of system prices, it is a useful side effect as it reduces the necessary level of taxation. Increased revenue provides an additional motivation for pricing as many services and externalities as can be priced with reasonable efficiency.

**Efficient Taxes.**  Efficient pricing will raise too little revenue to pay for the efficient level of services. Fixed costs make this nearly inevitable. For example, if a power line costs $100 million plus $1 per watt of line capacity, the optimally sized line would collect congestion revenues to cover the $1/W, but would not pay the $100 million fixed cost. Because the ISO must meet its total costs, it will need taxes to raise revenues.

Because every service should be priced as efficiently as possible before taxes are considered, there is no possibility of increasing efficiency with a well-designed tax. If there were, it would be called a price. Attempts to improve on efficient pricing can only result in reduced efficiency. For example, if efficient congestion pricing is already in place, time-of-use charges can only interfere with it and reduce the efficiency with which the grid is used.

Having placed all of the burden for increasing efficiency on prices, the job of designing taxes is greatly simplified. The efficient tax is the one which causes the least inefficiency. Taxes change behavior, and in a market where behavior is as efficient as realistically possible, induced changes in behavior must work in the wrong direction. Such changes are called distortions of behavior.

**Techniques for Minimizing Distortions.**  The most sophisticated approach to minimizing the distortion of behavior is called Ramsey taxation (MasCollel et al., 1995). This approach taxes most heavily those who will change their behavior the least. If the tax is on electricity, those with the lowest demand elasticity are charged the most. This is a complex and controversial approach and will not be pursued.

An unsophisticated but effective approach to minimizing distortion spreads the tax as thinly and evenly as possible. This is based on the observation that tax distortions are proportional to the square of the tax rate. Thus if a tax is collected only on energy sold on Thursdays instead of being collected evenly over the week it will cause seven times more distortion than necessary. The Thursday distortion is 49 times greater with the uneven tax, but the even tax causes distortion seven days per week. This effect can be understood by studying the deadweight-loss triangle in Figure 4-2.2. Like the inefficiency of monopoly, the inefficiency of taxing is caused by raising total charges to consumers. Taxes increase proportionally the height and width of the welfare-loss triangle which increases its area in proportion to the square of the price change. The deadweight loss is given by the triangle's area.

If taxes amount to 10% of the retail price, this would be a tax of about $8/MWh and probably more than sufficient to cover the fixed costs of the grid and other

---

1. There may be rare exceptions when an auction price is ambiguous and the higher price will raise more revenue without causing inefficiency, but such choices should be checked for long-run implications.

system services. As a plausible and not incautious estimate, long-run elasticity of demand can be taken as 1. If total system taxes are spread as thinly as possible, the area of the tax-loss triangle is then $10\% \times 10\%/2$, or $\frac{1}{2}$ of 1% of retail costs.[2]

No taxing scheme can reduce this to zero, and most likely no realistic scheme can reduce it much at all. All schemes for improving on this would necessarily spread the tax less evenly and would risk increasing the tax inefficiency in doing so. Given the difficulty of the task of designing better electricity taxes, the scarcity of trained tax designers, the extremely small room for improvement, and the relatively larger possibility of loss, the recommendation must be to avoid all taxing innovations. When the goal is to raise revenue, simply place a flat charge on energy use.

---

| Result | 5-6.1 | **Price for Efficiency and Not to Raise Revenue** |
|---|---|---|

System prices should be designed with the sole purpose of improving the efficient allocation of resources. Price as many externalities and services as can be efficiently priced.

---

| Result | 5-6.2 | **Tax for Revenue and Not to Improve Efficiency** |
|---|---|---|

System taxes should be designed to raise revenue while minimizing the distortion caused by their affects on prices.

---

## 5-6.2   ENERGY TAXES

In addition to being simple and causing little distortion, a tax on energy has one more extremely convenient property. It has the same affect whether collected from generators or loads. Either way, load pays the tax. (Actually, customers pay the whole cost of power no matter how the charges are shuffled, so this should not be a surprise.) Throughout this discussion an energy tax is defined as being flat, that is, every megawatt hour of energy is taxed at the same rate regardless of time of day, level of demand or any other characteristic. The tax rate is expressed in dollars per megawatt-hour.

An energy tax of $X$ on suppliers increases their marginal costs by $X$. If the marginal cost of producing a certain megawatt-hour is $30 and the tax rate, $X$, is $2/MWh, the marginal cost of producing this MWh becomes exactly $32/MWh. Consequently an energy tax shifts every supplier's supply curve up by $X$.

Similarly, a negative energy tax, an energy refund, of $$X$/MWh to consumers shifts their demand curves up by $X$. If they would purchase $Q$ MWh at a price of $30/MWh and they are promised a refund of $2/MWh, they will now be willing to purchase $Q$ MWh at a price of S32/MWh. This causes a $2 vertical shift in their demand curve.

---

2. As a percentage of net social benefit, the tax loss is much smaller; and if the tax rate were 5% instead of 10%, the tax inefficiency would be 1/8 of 1%. If the tax were placed on top of other taxes that effectively raise the price of energy, its inefficiency would be greater.

In a competitive market, if both competitive supply and demand curves shift up by $X$, then the intersection of the two curves will increase by $X$, but the horizontal point of intersection, the quantity bought and sold, remains unchanged. The vertical level of intersection of the demand curve and the competitive supply curve is the competitive price.

---

**Result    5-6.3**    **An Energy-Based Transfer from Generators to Loads Has No Net Effect**
An energy-based tax on generation and refund to load of $\$X$/MWh increases the market price by $X$ but leaves unchanged the price net of taxes received by generators and the price net of refunds paid by load. Consequently the same quantity is sold, and all parties experience the same net costs and receipts.

---

The combination of an energy tax on suppliers and an energy-based refund to loads is an energy-based transfer from generators to loads. An increase in the competitive energy price with no change in the quantity sold is an energy-based transfer from loads to generators. Because the price increase equals the tax rate, these two cancel exactly. If the transfer is negative, this result means a transfer from loads to generators also has no effect.

Result 5-6.3 is stated without reference to the competitiveness of the market. If the market is not competitive, the generator's supply curve may not be its marginal-cost curve. Similarly, **oligopsony** power may affect consumers' demand curves. In spite of this, the above result holds because the actual supply and demand curves are still shifted equally by the transfer.[3]

## The Incidence of Energy Taxes

So far, only transfers have been considered, and these have been shown to have no effect. This result helps to establish the fact that energy taxes have the same effect whether placed on generators, loads, or any combination of the two.

An energy tax increases the cost of power and causes less to be consumed. This reduction is uneven in its effects on customers and on different uses of power. In spite of this, it has the same impact whether levied on generators, loads, or a combination of the two.

Say an energy tax with tax-rate $X$ on generators has a certain impact, $\mathcal{M}$, on the market. Now consider such a charge in combination with an energy-based transfer from loads to generators having transfer rate $X$. Result 5-6.1 guarantees that the transfer will change nothing, so the combination must still have impact $\mathcal{M}$ on the market. The combination is simply an energy tax on loads with tax rate $X$. This proves that an energy tax on generators has the same impact as an equal energy tax on loads.

---

3. With market power, there is still a market price, $P$, and a set of traded quantities, $q_1 \ldots q_N$. Each combination yields a certain profit to each supplier or a certain utility to each demander. Under an energy-based transfer of $\$X$/MWh, $(P + X, q_1 \ldots q_N)$ yields exactly the same set of profits and utilities. This shows that the transfer simply creates a new game that is the same as the old game in every respect except that $P + X$ takes the place of $P$ for any set of strategies employed by the players. This game will necessarily have the same outcomes except that $P + X$ will replace $P$. All of the quantities will remain unchanged as will the net costs and receipts under the new market price, $P + X$.

Now consider changing the transfer to $Y$/MWh. Again the impact on the market remains $M$. This time the result is a tax with rate $X - Y$ on generators and tax with rate $Y$ on loads. This proves that as long as the sum of the tax rates is $X$, the impact on all market participants is the same.

---

| | | |
|---|---|---|
| **Result** | **5-6.4** | **An Energy Tax on Load or Generation Will Be Paid by Load** |

**An Energy Tax on Load or Generation Will Be Paid by Load**

Assume an energy tax with tax-rate $X$ is placed on generators and one with tax rate $Y$ is placed on load, where $X + Y = T$. The combined taxes will have the same incidence as an energy tax with tax rate $T$ placed on loads.

---

# Pricing Losses on Lines

*Reason and Justice tell me that there is more love of man in electricity and steam, than in chastity and refusal to eat meat.*

Anton Chekhov
letter, concerning Tolstoy
1894

Losses ARE MORE THAN THE SUM OF THEIR PARTS. If one power flow loses 10 MW to heating power lines and another loses 20 MW, the two flows together will lose 58.3 MW. Billing for the extra 28.3 MW of losses has caused much controversy. To make matters worse, economists suggest charging the first power flow the cost of replacing 38.9 MW and the second the cost of replacing 77.7 MW. Together this is double the actual cost of losses.

Though charging transmission users for marginal losses often seems unfair to those being charged, it is what a competitive market in transmission would do. As always, the competitive approach increases market efficiency relative to the regulatory average-cost approach. The fact that it causes an "overcollection" is also an advantage as the revenue can be used to reduce the inefficient taxes that are needed to cover various fixed costs (see Chapter 5-6).

**Chapter Summary 5-7:** If transmission were provided by many small competing line owners, each with an unconstrained line, the price of transmission would equal the marginal cost of losses. This price would minimize the total cost of power.

**Section 1: The Competitive Price Is Twice the Average Cost.** Marginal-cost pricing means charging a power flow for the losses caused by a 1-MW increase of the flow times the amount of the flow. It collects about twice the cost of replacing total losses.

**Section 2: Competitive Loss Pricing.** A decentralized (bilateral) competitive transmission market would price losses at marginal cost thereby collecting about double the cost of the losses.

**Section 3: Inefficiency of Average-Loss Pricing.** Average-cost pricing of losses causes an inefficient dispatch and thereby increases the total cost of power. When generation is not otherwise constrained, the cost increase is approximately equal to the total cost of losses under an efficient dispatch.

## 5-7.1    THE COMPETITIVE PRICE IS TWICE THE AVERAGE COST

This chapter assumes that power lines are not congested. Section 5-7.2 models a competitive market, but this section simply assumes that the competitive price of losses equals their marginal cost.

### Finding the Marginal Cost of Delivered Power

Result 5-1.4 states that the transmission losses are proportional to the square of the power delivered to the load. This is true on a DC power line and approximately true on an AC line. According to Chapter 5-1, and as shown in Figure 5-7.1, losses, $L$, equal $a W^2$ where $a$ is a constant determined by the line voltage and resistance, and $W$ is the power transmitted to the load. (Losses are termed "quadratic" because $W$ is squared.)

Since the losses from a power flow of $W$ equal $a W^2$, the average loss per megawatt of low is given by $a W^2 / W = a W$. Marginal losses are the increase in losses that would be caused by a small increase in power flow. According to calculus, they are the derivative to total losses or $d\, a W^2 / dW$, which equals $2\, a\, W$, which is exactly twice average losses. Without calculus, the result can be found by increasing $W$ by one watt.[1] This will increase total losses $a (W + 1)^2$. Subtracting the original value of $a W^2$ gives an increase of $a (2\, W + 1)$. Because 1 is much smaller than $W$, ignoring it makes essentially no difference, so again marginal loss equals $2\, a\, W$.[2]

---

| Result | 5-7.1 | **Marginal Losses Are Twice Average Losses** |
|---|---|---|

**Marginal Losses Are Twice Average Losses**
If $W$ is the power received from transmission, the power lost in transit is $a W^2$, and the average loss is $a W$. Marginal losses are the derivative of $a W^2$ with respect to $W$ and thus equal to $2\, a\, W$, where $a$ is defined in Result 5-1.4.

---

### Pricing at Marginal Cost

Assume the transmission customer injects $W$ at one end, pays for delivery, and $W$ is delivered at the other end. The line owner physically replaces any losses by purchasing power at the generation bus at the price $P_E$. The cost of replacement power, $C_T$, is $P_E a W^2$. In this analysis, the only cost of transmission is the cost of replacing the losses. This is the physical variable cost of power transmission, and opportunity costs need not be considered as they are handled by congestion charges.

The competitive market <u>price</u> of transmission is the marginal cost of providing transmission. This is simply the price of replacement power, $P_E$, times the marginal losses, $2\, a\, W$, so the price of transmission will be $P_T = 2 P_E a W$. This is the price

---

1. Assume for the moment that the constants in our formula have been adjusted to work with watts instead of megawatts, and that W is a very large number of watts.

2. If this calculation is performed using watts for units instead of megawatts, then 1 watt is truly negligible.

**Figure 5-7.1**

A competitive market for transmission and losses.

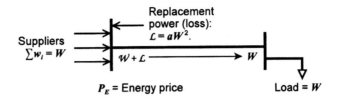

per megawatt of flow on the line, $W$, so the total charge for transmission is $G_T = 2 P_E a W \times W$. This is twice the cost of replacement power which was just computed to be $C_T = P_E a W^2$. The total charge for transmission is twice the total cost of replacing transmission losses when transmission is priced at the marginal cost of losses.

---

**Result 5-7.2**   **The Competitive Charge for Transmission Is Twice the Cost of Losses**
The cost of losses is $P_E a W^2$, where $W$ is the power delivered and $P_E$ is the cost of energy at the transmitting bus. The competitive price of transmission times $W$ is exactly twice this amount in a DC system and typically slightly less in an AC power system due to reactive power consideration.

---

## 5-7.2   COMPETITIVE LOSSES PRICING

A decentralized competitive transmission market would, as previously assumed, price losses at marginal cost. First a theoretical argument is given and then a decentralized market is modeled.

There is no market for losses; instead, the replacement of losses is a cost of providing transmission. Imagine that transmission is provided by many small competitive line owners who replace the losses on their lines and thereby, in effect, deliver all of the power injected by a trader using the line. Losses are essentially the only variable cost of providing this service.

According to the Marginal-Cost Pricing Result (1-5.2), a supplier will choose its level of output so that its marginal cost of production equals the market price. Because loss replacement is the variable cost of transmission production, the marginal cost of losses will be kept equal to market price by adjustments of individual supply levels. Price will be adjusted to clear the market, and marginal cost will follow price. In equilibrium, transmission will be priced at the marginal cost of loss replacement. As a consequence, every bilateral trade of energy will be charged the marginal cost of losses for its trade times the amount of power traded.

An alternative decentralized market structure would specify that the line owners did not replace losses. Instead, they would require traders to replace losses on their line in proportion to their use of the line. Traders would have to "self-provide" losses. With this system, the traders would pay half as much to replace their proportionate share of losses as they would pay for transmission in the previous market structure. They would still be charged for transmission and the competitive price of transmission would also be exactly half of the price paid under the previous

**Figure 5-7.2**

A competitive
transmission market
(Example 1).

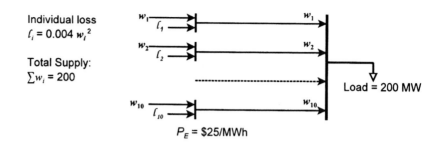

arrangement. This charge would be a scarcity rent. When the cost of loss replacement is added to the price of transmission, traders who "self provide" losses in a competitive market would pay as much for losses as traders in any other competitive market.

## Example 1: Competition

**Assumptions.** Consider a market with 10 perfectly competitive transmission owners, each owning a separate power line as shown in Figure 5-7.2.[3] They compete for business from many small generators who wish to ship a total of 200 MW of power from A to B. The lines have no capacity limit. Losses on each line are given by $\ell = 0.004\,w^2$, where $w$ is the power flow on the individual line. The price of energy at the generation bus is $P_E = \$25/\text{MWh}$.

**The Competitive Supply Curve.** A price-taking supplier will adjust its output to maximize profit given the market price of transmission, $P_T$. The competitive supply curve is the sum of these individual outputs.

*One supplier's short-run profit:* $\qquad\qquad \text{SR}_\pi(w) = P_T \times w - P_E \times \ell$

$$\text{SR}_\pi(w) = P_T \times w - P_E \times 0.004\,w^2$$

*Profit-maximizing condition:* $\quad d\,\text{SR}_\pi(w)/dw = 0 = P_T - P_E \times 2 \times 0.004\,w$

$$P_E = 25$$

*Individual supply of transmission:* $\qquad w(P_T) = P_T/0.2$

*Total supply of transmission:* $\qquad W(P_T) = 10 \times P_T/0.2 = 50 \times P_T$

Short-run profit is revenue from selling transmission $w$ at transmission price $P_T$ minus the cost of replacing losses. (Readers wishing to avoid calculus may check the competitive equilibrium by testing values of supply near $w = 20$ to see that this level of supply does maximize profits at the equilibrium price level of $P_T = \$4/\text{MWh}$.) The profit-maximizing condition of a price-taker is solved to find that the competitive supply curve of an individual transmission provider is $w(P_T) = P_T/0.2$, and the aggregate supply is 10 times greater.

**Supply Equals Demand.** In the competitive equilibrium, total competitive supply equals total demand which is constant at 200 MW.

$$\text{Supply } (W) = \text{Demand}$$

---

3. The only role played by the 10 lines is to make price-taking behavior plausible. In fact, because demand for transmission is elastic, 10 competitors would act very much like true price-takers.

$$50 \times P_T = 200 \text{ MW}$$

$$P_T = \$4/\text{MWh}$$

This is the competitive price of transmission, and it is what transmission suppliers charge for losses because that is their only variable cost. (This price may or may not cover their fixed costs, that depends on the number of transmission suppliers. The short-run equilibrium is not affected by this long-run consideration; rather, the long run will be determined by the short-run equilibrium.)

**Marginal Cost.**   In the competitive equilibrium, 20 MW flows on each line and the marginal cost of loss replacement is $2 P_E \, a w$ (see Result 5-7.1).

$$MC = 2 P_E \, a \, W = 2 \times 25 \times .004 \times 20 = \$4/\text{MWh}$$

That the competitive price equals marginal cost is neither coincidental nor surprising but is, once again, the standard result of competitive economics. Those who argue against marginal-cost pricing of losses are arguing against competitive pricing. When they argue in favor of average-cost pricing, or some form of "adjusted" or "scaled" marginal-cost pricing, they argue in favor of old-fashioned regulatory pricing. If the market price of transmission were set by regulation to the average cost of losses ($2/MWh), competitive transmission providers would maximize profits by suppling only half of the needed transmission.

---

**Result      5-7.3          Competitive Bilateral Loss Prices Equal Marginal Cost**
A perfectly competitive market for transmission would charge transmission customers the marginal cost of their losses.

---

## 5-7.3   INEFFICIENCY OF AVERAGE-COST LOSS PRICING

Electricity is produced more cheaply with competitive transmission prices than with any other prices. As an example, this section compares the cost of generation under average-cost pricing to the cost of generation under marginal-cost pricing of losses.

### Example 2: Inefficiency of Average-Cost Pricing

**Assumptions.**   Assume that the network looks as shown in Figure 5-7.2. Let the total transfer capability from A to B be the same as in Example 1, so that losses are given by $L = 0.0004 \, W^2$.[4] The price of energy at the generation bus, $P_E$, is again $25/MWh, and the price of energy at the load bus is $35/MWh.

**Equilibrium.**   Whether transmission is priced at the average cost or the marginal cost ($MC$) of losses, trade will take place until the price of transmission is $10/MWh. At any higher price, it makes no sense for the load to buy power from the remote bus. To find the amount of trade under the two systems, solve the two equilibrium conditions: average cost = $10/MWh, and $MC$ = $10/MWh. Average

---

4.  With 10 lines, the total resistance is 1/10 as great as with one line and the losses are also 1/10 as great.

**Figure 5-7.3**

Dispatch with average-cost and marginal-cost loss pricing.

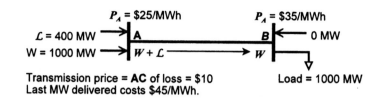

$P_A = \$25/MWh$          $P_A = \$35/MWh$

L = 400 MW → A          B ← 0 MW
W = 1000 MW → W + L ————→ W

Transmission price = **AC** of loss = $10          Load = 1000 MW
Last MW delivered costs $45/MWh.

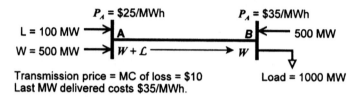

$P_A = \$25/MWh$          $P_A = \$35/MWh$

L = 100 MW → A          B ← 500 MW
W = 500 MW → W + L ————→ W

Transmission price = MC of loss = $10          Load = 1000 MW
Last MW delivered costs $35/MWh.

cost is total cost ($P_E\ 0.0004\ W^2$) divided by $W$, which, as demonstrated in Section 5-7.1, is half of marginal cost. Solving for $W$ in these two cases gives $W = 1000$MW under average-cost pricing, and $W = 500$ MW under marginal-cost pricing. The following table compares the system costs under the two pricing systems.

**Table 5-7.1** The Cost of Serving 1000 MW of Load at Bus B

|  | Average-cost pricing | | Marginal-cost pricing | |
|---|---|---|---|---|
|  | Power | Cost (/h) | Power | Cost (/h) |
| Power arriving at B from A | 1000 MW | $25,000 | 500 MW | $12,500 |
| Loss replacement at A | 400 MW | $10,000 | 100 MW | **$2,500** |
| Supplemental generation at B | 0 MW | $0 | 500 MW | $17,500 |
| Total generation | 1400 MW | $35,000 | 1100 MW | $32,500 |
| Increased cost due to average-cost pricing: | | **$2,500** | | |

Notice that the total cost of losses under marginal-cost pricing is $2,500, as is the waste due to the use of average-cost pricing. This equality is not coincidental but holds for any pair of energy prices and any value of line losses. It holds when power flow from remote generators is determined by the charge for losses. But if a power flow is limited by congestion or the amount of available generating capacity, it does not hold. When loss limits the dispatch of cheap generation, average-cost pricing quadruples losses and wastes as much money as the cost of efficient losses. When flows are determined by other constraints, loss charges may still send long-run signals that matter, but they do not affect short-run efficiency.

As an example of such inefficiency, consider the California ISO which computes the "full marginal loss rate" for each location.[5] This is the competitive price, but it does not charge generators this rate. Instead, it scales it down by the "loss scale

5. Attachment D of the Amendment to the ISO Tariff, filed with FERC by the California ISO on September 27, 1999.

factor" to find the "*scaled marginal* loss rate."[6] This euphemism for the *Average Loss Rate* was invented to pay lip-service to economics while pursuing a regulatory approach. As a consequence, whenever marginal-cost loss prices would limit the flow of power from remote locations, California wastes approximately as much money on excess losses as it costs to replace losses in an optimal dispatch. This is not a transfer of wealth; it is a deadweight loss.

| | |
|---|---|
| **Result**     **5-7.4** | **Average-Cost Loss Pricing Raises the Cost of Production** |

**Result**    **5-7.4**    **Average-Cost Loss Pricing Raises the Cost of Production**
In a power system in which losses would impose limits on competitive trade, the extra cost of production caused by average-cost pricing equals the cost of losses under marginal-cost pricing. When transmission or generation capacity would impose the limits on competitive trade, and not the cost of losses, under-charging for losses will not cause inefficiency.

---

6. California uses load as the reference bus (explained in the next chapter) to calculate nodal marginal losses. Although nodal marginal losses from generators are not necessarily double average losses, they are nearly so when the reference bus is "Demand distributed on a pro-rata basis throughout the ISO Controlled Grid."

# Pricing Losses at Nodes

*Her own mother lived the latter years of her life in the horrible suspicion that electricity was dripping invisibly all over the house.*

James Thurber
*My Life and Hard Times*
1933

Loss PRICING, LIKE CONGESTION PRICING, CAN BE SIMPLIFIED BY USING NODAL PRICES. By charging each generator and load the loss price at its bus, all trades will be charged properly for their losses. Also, if politics prevents locational pricing for loads, generators will still receive correct price signals.

It is easiest to charge for losses by including loss prices in the competitive locational prices (CLPs), but if they are charged separately and billed after the fact, generators will learn to adjust their bids and the market will still handle losses quite efficiently. Even the restrictions that total loss charges not exceed the value of lost power and that losses be paid for in kind can be handled quite efficiently by shifting nodal loss prices.

**Chapter Summary 5-8:** By choosing a reference bus, marginal loss prices can be computed at every bus. These are relative prices, but that is all that is needed. If losses are not collected from loads, then the total collection of losses can be controlled by choosing the reference bus. One nearer to load increases collections, and one nearer to generation decreases it. Shifting the reference bus shifts all loss prices uniformly, which has no economic impact for reasons discussed in Chapter 5-6. In particular, loss charges can be adjusted so that generators pay marginal loss prices, but total collections equal the total cost of loss replacement.

**Section 1: Nodal Loss Prices.** Unlike bilateral loss prices, which have an absolute value, nodal prices are relative. System physics cannot attribute losses to a power injection but only to a point-to-point power transfer. Similarly, only the differences between nodal loss prices are meaningful.

Choosing a reference or "swing" bus allows both losses and loss prices to be computed on a nodal basis, but both give absolute answers only when used in pairs corresponding to an injection and a withdrawal.

**Section 2: Full Nodal Pricing with Losses and Congestion.**  The concept of competitive locational price can be extended to include both loss prices and congestion prices at the same time. If loss pricing is separated from the central calculation, generators are forced to estimate it and adjust their bids accordingly. They cannot simply bid their true marginal costs.

**Section 3: Three Common Restrictions on Loss Pricing.**  If loads cannot be charged according to their location, loss pricing will not be fully efficient, but generators can and should still be charged their marginal loss prices. Restricting loss charges to collect only the total cost of losses need not interfere with marginal-cost pricing of generation losses, provided loads are not charged for losses. The restriction that generators must be allowed to pay "in kind" need not interfere with marginal-cost pricing if all charges are shifted down uniformly.

## 5-8.1   NODAL LOSS PRICES

In the previous chapter, the charge for losses was associated with a particular line or path rather than a single location. That path was defined by the origin and destination of a bilateral trade. While the bilateral approach is natural, a simplification in computing and listing loss charges can be achieved by associating losses with particular locations. This is completely analogous to the simplification in congestion prices that result from using nodal energy prices. Just as the congestion charge on a path is the difference between the energy prices at the path's ends, so the loss charge on a path is the difference between the loss prices at the path's ends.

Typically, the market for transmission services is not competitive but monopolistic, and competitive loss prices are centrally computed. A set of nodal loss prices is computed, and from them bilateral prices are found by subtraction. For instance, if the loss price at A, $P_A^L$, is \$1 per megawatt-hour injected, and the loss price at B, $P_B^L$, is \$3 per megawatt-hour injected, then a bilateral trade from A to B is charged $(P_A^L - P_B^L)$ which is negative, so it receives a \$2/MWh loss credit. In a 1000-bus network, half a million bilateral loss charges can be computed from 1000 nodal loss prices.

### Using a Reference Bus

The definition of nodal loss prices makes use of the standard engineering technique of picking a "reference" or "swing" bus. All losses and loss charges are then computed relative to that bus. The choice of reference bus is irrelevant because all meaningful economic calculations involve a subtraction that causes the effects of the choice to cancel out. This is the same type of cancellation that occurs in electrical calculations, and it makes the choice of reference bus just as irrelevant to the calculation of loss charges as it is to physical losses. Calculations at all buses are made relative to flow from the bus in question to the reference bus.

**Figure 5-8.1**

Market equilibria with
marginal-cost pricing.

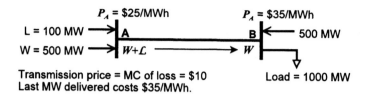

## Computing Nodal Loss Prices

The computation of nodal prices will be illustrated using Example 2 from the previous chapter and illustrated again in Figure 5-8.1. There are two ways to compute nodal loss prices in this example. Either Bus A can be taken as the reference bus, in which case the loss price at Bus A is zero, or Bus B can be taken as the reference bus, in which case the loss price at Bus B is zero.

If Bus B is selected as the reference bus, the loss price at Bus A is $10/MWh because the marginal cost of losses from transmitting a megawatt from A to B is $10/h. This means that if a generator injects one additional megawatt at Bus A, it will be charged $10/h. In contrast, a load will be paid $10/h for withdrawing a megawatt from Bus A. Consequently, there is no loss charge for shipping power from Bus A to Bus A. If a generator at A sells power to a load at B, the generator will be charged $10/MWh and the load will be paid nothing. Consequently, the net charge on this bilateral transaction is $10/MWh.

**Table 5-8.1**  The Choice of Reference Bus Is Irrelevant

| Nodal loss prices: (charge for power injected) | Price at Bus A | Price at Bus B | Charge from A to B |
|---|---|---|---|
| Reference bus = B | $10 | $0 | $10 |
| Reference bus = A | $0 | – $10 | $10 |

Picking Bus A as the reference bus results in a loss charge that is $10 less at both buses, so the difference is unchanged as is the loss charge for shipping power from A to B or the reverse. Nodal loss prices apply to any power injected or withdrawn at a bus. Injections are charged the loss price, and withdrawals are paid the loss price. If the price is negative, then the generator is paid, and the load is charged.

## Changing the Reference Bus

Fortunately, the effect of changing the reference bus is quite simple; all nodal loss prices are shifted by a constant amount. This property of loss prices holds even in a general meshed network.

| | | |
|---|---|---|
| **Result** | **5-8.1** | **Changing the Reference Bus Changes Loss Prices Uniformly** |

Locational loss prices are computed relative to a reference bus. If they are recomputed relative to a different reference bus, the new prices will equal the old prices plus a shift term that can be positive or negative.

Because a change in reference bus shifts all nodal loss prices by the same amount, and because bilateral loss prices are just nodal price differences, bilateral loss prices are unaffected by a change in the reference bus.

| | | |
|---|---|---|
| **Result** | **5-8.2** | **Changing the Reference Bus Does Not Effect Bilateral Trades** |

Changing the reference bus used to compute the nodal loss prices does not change $(P_A^L - P_B^L)$ and consequently leaves the bilateral loss price from A to B unchanged.

## 5-8.2   FULL NODAL PRICING: LOSS, CONGESTION AND REFERENCE PRICES

Competitive locational prices, CLPs, were defined in Chapter 5-3 to include only the effects of congestion. Having introduced losses and seen that they determine their own competitive locational prices, a unified approach is the logical next step.

The fundamental principle of competitive markets can be applied once again to the combined problem of losses and congestion and provides the most convenient and secure foundation for the synthesis. A competitive market minimizes the cost of power, so the first step is to find the efficient dispatch accounting for line limits and losses. Barring coincidence, this is unique. Using the optimal dispatch, the marginal surplus of an additional free megawatt can be evaluated at every node in the system to find all CLPs. (Recall that the additional megawatt is put to the best use which means completely re-solving for the optimal dispatch with the free megawatt included. See Sections 3-3.1 and 5-4.1.)

Having found the CLPs, there is no need to decompose them into a reference-bus energy price and separate loss and congestion prices relative to that bus. This can be done, but the CLPs provide the only prices that are needed.

Some systems compute congestion and losses separately, perhaps for practical reasons. California and PJM both compute congestion-based locational prices as a first step. This can be done simply by modeling the system as if it did not have losses. A better approximation to congestion prices can be obtained by computing the losses and balancing the market with losses included even though they are not taken into account when computing locational prices. The CLPs computed in this way can be thought of as the CLPs discussed in Chapter 5-3 which were computed for a network without losses but with congestion. If a swing bus is chosen and defined to be bus zero and have price $P_0$, the congestion CLPs allow the definition of locational congestion prices (as opposed to energy prices) in analogy to locational loss prices.

***CLPs with congestion:***               $$\text{CLP}_i^C = P_0 - P_i^C$$

Having dispatched the system based on congestion prices, all power flows are known. This allows the computation of marginal losses. These are computed relative to the same swing bus in the manner described in the previous section and will be denoted by $P_i^L$. When computed separately, the loss price is usually not incorporated into the CLP but charged separately. This has an impact on the bidding strategies of generators. They can no longer bid their marginal cost.

## Separate Loss Pricing

Suppose remote and local generators have marginal costs of $30/MWh and $31/MWh respectively, and they bid their marginal costs. If there is no congestion, the market will clear at $31/MWh. Next, the system operator computes loss charges, which could be $3/MWh for the remote generators (marginal losses are double average losses). This gives the remote generators an effective price of $28/MWh, $2/MWh below their cost. Clearly, honest bidding does not pay under the scheme of separate loss pricing.

**How Generators Compensate.**  This does not prevent an efficient competitive outcome; it just makes the generators game their bids. On the next round of bidding, remote generators will recognize the cost of losses as part of their cost of selling power and will add it to their bids—they will bid $33/MWh. Relative to centralized loss-included CLPs, there will be some additional error in this process as they cannot estimate their loss charge in advance as well as the system operator can estimate it during the calculation of the optimal dispatch.

The net result of this system of including congestion prices in the CLP but billing losses separately is that generators bid their marginal cost plus their estimated locational loss price and this sum, rather than their marginal cost, ends up equaling the CLP which includes only congestion prices. That produces the following equality:

$$P_0 - P_i^C = MC + P_l^L$$

which implies

$$P_0 - P_i^C - P_l^L = MC$$

which is

$$CLP_i = MC$$

So the result of charging for losses separately, and having the generators guess these charges and adjust their bids accordingly, is to equate the true CLP, with both congestion and loss prices included, to the marginal cost. This is approximate both because generators must guess their loss charges and because the two step procedure for finding the congestion and loss prices is an approximation. Nonetheless, this approximation is probably good enough in most circumstances. Certainly it is better than using average instead of marginal loss prices or simply adding losses into uplift. The strongest argument for the greater accuracy of loss-included CLPs is that it is simpler for generators, who do not have to guess loss charges but can instead bid their marginal cost into a central market.

## 5-8.3   THREE COMMON RESTRICTIONS ON LOSSES PRICING

Three restrictions are commonly imposed on loss pricing. None makes sense from an economic point of view, but once imposed, it is sensible to ask how to accommodate them most efficiently. The restrictions are:

1. Consumers must not be charged directly for losses.
2. Loss collections must equal the total cost of losses.
3. Loss charges must be assessed in electrical energy, not in dollars.

In each case the most efficient response is to drop loss charges on loads and charge generators marginal-cost loss prices.

**Consumers Must Not Be Charged for Losses.**   Restriction 1 will inevitably cause inefficiency on the demand side but need not do so on the supply side which is the source of most efficiency gains from competitive loss pricing. Demand-side problems consist of overconsumption by remote (downstream) consumers (who face prices below the cost of their power), and underconsumption by consumers nearer to generation. Given the restriction, the best pricing strategy is to use competitive loss prices on the generation side.

**Loss Collections Must Equal Loss Costs.**   If both generators and loads are charged marginal-cost loss prices, these will collect approximately double the cost of the losses. Moving the swing bus will shift the relative contribution of generators and loads, but the total will not change. Moving the swing bus, because it shifts all loss prices uniformly is equivalent to flat energy-based transfer between loads and generators. As Chapter 5-6 showed this has no effect because the market price exactly compensates for such a transfer.

   The appropriate response to this restriction is to implement efficient loss pricing only for generators. In the one-line example of Figure 5-8.1, if Bus B is selected as the reference bus, the generators at A will pay $10/MWh in loss charges, and the generators at B will pay nothing. If Bus A is selected, generators at A will pay nothing and the generators at B will *be paid* $10/MWh. In the first case, generators are "overcharged," and in the second case they are "undercharged," but in both cases the loss prices are efficient. By selecting the right point between A and B for the reference bus, the average charge to generators can be made to equal the average cost of losses.

   In general the reference bus can be moved, or the loss prices can be shifted up or down to adjust the total payments of generators.[1] When the loss prices are shifted down the system operator collects less from generators and so it must increase the uplift—tax load more. Shifting the loss prices down uniformly is an energy-based transfer from load to generation of the type discussed in Chapter 5-6. Generators pay less for losses; loads pay more for uplift. Result 5-6.3 ensures that such transfers have no effect. Generators pay less; this lowers their marginal cost; competition

---

1. Shifting the reference bus shifts all loss prices uniformly, and any shift in a wide range can be achieved by choosing the right swing bus. Rather than look for the right swing bus for a particular adjustment of prices, any bus can be chosen and all loss prices adjusted with an arbitrary adder.

forces the price down; and loads pay less for energy. The price change exactly cancels the system operator's transfer.

Typically when loss prices are found to "overcollect" because the swing bus is located too far toward the load end of the system, the remedy is to divide the prices down, often by nearly a factor of two. Instead, they should be adjusted by subtraction and not by division. Dividing the charges by a constant is not equivalent to moving the swing bus, so it reduces incentives and causes inefficiency.

For instance, if Generator X has a loss price of $5 and Generator Y has a loss price of −$5/MWh, increasing B's and decreasing A's output by 1 MW saves $10/h in losses. But if these prices were divided by two, such a change in output would save only $5/h. If the two generators are owned by one company, it will receive an incorrect pricing signal and will not optimize its dispatch. Less obvious, the price signals will cause the same problem if the generators are separately owned. Shifting the prices causes no such distortion of the price signals.

**Loss Charges Must Be Paid for with Energy.**   The most peculiar restriction on losses pricing is that all charges must be payable "in kind," in other words, with energy. Because marginal losses are greater than average losses, marginal-cost charges would cause more power to be collected than is lost, which would cause a system imbalance. (Again this is only true if the swing bus is placed too near the load.) This is equivalent to arguing that bus fares should be payable in gasoline, and because this would amount to more gasoline than the bus can use, the bus fair must be reduced.

If loss charges on generators are adjusted by subtracting a constant, the total collections from generators can be brought in line with total losses. In this way marginal-cost pricing can be made compatible with a requirement for barter.

# Transmission Rights

*Today's scientific question is:*
*What in the world is electricity and where does it go after it leaves the toaster?*

Dave Barry

FINANCIAL RIGHTS REFLECT ELECTRICAL REALITY; PHYSICAL RIGHTS REFLECT AN ILLUSION—THE NOTION THAT SUPPLIERS ACTUALLY DELIVER *THEIR* PRODUCT TO *THEIR* CUSTOMERS.[1] If supplier A sends power to load B and supplier B sends power to load A, their shipments may physically cancel each other on the connecting power line with the result that no power flows from A to B or from B to A. Instead supplier A's power goes to supplier B's customer and vice versa.

Suppose that instead of selling in their own regions, the Northern California generators decide to sell to Southern California and vice versa. Nothing physical changes. The same generators produce, the same loads consume, and the same amounts of power flow over the same paths. But with the new contracts, traders wish to own 10 GW of north-south rights and 10 GW of south-north rights. If the rights are financial, they just cancel out for the issuer. Whatever they pay to one set of rights they collect from the other. This calculation so perfectly mirrors physics that when financial rights are summed to find out if the total set is feasible, they are first converted to power flows and then summed by the engineers.

Issuing 10 GW of physical rights in each direction is next to impossible. The physical path may be limited to 2 GW. What if 4 GW of south-north rights were not exercised? The path would be burned out if the 10 GW of north-south rights were exercised, so 2 GW of north-south flow would be cancelled. Such rights are not very firm. To ensure that physical rights are firm, the issuance of such rights is limited to 2 GW in each direction on a 2 GW path. This forces trade to fit the limited concept of goods moving from supplier to customer without the possibility of automatic rerouting according to the far more efficient laws of power flow. Financial rights automatically cancel and reroute just as do power flows.

---

1. This is not the contract-path fiction. It accounts for paths but not for counterflow cancellations.

**Chapter Summary 5-9:**   Classic financial transmission rights, called transmission congestion contracts (TCCs), pay their owner the price difference between destination and origin for a specified power flow. Any association the owner may have with actual power flows is irrelevant. If they match the actual trade, these rights hedge their owners against all congestion costs. By bidding extremely aggressively in the day-ahead (DA) market, the trader can ensure dispatch and thereby synthesize a physical transmission right.

**Section 1:  Using Financial Transmission Rights.**   The classic financial transmission right is a TCC that pays the price difference between two locations for a fixed amount of power. This is the perfect hedge for an actual power flow of this quantity between these locations. In spite of the removal of risk by the hedge, traders will still be exposed to the full force of locational price signals and should still make efficient choices for production and consumption.

**Section 2:  Revenues from System-Issued Financial Rights.**   TCCs in opposite directions cancel out. If the remaining "net" TCCs are "feasible" because they correspond to a power flow that violates no line limits, the revenue from congestion rents will cover the system operator's cost of the TCC hedge payments.

**Section 3:  Physical Transmission Rights.**   Physical rights confer on their owners the right to transmit power. Physical rights in opposing directions cannot be netted out when determining the quantity of rights to be issued. This unnecessarily restricts the set of bilateral trades that can be based on a set of physical rights. TCCs in combination with extreme bids in the DA market can act as synthetic physical rights.

## 5-9.1   THE PURPOSE OF TRANSMISSION RIGHTS

Transmission rights may be awarded to the builder of a new power line and thereby serve the long-run purpose of encouraging investment. But, primarily, they are used to facilitate trading in the power market, and only that purpose is considered here.

**Trade without Transmission Rights.**   All power markets have centralized RT balancing markets. Consider one with competitive locational pricing that facilitates bilateral trades. If a bilateral generator-load pair registers their trade of 100 MW from A to B, the load will not be charged for taking power at B, and the generator will not be paid for delivering power at A. Instead the trade as a whole will be charged $100 \times (P_B - P_A)$, the RT price of transmission from A to B.[2] Assume this is the only centralized market and that all other trade takes place in bilateral forward

---

2. Of course this amounts to charging the load and paying the generator, but such psychological differences have played a major role in the bilateral-nodal debates.

markets. There are no transmission rights, financial or physical. How would such a market perform? Would transmission constraints be ignored?

First consider a system with constant supply and demand functions at every location but with a transmission grid that cannot support all economic trades that would take place in an unrestricted market. As a model, imagine $20/MWh generation at A, $40/MWh generation at B, and enough load at B to congest the line from A to B. For the opening day of the market, load at B would buy all its power from A in bilateral forward markets for $20. This would keep the system operator very busy in real time accepting $40 incremental bids at B and $20 decremental bids at A, but the line would be safe and there should be no physical problems. Power marketers who bought power at A for $20 and sold it for $20.10 to load at B would be chagrined to receive a bill for $20/MWh from the system operator—the price of transmission from A to B.

From then on traders would buy power at A for $20 and sell it for $40 at B, anticipating the $20/MWh cost of transmission. If the supply curves had some upward slope similar to the ones specified in Figure 5-3.1, forward trading would arrange a set of trades that neither over- nor underutilized the transmission line. No transmission rights, financial or physical, would be needed.

It is not always recognized that the pattern of trade might look quite different from the one-way trade from A to B that is usually depicted in this model. Consider a $40 generator at B that decides to sell to loads at A that are paying only $20 for power. Of course the load will pay only $20, but the generator registers the trade with the system operator and receives a transmission bill for *minus* $20/MWh—a transmission refund. There is no more reason for the generator to sell power in the high-priced local market than in cheap remote regions.

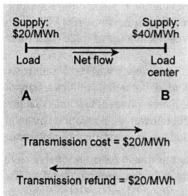

Supply:
$20/MWh

Supply:
$40/MWh

Load          Net flow          Load center

A                               B

Transmission cost = $20/MWh

Transmission refund = $20/MWh

---

**Result   5-9.1**

**Trading Opportunities Are Not Blocked by Congested Lines**
Although the cost of transmitting power from A to B may be high due to congestion, this will be offset precisely by a higher price at B. Similarly, generators will find that selling to low price locations is offset by negative congestion prices. Thus trades from any region to any other can be profitably arranged, just as in a market without congestion. (If congestion prices are unknown at the time of the trade, this introduces risk, but on average, trades across congested lines are just as profitable as local trades.)

---

In the long run, trades would be arranged quite differently from net power flows in a market with fully competitive transmission pricing. Some generators would sell from A to B while others sold from B to A. The set of generators producing would be determined by economics, but within this constraint, the matching of loads to generation could follow any pattern.

**Risk.**  Supply and demand curves are neither constant nor completely predictable, so prices are risky. If a generator trades with load at the local bus, there can be no charge for transmission and, as Chapter 3-5 showed, contracts for differences (CFDs) hedge such trades perfectly. If a generator trades with a remote load, and there is a chance of congestion, the trade is exposed to transmission-price risk which discourages trade. In the present example, the $20/MWh price difference between A and B would overcome this impediment and guarantee sufficient trade from A to B. But trades across a congested path in either direction will be risky. The result will be to minimize this type of trading. This results in a set of trades that mimic the physical flows on the network.

The minimization of long-distance trade means loads will have a narrowed choice of suppliers. Local purchases are unaffected, but remote purchases impose the cost of a risk premium. The risk premium will raise prices at locations that must make risky purchases and it may increase market power at all locations by acting as a barrier to the competition normally provided by remote trading opportunities.

**Hedging Transmission Costs.**   If the impact of transmission-price risk on the market place must be significant if it can be gauged by the reaction of traders. Two approaches have been proposed for reducing it: financial transmission rights, and physical transmission rights. The former pays its owner the RT cost of transmission on a given path, while the latter gives its owner the right to use the path.[3] Normally these two approaches would be essentially interchangeable, but transmission is an unusual commodity. Every line provides two types of transmission; call them "forward" and "backward." As is explained in Section 5-9.3, an unused line provides a certain amount of transmission of both types, but consuming forward transmission produces backwards transmission and vice versa. This property is very useful for facilitating trade but is very difficult for physical rights to take into account.

Ignoring this limitation of physical rights, it can be seen that both types of rights make good hedges. If a 100-MW trade from A to B is accompanied by a matching financial right, it can proceed with confidence that its transmission costs will be paid in full. All risk is eliminated. If the same trade is accompanied by a matching physical right, the trade is permitted to execute without charge.

A second motivation for physical rights is to help schedule power flows on potentially congested paths. The notion is that these physical controls are needed to protect the system because prices and markets are unreliable for this purpose. This approach would be called "command and control" by bilateralists were it not their own suggestion. Are these physical controls really needed? The answer is simple. There really are no physical controls, so physical controls must not be needed. There are only regulatory limits, violations of which are occasionally punished financially, but the existence of "physical rights" gives the system operator no telemetry equipment hooked to the governors of the generators in question. The only difference between this approach and a market approach is that the financial signals are nonmarket penalties instead of market prices. Are nonmarket penalties needed? Perhaps in extraordinary circumstances, but the daily functioning of markets such as PJM, which faces frequent congestion, does not depend on such penalties; these markets run on prices.

---

3. Financial transmission rights are more often settled in the DA market, but that does not affect the present line of reasoning.

**Ideal Transmission Rights.** Transmission rights are needed to hedge long-distance forward trading but not to protect power lines. If they are well designed, they will minimize forward-trading risks and the market will work much like the above example that has transmission costs but assumed no transmission-price risk. Trading at a distance, in any direction, will be uninhibited by price risk. Within the optimal set of generators, the matching of generators to loads is quite random. This will result in many counterflows between loads and generators, but because of physics, these flows will be netted out before they happen and the same optimal power flow will result as if local trading had been maximized. Similarly, transmission costs will net out and every generator and load will pay and be paid as if it had traded locally. Financial arrangements will reflect the physical properties of electricity.

## 5-9.2   USING FINANCIAL TRANSMISSION RIGHTS

### Transmission Congestion Contracts

The classic financial transmission right, defined by Hogan (1992) and called a **transmission congestion contract** (TCC), is one of the simplest rights.[4] It is financial and not physical because it entitles its owner to a certain sum of money but does not confer a right to transmit power. However, as explained in Section 5-9.3, it can be used to synthesize a physical right.

TCCs are defined in terms of locational prices which need not be competitive prices though they are preferred. The prices should depend on congestion and, if possible, losses. A TCC from A to B for 100 MW is defined to pay as follows while it is in effect.

$$\text{Value of TCC}_{A,B} = 100 \times (P_B - P_A)$$

Prices will change during the time interval specified by the TCC so the total payment is calculated by summing a sequence of fluctuating values. Typically, TCCs are long-term contracts and are settled in the DA market; that is, $P_A$ and $P_B$ are DA prices.

The TCC value is unrelated to power that the owner of the TCC may or may not have transmitted over the path in question. TCCs are directly connected only to prices, not to power flows. Their value may be positive or negative; if the value of $\text{TCC}_{A,B}$ is positive, the value of $\text{TCC}_{B,A}$ is negative. TCCs are not related to particular wires but only to a starting and an ending bus. The lack of connection to anything physical other than these two end points gives TCCs their simplicity.

### Hedging Congestion Prices

In a market with physical transmission rights, forward trades are hedged by purchasing the physical right to flow the traded power. If the price of such rights changes, the trade is unaffected. In a market in which physical rights are neither available

---

4. See Hogan (2001c) for a nontechnical explanation of financial transmission rights.

nor required, traders can schedule their power in the DA market, or (depending on market rules) just execute the trade in real time. In the first case they are charged the DA congestion price, and in the second, the RT congestion price. If they make a bilateral trade a month in advance using a contract for differences (CFD), they can hedge their trade against uncertain DA and RT prices by purchasing a TCC at the time of the trade.

Just as CFDs hedge temporal uncertainties in energy prices, TCCs can hedge spacial uncertainties. A 100-MW TCC from A to B hedges a purchase of 100 MW of power at bus A by a load at bus B. Because the TCCs settle in the DA market, the trader will need to purchase an identical hedge in that market as the TCC expires. The full transaction works as follows:

**Table 5-9.1**  Sequential Hedges: TCC Followed by Day-Ahead Trades

| Transaction | Revenue | Cost |
|---|---|---|
| Purchase of 100-MW $TCC_{AB}$ | | Cost of TCC |
| Realized value of TCC | $100 \times (P_B - P_A)$ | |
| Purchase/sale of 100 MW in the DA market | $100 \times P_A$ | $100 \times P_B$ |
| RT 100-MW flow of power from A to B | $P_{A0} \times$ deviation | $P_{B0} \times$ deviation |

The traders sell 100 MW at Bus A and purchase 100 MW at Bus B in the DA market, which exactly matches the trade they hedged with the TCC. The costs and revenues from these two trades exactly cancel the payoff of the TCC. As long as they trade as planned in real time, they will owe nothing in the RT market which charges only for deviations from DA contract positions. The net cost is the initial cost of purchasing the TCC, and the unpredictable DA and RT congestion prices have no effect unless the traders choose to deviate from their contract in real time.

## Efficient Dispatch

Traders who are fully hedged with a TCC are guaranteed full protection from the vicissitudes of congestion pricing. In spite of this, they feel the full incentive effects of RT pricing and make efficient choices regarding the production and use of energy. (This conclusion is identical to those concerning the two-settlement system in Chapter 3-5, which can be consulted for more detail.) Because the CFD and TCC are both purely financial, and thus unrelated to physical changes such as increased production or consumption, any deviations of the physical flows from the flows stipulated in the financial contracts affects only the settlement in the RT market. Consequently, the hedges have no impact on the decision to supply or consume an additional MW in real time. Only the RT locational price affects this decision, and a competitive price  leads to an efficient choice.

## 5-9.3   REVENUES FROM SYSTEM-ISSUED FINANCIAL RIGHTS

Financial transmission rights can be privately supplied or supplied by the system operator. The private market may fail to provide them because they are risky to provide, because there are many types, and because the market for each type is thin. The California, New York, and PJM ISOs have all provided financial rights quite similar to TCCs.

Typically, a system operator will provide them through an auction, the proceeds of which replace revenue from congestion charges covered by the issued TCCs. Generally the auction income will be more predictable than the congestion rent because the auction price of a TCC will be the expected or average congestion price. If the auction is efficient the revenue from the auction should, on average, equal the congestion rent that would have been collected had the rights not been issued.

**Revenue sufficiency** concerns a different question: whether the DA congestion rents collected in the DA market will be sufficient to cover the payments to TCC holders. They will be, provided the system operator does not issue too many TCCs. An understanding of this limitation, known as the *feasibility condition*, requires an understanding of counterflows.

### Counterflows

If generator A sells 1500 MW to load B while generator B sells 500 MW to load A, the actual flow on the line will be 1000 MW from A to B. Generator B's flow is a "counterflow" to generator A's flow. A **counterflow** is defined as a flow of power in the opposite direction to the predominant flow, but both the flow and the counterflow are fictitious. An engineer who is sent to measure the flow on the line will find only a 1000-MW flow; the two commercial flows will be undetectable as they have no separate physical reality. In spite of this, it does no harm to analyze the power flow as if it had two separate components.

Counterflows follow the same rules in meshed networks. If 1000 MW is sold from A to B, the power will flow over every line in the network. If another 1000 MW is sold from a generator at B to a load at A, it too will flow over every line in the network, and every power flow of this second trade will be equal and opposite to every flow caused by the first trade. As a result, there will be complete cancellation and no power will flow on any line. The actual power flows will go from the generator at A to the load at A and from the generator at B to the load at B. Physics reschedules the commercial flows so that they take the path of least resistance.

Physical transmission rights should not inhibit trades that produce counterflows, and financial rights should not be limited in ways that real power flows are not limited. Consider the two arrangements of trade in Table 5-9.2. Even if these trades take place in a meshed network with many lines, the actual power flows will not differ by even 1/1000 of a watt. The power flows are identical, not because of the principle of superposition or any law of physics but because all of the generators and loads are taking the same physical actions. All that differs are the words in the contracts. It would be absurd to allow one set of trades and not the other, yet many proposals for rights and many current restrictions on imports and exports interfere with trade arrangements analogous to the second arrangement of Table 5-9.2 while permitting the first.

**Table 5-9.2** Flows and Counterflows

| Power flows: | A | | B | |
|---|---|---|---|---|
| | In | Out | In | Out |
| **Trade Arrangement 1** | | | | |
| 500 MW: A ➜ A | 500 | 500 | | |
| 100 MW: A ➜ B | 100 | | | 100 |
| 400 MW: B ➜ B | | | 400 | 400 |
| Total | **600** | **500** | **400** | **500** |
| **Trade Arrangement 2** | | | | |
| 100 MW: A ➜ A | 100 | 100 | | |
| 500 MW: A ➜ B | 500 | | | 500 |
| 400 MW: B ➜ A | | 400 | 400 | |
| Total | **600** | **500** | **400** | **500** |

The cause of such restrictions, if they are not deliberate attempts to limit trade (exports for example), is a failure to understand that counterflows are an accounting fiction. Sometimes they are a fiction used by engineers in their calculations, and other times a fiction of trading (generators pretend their power flows to their customers). These are useful fictions as long as they are not used to prohibit trades that would be allowed if they were simply relabeled.

## The Feasible Set of Rights

With financial rights, there is no concern that a limited set of rights will limit trade, but it still makes sense to issue rights that do not favor one arrangement of trade over another when the two arrangements correspond to identical physical behavior. This is easily accomplished by restricting only the sum of the issued rights instead of by using a more detailed list of restrictions. This requires understanding the correspondence between TCCs and power flows. (As a simplification, a lossless network will be assumed.)

**Summing Power Flows.** A *power flow* on a network lists the power flowing on every line and its direction of flow. In a linear approximation of a network, two power flows can be added by adding the flows on every line and taking account of cancellations due to counterflows. In a real network, the net inflows to the grid at every bus must be summed and a new power flow computed. This is far more difficult, but still trivial by engineering standards, so any two power flows or any number of power flows can easily be added together.

**Feasible Power Flows.** Some power flows are considered "infeasible" because the flow on at least one line violates that line's power-transmission limit. A power flow that is not infeasible is *feasible*. All power flows fall into one category or the other.

**TCCs Correspond to Power Flows.**  A TCC specifies a power flow from A to B, but it does not specify the power that will flow on any line. Given the specification of the grid, an engineer can compute the power flow that corresponds to any TCC by assuming that its specified flow of power is injected at A and withdrawn at B. Similarly, every set of TCCs corresponds to the power flow that is the sum of all the power flows corresponding to TCCs in the set.

**Feasible Sets of TCCs.**  Any set of TCCs that corresponds to a feasible power flow is a feasible set. Other sets are infeasible. The set of TCCs to be issued is often restricted to be any feasible set. Restricting TCCs by restricting the sum of the corresponding power flows automatically accounts for counterflows, so trade will not be restricted except because of physical limitations.

If a full set of TCCs is to be issued, then traders are allowed to purchase any feasible set of TCCs. If a half set is to be issued (California issued a fractional set), then "half a feasible set" can be defined by cutting all of the line limits in half.

## Revenue Sufficiency for TCCs

The system operator collects congestion rent, $R_C$, in the DA market but refunds much or all of it to the owners of the TCCs. There is concern that $R_C$ cover the TCC hedge payments, $C_H$, which is a cost to the system operator. This condition ($C_H < R_C$) is termed "revenue sufficiency." While this is a sensible goal, the system operator will not lose money on TCCs as long as $C_H < R_C + R_{TCC}$, where $R_{TCC}$ is the revenue from selling TCCs.

Hogan (1992) has shown that for a DC approximation to a meshed network, the revenue sufficiency condition holds provided the set of issued TCCs is feasible.

---

| | | |
|---|---|---|
| **Result** | **5-9.2** | **Revenue from a Feasible Set of TCCs Will Be Sufficient** |
| | | If the set of TCCs distributed by the ISO is feasible (does not exceed the capacity of the network), the congestion rent will be at least sufficient to cover the payments owed to owners of TCCs. |

---

This result can be partially understood by considering a one-line network from A to B. If it has a capacity of 100 MW, and 200 MW of TCCs are sold from A to B, and the congestion price is $20, the congestion rent, $R_C$, will be $2,000/h while the cost to the system operator of the TCC hedge payments, $C_H$, will be $4,000/h. If, however, the sale of TCCs is restricted to the feasible level of 100 MW, the revenue from the congestion rent will be sufficient to cover the cost of the TCC hedge payments.

## Financial Transmission Options

Traders often prefer financial transmission rights (FTRs) that cannot have negative values. Such rights are options, and while reasonable, make revenue sufficiency impossible to guarantee without severely restricting the sale of FTRs.

Consider a 100-MW line on which 100 MW of options have been sold from A to B. With normal TCCs, another 200 MW of rights could be sold from B to A without violating the feasibility condition. But if this is done with options, the ISO will no longer be guaranteed revenue sufficiency. If $(P_A - P_B)$ is $10/MWh, the 200 MW financial option from B to A will be paid $2,000/h, and the owner of the 100 MW financial option from A to B will choose not to pay the $1,000/h that would have been owed had the option been a TCC. While the congestion rent is only $1,000, the ISO must pay $2,000 on the financial options that have been issued. The set of financial options that ensures revenue sufficiency must be far smaller than the feasible set.

## 5-9.4   PHYSICAL TRANSMISSION RIGHTS

Physical transmission rights confer a right to schedule and flow power. Unlike financial rights, they do not provide payments, and they are only useful to those actually trading power. The classic physical right gives its owner the right to transmit power on a particular power line and is required for the use of that line.[5] Usually a trade will require rights on a number of lines.

### The Feasible Set of Rights

| | | |
|---|---|---|
| **Result** | **5-9.3** | **The Feasible Set of Physical Rights Cannot Account for Counterflows**<br>On a power line with a capacity limit of $W$ MW, at most $W$ MW of physical rights can be issued in each direction. |

Like financial options, the feasibility limit on the set of physical rights is much stricter than on the set of TCCs because netting out of counterflows is impossible. This is because physical rights are also options; they can be used or not. Only 100 MW of physical transmission rights can be sold in each direction on a 100-MW power line because physical rights do not commit their owner to flow power on the line. Suppose 100 MW were sold from A to B and 200 MW from B to A, as is possible with TCCs. The owner of the rights from A to B might choose not to transmit power with the result that either the 200-MW flow from A to B would overload the line or the owner of the 200-MW right would be curtailed, in which case it was not a true 200-MW physical transmission right. If counterflow rights were netted out, someone wishing to buy a right from B to A after all such rights

---

5. Power lines are often built in groups for reliability purposes, and the entire "path" is used almost as a single line and is given a single flow limit. Physical rights would treat such a "path" as a single line.

had been sold could simply buy a counterflow right from A to B with no intention of using it, and that would make the desired additional right from B to A available.

## Synthetic Physical Rights Using TCCs

A TCC can be used to mimic a physical transmission right. Suppose a trader has purchased a TCC for 10 MW from A to B months in advance. To convert the TCC into a physical right, the trader can enter an offer to sell 10 MW at A in the DA market for a price of minus $1,000/MWh and to buy 10 MW at B for a price of $1,000/MWh. Such bids will certainly be accepted. Suppose the selling price at A is $P_A$ and the purchase price at B is $P_B$. In this case the trader will pay $10 \times (P_B - P_A)$ for its purchase and sale of energy and will be paid $10 \times (P_B - P_A)$ on its TCC. (TCCs are generally settled in the DA market.) These sums cancel exactly regardless of energy prices. The bilateral trade can now proceed as planned with no consequence from RT price fluctuations because it will just follow the schedule it established in the DA market. The TCC has protected it from the financial consequences of both the DA and RT markets, and its extreme bids have guaranteed its acceptance for DA scheduling. This is exactly what a physical right would have accomplished.

There are two unlikely possibilities that the TCC owner's synthetic physical right will fail to guarantee physical access. First, the rights from A to B may be oversold because some of them have been netted against counterflow rights from B to A. In this case, if all of the A-to-B rights holders place extreme bids, some could lose. But then the prices at A and B would be −$1,000, and $1,000, respectively, and those whose bids were not accepted would be paid $2,000/MWh for their TCC and would have no offsetting costs from the energy market. Surely this would be enough compensation, but such a windfall is extremely unlikely because no one will buy a counterflow TCC and fail to exercise it when the value of flowing in the forward direction is so great.

The second possibility is a transmission line outage. If a line is out, neither financial nor physical rights can guarantee the trade will go through. With a TCC, the fact that the trade is not executed will have no impact on the value of the TCC. So in principle the owner should still be compensated, but this right may be overridden by a "force majeure" clause in the TCC. In any case the TCC owner will be no worse off than the owner of a physical transmission right and may be much better off.

# Glossary

**ACE.** Area control error is the instantaneous difference between actual and scheduled *interchange*, taking into account the effects of *frequency bias* (NERC).[1]

**adequate installed capacity.** *See* reliability.

**adequacy.** *See* reliability.

**AGC.** *See* operating reserve.

**aggregate price spike.** *See* price spike.

**ampere (amp, A).** The unit of electrical current flow. One amp flowing from a 120-V outlet delivers 120 W of power.

**area control error.** *See* ACE.

**arbitrage.** A zero-risk, zero-net-investment strategy that still generates profits. One type of arbitrage involves the transfer of a commodity from a high-priced location to a low-priced location; a second type involves the transfer of demand from a high-priced product to a low-priced product.

**architecture.** *See* market architecture.

**auction market.** A market where all traders in a commodity meet at one place or communicate with a central auctioneer to buy or sell an asset, e.g., the NYSE. Auctions require bids of buyers, sellers or both. The following are types of auctions:

**English:** buyers start bidding at a low price. The highest bidder wins and pays the last price bid

**Vickrey:** buyers submit sealed bids, and the winner pays the price of the highest losing bid. Also known as a **second-price** auction and, confusingly, as a **Dutch** auction.

**Dutch:** the auctioneer starts very high and calls out progressively lower prices. The first buyer to accept the price wins and pays that price.

**Sealed-Bid:** buyers submit sealed bids (as in a Vickrey auction), and the winner pays the price that is bid. Also known as a **first-price** auction, a **pay-as-bid** auction, and a **discriminatory** auction.

**Reverse auction:** used to purchase instead of sell. The lowest bid wins. All auction types can be used in reverse.

**Double auction:** both buyers and sellers submit bids. It can be run as a first-price auction, a second-price auction, or as a bid-ask market which trades continuously as the NYSE does after its opening second-price auction.

**augmented load.** Load plus *installed generating capacity* that is out of service.

**automatic generation control.** *See* operating reserves.

**baseload.** The minimum load for a given control area. This part of load is constant.

**baseload generating capacity.** Generators normally operated around the clock, also referred to as baseload generators. *See also* midload generator; peaker.

**bilateral contracts.** Contracts used to make trades between two private parties.

**bilateral market.** A market in which private parties, generators and loads, trade directly at negotiated prices. Neither an *exchange market* nor a pool. Trades may be arranged by brokers or dealers.

**black-start capability.** The ability of a generator to start without taking power from the grid. This allows it to help restart the power system in case of a complete failure.

**broker.** An intermediary in a *bilateral market* who arranges trades but does not "take a position," i.e., does not buy or sell the commodity. *See also* dealer.

**capacity factor.** The ratio of the total energy generated by a generating unit for a specified period to the maximum possible energy it could have generated if

---

1. Definitions followed by "(NERC)" are from NERC (1996) and those followed by "(DOE)" are from DOE (1998b).

operated at its maximum capacity rating for the same specified period, expressed as a percent (NERC).

**capacity requirement.** A control area's requirement that all load-serving entities own, or have under contract, a certain amount of installed capacity. The amount required is typically near 118% of their peak load.

**capacity-requirement market.** A capacity-requirement market relies on its *capacity requirement* and the associated penalty to induce sufficient installed capacity for reliability purposes. A capacity-requirement market does not need price spikes to induce investment.

**classic competitive equilibrium.** *See* equilibrium.

**cogeneration.** Production of electricity from steam, heat, or other forms of energy produced as a by-product of another process (NERC).

**combustion turbine.** *See* gas turbine.

**commitment.** *See* unit commitment.

**competition.** The process by which suppliers attempt to maximize profit or demanders attempt to maximize net value. This book considers only price competition. **Perfect competition** is a market process in which neither suppliers nor demanders exercise market power and suppliers still earn a normal rate of return on their investment. *See also* natural monopoly; market power.

**competitive equilibrium.** *See* equilibrium.

**competitive market.** One in which perfect *competition* is approximated.

**competitive price.** The price that equilibrates supply and demand in a competitive market. The *market-clearing price* in a competitive market.

**congestion.** A line that would be overused if its flow limit were not enforced is congested. In a market, the path from A to B is congested if price of power at B is greater than the price at A.

**congestion management.** In the United States the term refers to controlling the dispatch to prevent violations of power-line limits. In Alberta and Ireland it also refers to building lines or generators to reduce congestion. Locational pricing of energy is the market approach to short-run congestion management.

**contingency.** The unexpected failure or outage of a system component, such as a generator, transmission line, circuit breaker, or switch. May also include multiple components, which are related by situations leading to simultaneous component outages (NERC).

**contract for differences.** A trade contract in which the purchaser pays the seller the difference between the contract price and some market price, usually the spot price.

**control area.** An electric system or systems, bounded by interconnection metering and telemetry, capable of controlling generation to maintain its interchange schedule with other control areas and contributing to frequency regulation of the Interconnection.

**cost of production.** Includes fixed costs, variable costs, startup costs, and no-load costs. Also called total cost of production or total cost.

**Fixed costs** are those such as debt and equity costs associated with the capital invested in the plant that do not depend on its level of operation.

**Variable costs** are those which vary with output. For its use in simplified examples, *see* variable cost.

**Average cost** is total cost divided by total output.

**Marginal cost** is the *derivative* of total (or variable) cost with respect to output. *See also* marginal cost.

**Startup costs** are incurred only because of starting up a generator. There are also shut-down costs, but they are less.

**No-load cost** is the cost of running a generator while producing no output. As with *startup costs* these may or may not be counted as variable costs depending on the problem under consideration. No-load cost is the same at all output levels and is measured in $/h. *See also* nonconvex costs.

**cost-of-service regulation.** Setting prices so that the regulated firm earns a normal rate of *profit*.

**counterflow.** A flow of power in the opposite direction to the predominate flow. Both the flow and the counterflow are useful fictions; only the net flow is measurable.

**current.** Electrical current is measured in amps and corresponds to the motion of electrons through wires that is fully analogous to the flow of water through a pipe. Benjamin Franklin, not knowing that electrons were negative, defined current as flowing from a positive voltage to a negative voltage, while electrons flow from negative to positive.

**dealer.** An intermediary in a *bilateral market* which buys and sells for its own account. It attempts to profit from taking speculative positions. *See also* broker.

**decremental bid.** An offer to buy back power that has been sold, usually at a price below the selling price.

**demand.** The amount of power that would be consumed by loads if system frequency and voltage were equal to their normal operating values for all consumers. Shed load is included as part of demand. This text uses an economic definition of demand and not the engineering definition which is "the power received by load (NERC)." Under normal conditions, the two coincide—supply equals demand. The economic definition allows supply and demand to differ, while the engineering definition does not. (Losses are included in both definitions of demand.)

**demand charge.** That portion of the consumer's bill for electric service based on the consumer's maximum electric capacity usage and calculated based on the billing demand charges under the applicable rate schedule (DOE).

**demand elasticity.** *See* elasticity of demand.

**demand-side flaws.** The first flaw is lack of metering and real-time billing, which prevents consumers from responding to the real-time price and makes very-short-run demand inelastic. The second is lack of real-time control of power flow to specific customers. This prevents physical enforcement of bilateral contracts in real time.

**derivative.** A mathematical concept that measures the change in one variable caused by the change in another. For example, if total cost changes by $20 when output changes by 1 MWh, the derivative of total cost with respect to output is $20/MWh. Economics calls this the *marginal cost*.

**dimensionless.** A variable whose value can be expressed as a pure number; for example, a probability is dimensionless. If the dimensions of units are the same, one can be converted to another by multiplying by a pure number, e.g., a year is 8760 hours. Time, energy, and money are the dimensions used in power economics. *Duration* can be measured in units of hours per year, but

time per time is dimensionless, so such a value can be converted to a pure number by dividing by 8760.

**dispatch.** To operate and control a power system, especially with respect to determining the outputs of the system's generators. Also the set of generators that are providing power at any point in time and their output levels. If an inappropriate set of generators is dispatched, the dispatch is said to be inefficient. An **economic dispatch** is one that minimizes production cost given transmission constraints.

**duration.** The fraction of time that a particular *load* takes power. This can be expressed in hours/year or as a percentage or a fraction. It can be usefully interpreted as the probability that the load is taking power.

**efficiency.** The state of having maximized the sum of consumer and producer *surplus*. **Productive efficiency** means production costs have been minimized (producer surplus maximized) for the given level of output.

**elasticity of demand.** The responsiveness of the quantity demanded of a good to its own price. Short for price elasticity of demand. The percentage change in demand that occurs in response to a percentage change in price. $(dQ/dP)(P/Q)$.

**emergency operating range.** Generation operating levels in excess of a generator's normal rating and below its emergency rating (NERC's terminology) or in excess of a normal maximum generation limit and below a maximum generation emergency limit (in PJM's terminology).[2] In PJM, just over 3% of total installed capacity falls into this range.

**emergency output.** Output produced when operating in the *emergency operating range*.

**energy.** The capacity for doing work; thus, energy is a *power* flow (MW) over a period of time (h). Electrical energy is usually measured in kilowatt-hours (kWh) or megawatt-hours (MWh), while heat energy is usually measured in British thermal units.

**entire market.** *See* market.

**equilibrium.** A state in which forces balance and a system has no tendency to change. In economics this is often characterized by supply equaling demand. Confusingly, a *competitive equilibrium* is not necessarily an

---

2. See PJM (1999)

equilibrium; instead, it is a state that would be an equilibrium were market participants to behave competitively. Thus a monopolized market typically has a competitive equilibrium although it does not operate at that point. Instead it operates at the monopolistic equilibrium, a true equilibrium, at which supply equals demand.

A **competitive equilibrium**, also called a **Walrasian equilibrium**, is a price-quantity pair $(P^e, Q^e)$ that satisfies two conditions: (1) a price-taking supplier would supply $Q^e$, and (2) a price-taking consumer would demand $Q^e$.

A **monopolistic equilibrium** is a price-quantity pair $(P^e, Q^e)$ that satisfies two conditions: (1) the supplier, a monopolist, would supply $Q^e$, and (2) a price-taking consumer would demand $Q^e$.

A **Nash equilibrium** is a game-theoretic equilibrium consisting of a set of strategies, one per player. Each player can do no better by changing strategy, provided the other players continue to play their designated Nash strategies. A market without a *competitive equilibrium* can have a Nash equilibrium in which the players (suppliers) exercise no market power. This is a competitive Nash equilibrium.

A **classic competitive equilibrium** is simply a *competitive equilibrium*. "Classic" is used to distinguish it from a competitive *Nash equilibrium*.

**exchange market.** Also called simply an exchange, or, in power systems, a power exchange. This is a centralized market that does not make side payments. Typically the bids would specify price and quantity or a supply or demand curve. Occasionally they also specify a minimum supply time or quantity or a startup cost. The bids are typically much less complicated than multipart bids used in *pools*.

**expected.** The expected value of a random variable is the average of the values that would be observed for that variable if it were sampled many times (technically an infinite number of times). The expected value of the role of a die is 3.5.

**externality.** An externality is present whenever the well-being of a consumer or the production possibilities of a firm are directly affected by the actions of another agent in the economy. "Directly" means the effect is not mediated by price. Externalities are termed negative if they are detrimental and positive if they are beneficial.

They occur because there is no market for the direct effect.

**Federal Energy Regulatory Commission (FERC).** The branch of the federal government that regulates the price, terms, and conditions of most power sold in interstate commerce and most transmission services. Municipal utilities and the great State of Texas are not under FERC's jurisdiction.

**financial withholding.** *See* withholding.

**fixed costs.** *See* cost of production.

**fixed-cost recovery.** Revenues minus variable costs. Variable costs are all costs that can be avoided by not producing. *See also* profit.

**forced outage.** The removal from service availability of a generating unit, transmission line, or other facility for emergency reasons or a condition in which the equipment is unavailable due to unanticipated failure (NERC).

**forward contract.** An agreement calling for future delivery of a commodity at an agreed-upon price. *See also* futures contract.

**forward markets.** The decentralized market in which a commodity (energy) is sold using forward contracts. Delivery times can range from days to years in the future.

**frequency.** The rate at which alternating current completes a cycle of two reversals of direction. It is measured in *hertz* (Hz), which are cycles per second. The scheduled frequency in North America is 60 Hz, in Europe it is 50 Hz. There is a single frequency in an *interconnection*—individual *control areas* do not have individual frequencies.

**frequency bias.** A value, usually given in megawatts per 0.1 Hertz, that relates the difference between scheduled and actual *frequency* to the contribution a specific *control area* is required to make to the correction of that difference.

**futures contract.** Obliges traders to purchase or sell a commodity at an agreed-upon price on a specified future date. The long position is held by the trader who commits to purchase. The short position is held by the trader who commits to sell. Futures differ from forward contracts in their standardization, exchange trading, margin requirements, and daily settling (marking to market).

**gaming.** An attempt by a market participant to profit by exploiting imperfections or loopholes in the market rules.

**gas turbine (GT).** Short for a generator powered by a gas (combustion) turbine, which is essentially a jet engine.

**giga.** 1,000,000,000. *See also* kilo.

**grid.** The transmission network.

**hedge.** A financial instrument that insures the purchaser against uncertainties in the spot price. Also the act of purchasing such an instrument to reduce risk. The opposite of speculation.

**Herfindahl-Hirschman index (HHI).** More commonly called the Herfindahl index, it is a measure of supply concentration found by summing the squares of the *market shares* of suppliers. Economists measure shares as fractions and compute HHIs between 0 and 1, while lawyers use percentages and compute HHIs between 0 and 10,000. A supplier's *market share* is its output divided by total industry sales.

**hertz.** *See* frequency.

**ICap.** *See* installed capacity.

**impedance.** The complex resistance to AC power flow. It includes ordinary resistance and reactance. Reactance comprises inductance and capacitance. Power lines tend to have impedance that is mostly inductance.

**inadvertent interchange.** The difference between a control area's net actual *interchange* and net scheduled *interchange* (NERC).

**incentive compatible.** An economic mechanism is incentive compatible if it induces participants to reveal their true costs or preferences. A second-price *auction* is incentive-compatible because bidding lower does not reduce what you pay if you win, and bidding higher improves your chance of winning. Hence the best strategy is to bid as much as winning is worth but no more.

**incremental cost.** Avoidable cost of producing output. Includes startup cost and variable costs. Also includes any cost of redispatch.

**independent system operator.** A system operator that utilizes markets for scheduling and dispatch and that has been approved by FERC as meeting its criteria for independence and market design. An ISO must be nonprofit and have no affiliation with any market participant.

**inframarginal rent.** *See* profit.

**installed capacity (ICap).** Generating capacity that has been operational and will be again although it may currently be experiencing a planned or unplanned *outage*. ICap refers to the total installed capacity of a control area.

**integer problem.** *See* lumpiness problem.

**interchange.** Power or energy that flows from one *control area* to another.

**Interconnection.** When capitalized, any one of the five major electric system networks in North America: Eastern, Western, Texas (ERCOT), Québec, and Alaska. An Interconnection is the largest *synchronized* part of an electrical network. Only DC lines connect the Interconnections.

**interconnection.** The facilities that connect two systems or *control areas* or that connect a nonutility generator to a *control area*.

**interruptible load.** Load that has a contract stating that it can be interrupted no more than a set number of times per year in return for some compensation, generally a reduction in its rate.

**ISO.** An independent system operator is a nonprofit system operator, not under the control of any small group of market participants, that also runs a real-time balancing market and usually a day-ahead market of some type. FERC has officially designated some markets as ISOs.

**kilo.** A prefix meaning 1000 and abbreviated with a small k. Related prefixes come in multiples of 1000 and are *mega* (M), *giga* (G), and *tera* (T). Thus a megawatt (MW) is a million *watts*, and 1 GW = 1,000,000,000 watts. The symbol mW means milliwatt or 1/1000 of a watt and has no use in power economics.

**Kirchhoff's laws.** Two laws governing voltage and current in DC networks, discovered by Gustav Kirchhoff (1824-1887). Euphemistically used by engineers to describe similar laws governing lossless AC power flows.

Euphemistically used by non-engineers to refer to all physical laws governing power flows, especially to the generalization of Ohm's law.

**left-hand marginal cost.** *See* marginal cost.

**Lerner index.** The relative *markup*, $L_X = (MC - P)/P$. Also called the relative markup (Tirole 1997, 66). *Marginal cost, MC,* is measured at the actual *monopolistic equilibrium*, not the *competitive equilibrium*. Zero indicates no market power while 1 indicates infinite market power.

**load.** An end-use device or a customer that receives power from the electric system. Often *load* is used as synonymous with *demand*, as in the *Simple Model of Reliability*.

**load factor.** The ratio of the average load to peak load during a specified time interval.

**load-serving entity (LSE).** A load aggregator, power marketer, or electric distribution utility that purchases electricity at wholesale for resale to individual or groups of end-use customers.

**load shedding.** The process of deliberately removing (either manually or automatically) preselected customer demand from a power system in response to an abnormal condition to maintain the integrity of the system and minimize overall customer outages (NERC). Load shedding is involuntary and occurs outside of contracts with load customers.

**long run.** Shorthand for the assumption that the capital stock (total generating capacity) is in equilibrium.

**long-run profit.** *See* profit.

**loop flows.** The difference between the scheduled and actual power flow, assuming zero inadvertent interchange, on a given transmission path. Synonyms: parallel path flows, unscheduled power flows, and circulating power flows (NERC). It also refers to the existence of parallel power flows in a system in which flows have not been scheduled on any particular path.

**losses.** Power flowing on power lines heats the wires. This uses power, so less power reaches the receiving end than is injected at the sending end of the power line. The difference is the power loss.

**low operating limit.** Some generators cannot produce less than a certain amount of power, called the low operating limit, and still maintain stable operation.

**LSE.** *See* load-serving entity.

**lumpiness problem.** Some inputs to production may come in lumps, which may make it impossible to purchase the amount that would be optimal were this restriction lifted. For example gas turbines might come in only two sizes, 100 MW and 150 MW. In this case, if 275 MW of capacity is needed, it cannot be purchased. This problem can invalidate efficiency results. Generally if a large number of lumps are required the problem becomes unimportant. The **integer problem** is the special case in which lumps are of only one size and a whole number of them must be utilized.

**marginal.** A term left over from the pre-calculus days of economics usually meaning the derivative with respect to quantity.

**marginal cost (MC).** At all output levels where a supplier's variable cost curve is smooth, marginal cost is the slope of that curve. It is the rate of change of cost with output at a given level of output, or technically,

$$MC = dVC/dQ.$$

When the variable cost curve is not smooth at a given level of output because it ends or becomes vertical, marginal cost is not defined for that level of output. In such situations the following concepts usually will be defined and can be used to restate standard economic results concerning marginal cost. In short, left- and right-hand marginal cost are defined to be the left- and right-hand derivatives of $VC$ and often exist when the *derivative* itself does not.

   The following definitions are non-technical, and assume "one unit" changes are equivalent to *derivatives*.

   **Right-hand marginal cost ($MC_{RH}$)** is the cost of producing one more unit.

   **Left-hand marginal cost ($MC_{RH}$)** is the saving from producing one less unit.

   **Marginal-cost range ($MC_R$)** spans cost from $MC_{RH}$ to $MC_{LH}$.

   When the variable cost curve is smooth, $MC_{RH}$, $MC_{LH}$, and $MC_R$ equal marginal cost. In a competitive equilibrium, if marginal cost is not defined, price will still fall within the marginal-cost range. As long as this condition holds, market power has not been exercised.

**marginal-cost pricing.**    Originally the basic principle of economic regulation. In the context of power markets it refers to setting price equal to the system marginal cost based on multipart bids. Sometimes called marginal pricing. Many disagreements concern the details of determining system marginal cost.

**marginal value.**    The consumer's willingness to pay for an additional unit of consumption. This assumes a smooth demand curve. At points where the demand curve is vertical, marginal value is undefined. It is the demand-side analogue of marginal cost.

**market.**    Any context in which the sale and purchase of goods and services takes place, for example, an *exchange*, *pool*, or *bilateral market*. The trades in a market usually involve goods that are close substitutes, but the term power market includes services that are inputs to the production of delivered power, such as transmission. Many of the market design problems discussed in the present book concern such broadly defined markets.

An **entire market** is the complete set of *submarkets* that constitutes the scope of a market design or analysis project.

A **sub-market** is a component of an *entire market* that is itself a market. The day-ahead wholesale market is a sub-market of the power market.

**market architecture.**    A market's architecture is a map of its component *submarkets* including the type of each *submarket* and the linkages between them. Market type classifies markets as, for example, bilateral, private exchanges, or pools. Linkages between submarkets may be, for example, implicit price relationships caused by arbitrage or explicit rules linking rights purchased in one market to activity in another. The architecture of an *entire* market includes the list of both designed and naturally occurring *submarkets*. This concept is developed in Wilson (1999). *See also* market structure.

**market-clearing price.**    The price at which supply equals demand, which may or may not be the competitive price. If some suppliers are buying back power through decremental bids at a price of, for example, $20/MWh, while the rest are paid $30/MWh to produce, neither price is the market clearing price. If suppliers have *nonconvex* costs, such as startup and no-load costs, there may be no market-clearing price.

**market failure.**    A market flaw that prevents the market from operating efficiently. Usually the inefficiency must be fairly dramatic to warrant the use of this term. Also the state of a market in which demand cannot be brought in line with supply by adjusting the price, for example, when totally inelastic demand exceeds maximum physically-available supply.

**market power.**    The ability to alter profitably prices away from competitive levels (Mas-Colell et al., 1995, 383). When suppliers exercise market power, they do not act competitively as price takers. Instead they produce less than their competitive output causing an increase in market price. *See also* marginal cost; monopoly power.

**market price.**    Usually short for the *market-clearing price* but can apply to a prevailing price that fails to clear the market if there is some type of market failure.

**market share.**    The proportion of total market sales accounted for by one firm. *See also* Herfindahl-Hirschman index.

**market structure.**    Properties of the market closely tied to technology and ownership. The classic structural measure is a supplier concentration index. The cost structure of a power market describes generation and transmission costs. Concentration and elasticity of demand are two key determinants of the competitiveness of a market's structure. *See also* market architecture.

**markup.**    *See* withholding.

**mega.**    1,000,000. *See also* kilo.

**merit order.**    A ranking of generators from those with the lowest average variable cost to those with the highest. Also, the ranking by marginal cost of started generators.

**midload generator.**    In this book, the term is equivalent to "intermediate capacity" which is defined by NERC to mean plants that operate for more hours than peakers and fewer than baseload plants, approximately 20% to 60% of the time. A plant with intermediate fixed and variable costs. Midload plants are often started daily, but run for most of the day not just a few peak hours. Used in models of startup cost to contrast with peakers which have lower startup costs.

**mill.**    Short for milli-dollar, 1/1000 of a dollar. Also, casually used for "mills per kilowatt-hour," as in "a price of 30 mills," because this cost unit is always combined with kilowatt-hours. Any cost expressed in mills per kilowatt-hour is numerically equal to the same cost in dollars per megawatt-hour.

**monopoly.** A firm that is the sole supplier of its product for which there are no substitutes and many buyers. This is an absolute monopoly and situations that approximate this condition are also referred to as monopolies. *See also* natural monopoly.

**monopoly power.** The ability of a firm or group of firms to exercise *market power*, in other words, market power on the supply side. (*Monopsony power* is market power on the demand side.)

**monopsony power.** The ability of a buyer or group of buyers to exercise *market power*.

**multipart bids.** *See* pool; marginal-cost pricing.

**Nash equilibrium.** *See* equilibrium.

**natural monopoly.** An industry is a natural monopoly if the total output of the industry can be produced more cheaply by one firm than by more than one. If the cost advantage of one firm is very slight, the industry is a very weak natural monopoly, and a competitive market will be advantageous if it can be maintained. If the cost advantage is large enough, the best choice will be to allow a regulated monopoly.

**net interchange.** The net flow of power out of one control area into all other control areas.

**no-load cost.** *See* cost of production.

**nonconvex costs.** If the variable cost of producing two units is less than twice the cost of producing one, the cost of production is termed nonconvex. More generally if the average of the costs of producing $q_1$ and $q_2$ is less than the cost of producing the average of $q_1$ and $q_2$, the cost of production is nonconvex. *See also* market-clearing price.

**ohm ($\Omega$).** The unit of measurement of resistance to the flow of electrical current. Ohm's law states that current is voltage divided by resistance. If a light bulb has 120 ohms of resistance and is plugged into a 120 V outlet, 1 A of current will flow through it. *See also* volt; amp.

**Ohm's law.** DC current flow through a wire, or any electrical device, equals the (input voltage minus the output voltage) divided by the resistance of the device.

An analogous law concerning AC power flows relates power flow to impedance.

**oligopoly.** A small group of suppliers that produces the entire output of a product with no close substitutes and as a consequence have market power.

**oligopsony.** A small group of buyers that trades with large group of sellers.

**operating reserve.** Generation in excess of demand, scheduled to be available on short notice to ensure the reliable operation of a control area. Operating reserves vary by response time. **Regulation** is used constantly to balance load fluctuations and is controlled by AGC, automatic generation control. **Spinning reserve**, generators that are running and synchronized with the AC grid, and producing at less than full output, can begin supplying more power almost instantly after a *contingency* occurs. Non-spinning reserves must be started which may take anywhere from ten minutes to an hour.

**operating-reserve pricing.** *See* OpRes pricing.

**operating-reserve requirement.** A level of operating reserves measured in megawatts. If reserves fall below this level the system operator attempts to increase them by purchasing more reserves or by purchasing more power and restoring some generation to reserve status. There are several types of operating reserve and each has its own requirement, though this book often treats all types as equivalent.

**OpRes market.** A market that employs only OpRes pricing for the purpose of inducing adequate investment in installed capacity.

**OpRes pricing.** Setting a high price for power when operating reserves are short of their required level as opposed to *VOLL pricing* which waits until load has been shed to set the price of power to $V_{LL}$. OpRes pricing uses a lower price but uses it much more often than *VOLL pricing*.

**optimal power flow.** A program for determining the least-cost dispatch of a power system subject to transmission constraints. Also, the power flow discovered by such a program.

**outage.** The removal from service availability of a generating unit, transmission line, or other facility.

A **forced outage** occurs because of an emergency such as an unanticipated equipment failure.

A **maintenance outage** removes equipment to perform maintenance that can be deferred a week but not until the next **planned outage**.

**parallel path flows.** *See* loop flow.

**peaker.** A peak-load plant.

**peak load.** Load during periods of relatively high system demand. "Relatively" can mean relative to the day, week or season. Peakers are used for daily peak load even when this is much lower than the annual peak.

**peak-load plant.** A generator, having lower *fixed costs* and higher *variable costs*, designed to serve peak load. These are often gas turbines, diesels, or pumped-storage hydroelectric equipment.

**perfect competition.** *See* competition and *under* equilibrium.

**physical limit.** In contrast with a *security limit*, the violation of a physical limit is likely to cause damage to some part of the system, though not necessarily the limited part. Physical limits on lines include *thermal limits* and *stability limits*.

**PJM.** The PJM Interconnection, L.L.C. is the organization responsible for the operation and control of the bulk electric power system in parts of five Mid-Atlantic states and the District of Columbia. With a peak load of 52 GW, it is the third largest, centrally-dispatched control area in the world after EDF in France and Tokyo Electric. It began operating its nodal spot market on April 1, 1998.

**planning reserve.** The difference between a *control area*'s expected annual peak capability and its expected annual peak demand expressed as a percentage of the annual peak demand. (NERC)

**pool.** A centralized energy market that makes side payments. Pools allow or require **multi-part bids** describing aspects of a generator's costs and production limitations. A pool solves the unit-commitment problem to optimize the dispatch. The resulting market price is not a *market-clearing price* but, combined with *side payments*, does clear the market. Also referred to as a *power pool* in this book.

**power.** The rate of flow of *energy*. If power is not specified as *reactive*, *real* power is indicated. Power is measured in watts which are joules per second, where a joule is a unit of *energy*. *See also* watts.

**power exchange.** *See* exchange market.

**power pool.** Originally an organization of utilities that coordinated their dispatch or traded bulk power. Today's *pools* in the eastern United States are direct descendants of the original power pools and are often referred to as power pools in this book. The original power pools solved the unit commitment problem but did not implement nodal pricing. *See also* pool.

**power system.** All electrically connected power grids plus all generators and loads and control centers associated with them.

**price distortion.** *See* withholding.

**price spike.** A rapid increase and decrease in price. In power markets, their *duration* is typically less than a day and the term is generally used only when prices exceed the variable cost of a peaker.

The **aggregate price spike** is the part of the *price-duration curve* that exceeds the variable cost of the peaker that is most expensive to operate but still investment worthy. Also referred to as **the price spike**.

The **price-spice revenue** is the average hourly scarcity rent per MW derived from price spikes by the highest variable-cost investment-worthy peaker. It equals the area under the aggregate price spike curve and above the variable cost of high-cost peaker. When the price-spike revenue equals the fixed costs of a high-cost peaker, investment in these peakers is in equilibrium.

**price taker.** A supplier that optimizes its production as if it could not affect the *market price*. A price taker produces to the point where its *marginal cost* equals the market price. The term can also apply to a buy and the buyer's consumption.

**procurement auction.** Auctions to buy instead of sell.

**profit.** This book uses profit to mean economic profit which is different from "net profit" used in business. Profit is revenue minus all costs including the cost of capital which includes a normal rate of return and any risk premium. Thus if the return on equity equals the cost of equity, in other words if the firm has a normal ROE, it will have zero economic profit.

**Long-run profit** is economic profit as defined above.

**Short-run profit** is revenue less all operating costs including startup and no-load costs. Short-run profit is needed to cover fixed costs such as the cost of capital.

**Scarcity rent** is revenue less all operating costs that vary with the level of output after a generator is started. Thus *short-run profit* equals *scarcity rent* less *startup* and *no-load costs*. In Part 2, *startup* and *no-load costs* are ignored, so *short-run profit* equals *scarcity rent*.

**Inframarginal rent** is another term for scarcity rent which can be calculated as the area between the market price and the supplier's marginal cost curve.

**profit function.** The profit function computes *expected* short-run profits, usually for a peaker, as determined by the level of installed capacity and the averaged *augmented load* conditions. The lower the level of installed capacity the greater are short-run profits due to greater scarcity of generating capacity. The steepness of the profit function indicates the riskiness of the market.

**public good.** A commodity for which use of a unit of the good by one agent does not preclude its use by other agents.

**quantity distortion.** *See* withholding.

**radial network.** A network without loops.

**ramping.** Increasing or decreasing the output of a generator. Steam units can only ramp at roughly 1% of their output per minute, so it can take them more than an hour to reach full output.

**reactive power.** An AC power flow can be decomposed into two components: real power which always flows from generator to load, and reactive power which flows back and forth with no net transfer of power in either direction. Reactive power flows, despite the naming convention, are real and cause real power losses and heating of system components.

**real power.** *See* power; reactive power.

**real-time prices.** Wholesale prices computed on an hourly or five-minute basis. Also used to mean real-time rates for end-use customers.

**real-time rates.** Retail electricity prices that fluctuate with the real-time wholesale price.

**regulation.** *See* operating reserve.

**reliability.** The ability of a power system to deliver power with a voltage and frequency within their normal limits. A system will be reliable if it has **adequate installed capacity** and is operated within security limits.

**Adequacy** is the ability of the electric system to supply the aggregate electrical demand and energy requirements of the customers at all times, taking into account scheduled and reasonably expected unscheduled outages of system elements (NERC).

**Security** is the ability of the electric system to withstand sudden disturbances such as electric short circuits or unanticipated loss of system elements (NERC).

**reserve.** *See* operating reserve; planning reserve.

**risk.** Uncertainty of income or cost. In economics, risk is not simply the chance of loss. An income stream that is occasionally $100 higher than usual is just as risky as one that has an equal chance of being $100 lower than usual.

Individuals or firms are **risk neutral** if they consider only long-run average costs and income and not their uncertainty.

A **risk premium** is an expected return in excess of that paid on risk-free securities. The premium provides compensation for the risk of an investment.

To test if a certain effect is caused by risk, consider whether a risk-neutral agent would be subject to the effect. If so, the effect is not due to risk.

**scarcity rent.** *See* profit.

**security.** *See* reliability.

**security-constrained dispatch.** A *dispatch* of a control area that minimizes generation cost taking into account the *security limits* on transmission lines. When security limits are binding, the result is different marginal costs of generation in different locations.

**security limit.** The power-flow limit imposed on a line to protect it from increased flows caused by unexpected outages of other lines. When any line goes out of service, physics instantly reroutes its power to other lines. A security limit generally prevents a line from exceeding its *physical limit* under the worst single *contingency*.

**shadow price.** The shadow price of a constraint is the reduction in cost (more precisely the increase in total surplus) that would result from a small relaxation of that constraint. For example, if a 1 kW increase in the flow limit of a power line would reduce the cost of delivered power by 3¢/h by allowing more use of a cheap remote generator, then the shadow price of the line is 3¢/kWh, or $30/MWh. If a line is not congested, its shadow price is zero because the line constraint is not binding.

**short run.** A term indicating the assumption that adjustments in the capital stock (total generation capacity) are being ignored but that output has been adjusted to maximize profit.

**short-run profit.** *See* profit.

**side payment.** A payment made by a power *pool* to compensate a supplier for the fact that a pool sets price equal to marginal cost and this is too low to clear the market. The payment equals the difference between the supplier's revenue from marginal-cost prices and its total short-run costs of providing power as indicated by its *multi-part bid.*

**Simple Model of Reliability.** A model presented in Section 2-3.3 in which load-shedding exactly equals *installed capacity* minus *augmented load.*

**spinning reserves.** *See* operating reserve.

**spot market.** The real-time market. Sometimes, but not in this book, it includes the day-ahead and hour-ahead markets.

**spot price.** The price in the spot market.

**stability limit.** A line limit based on the stability of the AC power flow. The physical stability limit of a line is determined by the power flow that requires a 90° voltage phase angle difference between the sending and receiving ends. Security limits based on the physical stability limit are closer to 30°. Stability limits are more likely to be binding on longer lines and *thermal limits* are more likely to bind on shorter lines.

**sub-market.** *See* market.

**surplus.** Total surplus is the difference between a product's value to the consumer (utility) and its cost of production. Also, the sum of producer and consumer surplus. Always used in one of the following compound terms.[3]

   **Consumer surplus** (or sometimes net consumer surplus) is the area under the market's demand curve and above the price of the commodity. (Gross consumer surplus is the total area under the demand curve.)

   **Producer surplus** (or profit) is the area below the market price and above the market's supply curve.

**Total surplus** (or aggregate surplus) is the sum of consumer surplus and producer surplus.

   **Marginal surplus** is the rate of change of total surplus with a costless increase in supply. Marginal surplus differs by location and is equal to the locational competitive price.

**swap.** A trade of energy at one location for energy at another location, often used to hedge uncertain transmission charges.

**synchronization.** The process of bringing a generator up to speed, making sure its AC voltage is "in step" with the power system voltage, and then connecting it to the system so it can deliver power.

**synchronized.** Two separate alternating-current apparatuses are synchronized if their voltages are alternating in lock step. If a cycle is considered a step, then the two synchronized voltages can never get more than a quarter of a step ahead or behind one another.

**system marginal cost (SMC).** The cost of producing one more unit of output using the cheapest available generator. As with *marginal cost,* SMC may be undefined, but in this case an analogous SMC range may be defined.

**system marginal value (SMV).** The increase in value to consumers of a small increase in consumption of power. Just as marginal cost determines the competitive supply curve, so marginal value determines the competitive supply curve.

**tariff.** The body of regulations governing a power market and approved by FERC.

**tera.** 1,000,000,000,000. *See also* kilo.

**thermal limit.** A power flow limit based on the possibility of damage by heat. Heating is caused by the electrical losses which are proportional to the square of the *real power* flow. More precisely, a thermal limit restricts the sum of the squares of real and *reactive power.*

**unit commitment.** The starting of a generator. Starting is expensive, so once a generator has been started, it is said to be committed. The **unit commitment problem** is the problem of finding the most economical times to commit and decommit all the individual generators in a control area. Because of the complex nature generation costs and constraints, solving the unit commit-

---

3. Tirole (1997), pages 8, 9, and 192.

ment problem requires advanced mathematics and enormous computations.

**uplift.**  A charge imposed on all customers, usually per MWh, that covers costs not covered by prices. Examples of such costs are redispatch costs when congestion is not priced, side payments in pools, and fixed costs of transmission. Often referred to as a tax in this book.

**value of lost load (VOLL).**  The average cost to customers per megawatt-hour of unserved load when they are disconnected during involuntary load shedding.

**variable cost.**  A cost that varies with the level of output. In examples, generator supply curves are often modeled as having constant variable cost up to full output. This is the generator's marginal cost until full output is reached; then marginal cost becomes undefined or may increase rapidly in some narrow *emergency operating range*. In this book, the constant variable cost below full output is referred to simply as the generator's "variable cost" rather than its *marginal cost* which could be undefined. *See also* cost of production.

**vesting contract.**  A contract signed by the purchaser of generating units, divested by a regulated utility, that generally specifies the price of a long-term power sale from these units to the regulated utility.

**VOLL.**  *See* value of lost load.

**VOLL market.**  A market that employs only VOLL pricing for the purpose of inducing adequate investment in installed capacity.

**VOLL pricing.**  A pricing policy that sets the spot market price to VOLL whenever load must be shed and there is a partial blackout.

**volt (V).**  The unit of electrical pressure. One *amp* of current forced through an appliance by 120 V of pressure delivers 120 W of power to the appliance.

**Walrasian equilibrium.**  *See* equilibrium.

**watt (W).**  The unit of *power* (electrical energy flow). One watt is the power delivered by 1 A of *current* flow under 1 V of pressure.

**withholding.**  Reducing output below the competitive, price-taking level at the market price. Withholding is termed **financial** if it is accomplished by asking a price above marginal cost and **physical** if it is accomplished by simply not producing.

The **quantity withheld** is the difference between the competitive supply at the market price and actual supply. There are three other measures of the effect of *market power* on prices and quantities. **Markup** is the price analog of quantity withheld and is the difference between actual quantity supplied and the competitive supply price (marginal cost) of the quantity.

**Quantity distortion** is the difference between competitive supply and the actual supply, while **price distortion** is the difference between the market price and the competitive price.

# References

Baldick, R. and W. Hogan. 2001. Capacity constrained supply function equilibrium models of electricity markets: stability, non-decreasing constraints, and function space iterations. PWP-089, University of California Energy Institute, University of California, Berkeley.

Bodie, Z., A. Kane, and A. Marcus. 1996. *Investments*. Chicago: Irwin.

Borenstein, S. 1999. Understanding competitive prices and market power in wholesale electricity markets. PWP-067, University of California Energy Institute, University of California, Berkeley.

Borenstein, S. 2001a. The trouble with electricity markets (and some solutions). PWP-081, University of California Energy Institute, University of California, Berkeley.

Borenstein, S. 2001b. Frequently asked questions about implementing real-time electricity pricing in California for Summer 2001. University of California Energy Institute, University of California, Berkeley.

Borenstein, S. and J. Bushnell. 1998. An empirical analysis of the potential for market power in California's electricity industry. PWP-044r, University of California Energy Institute, University of California, Berkeley.

Borenstein, S. and J. Bushnell. 1999. An empirical analysis of the potential for market power in California's electricity industry. *Journal of Industrial Economics* **47**(3), September: 285–323.

Borenstein, S. and J. Bushnell. 2000. Electricity restructuring: Deregulation or reregulation. *Regulation, The Cato Review of Business and Government* **23**(2): 46–52.

Borenstein, S., J. Bushnell, and C. Knittel. 1999. Market power in electricity markets: Beyond concentration measures. *Energy Journal* **20**(4).

Borenstein, S., J. Bushnell, and F. Wolak. 2000. Diagnosing market power in California's deregulated wholesale electricity market. PWP-064, University of California Energy Institute, University of California, Berkeley.

Brien, L. 1999. Why the ancillary services markets don't work in California and what to do about it. Working paper, National Economic Research Associates, San Francisco.

Bushnell, J. and S. Oren. 1994. Bidder cost revelation in electric power auctions. *Journal of Regulatory Economics* **6**: 5-26.

California Independent System Operator. 2000. Report on California energy market issues and performance, May-June 2000. Special Report, Department of Market Analysis, Folsom, Calif.

Cameron, L. and P. Cramton. 1999. The role of the ISO in U.S. electricity markets: A review of restructuring in California and PJM. *Electricity Journal* **12**(3): 71–81.

Chao, H. and R. Wilson. 1999a. *Design of wholesale electricity markets*. Book manuscript. Available at http://faculty-gsb.stanford.edu/wilson/E542/classmaterial.htm (accessed 7 February 2002).

Chao, H. and R. Wilson. 1999b. Incentive-compatible evaluation and settlement rules: multi-dimensional auctions for procurement of ancillary services in power markets. Forthcoming in *Journal of Regulatory*

*Economics*. Available at: http://faculty-gsb.stanford.edu/wilson/E542/classmaterial.htm (accessed 7 February 2002).

Cramton, P. and J. Lien. 2000. Eliminating the flaws in New England's Reserve Markets. Working paper, University of Maryland.

Cramton, P. and R. Wilson. 1998. A review of ISO New England's market rules. Working paper. Available at: http://www.market-design.com/files/ 98mdi-iso-ne-markets-review.pdf (accessed 6 January 2002).

Courant, R. *1937. Differential and integral calculus.* Vol. I, Second Ed. Translated by E .J. McShane. Bdlackie & Son Ltd, London and Glasgow.

DOE (U.S. Department of Energy). 1998a. Challenges of electric power industry restructuring for fuel suppliers. DOE/EIA-0623. Washington, D.C.

DOE (U.S. Department of Energy). 1998b. Electric trade in the United States 1996. DOE/EIA-0531(96). Washington, D.C.

DOE (U.S. Department of Energy). 2001a. Assumptions to the annual energy outlook 2001. Electricity market module. DOE/EIA-0554(2001). Washington, D.C., March 2.

DOE (U.S. Department of Energy). 2001b. Annual energy outlook 2001. Slide presentation by Mary J. Hutzler. Washington, D.C.

DOE (U.S. Department of Energy). 2001c. Electric power monthly June 2001. DOE/EIA-0226(2001/06). Washington, D.C.

DOJ (U.S. Department of Justice). 1997. 1992 horizontal merger guidelines (with April 8, 1997 revisions to Section 4 on efficiencies). Jointly issued by the U.S. Federal Trade Commission. Washington, D.C. Available at: http://www.usdoj.gov/atr/public/guidelines/horiz_book/hmg1.html (accessed 6 January 2002).

Federico, G. and D. Rahman. 2001. Bidding in an electricity pay-as-bid auction. Working paper, Nuffield College, Oxford University and University of California, Los Angeles. Available at: http://www.nuff.ox.ac.uk/Economics/papers/2001/w5/federico-rahmansept2001.pdf (accessed 6 January 2002).

Felder, F. 2001. "An island of technicality in a sea of discretion": A critique of existing electric power systems reliability analysis and policy. *Electricity Journal* 14(3):21–31.

FERC (U.S. Federal Energy Regulatory Commission). 1996a. Promoting wholesale competition through open access non-discriminatory transmission services by public utilities: Recovery of stranded costs by public utilities and transmitting utilities. Order no. 888. Final rule. Washington, D.C., April 24.

FERC (U.S. Federal Energy Regulatory Commission). 1996b. Inquiry concerning the commission's merger policy under the Federal Power Act: policy statement, Order no. 592. Docket No. RM96-6-000. Washington, D.C., December 18, 1996.

FERC (U.S. Federal Energy Regulatory Commission). 1999. Regional transmission organizations. Order No. 2000. Final rule: Docket No. RM99-2-000. Washington, D.C., December 20.

FERC (U.S. Federal Energy Regulatory Commission). 2000a. State of the markets 2000: Measuring performance in energy market regulation. Washington, D.C. Available at: http://www.ferc.fed.us /news/pressreleases/Perfpl_som2000.htm (accessed 6 January 2002).

FERC (U.S. Federal Energy Regulatory Commission). 2000b. Market order proposing remedies in California wholesale electrics. Washington, D.C., November 1.

FERC (U.S. Federal Energy Regulatory Commission). 2001a. Order on motion to implement hybrid fixed block pricing rule and requiring tariff filing, acting on related requests for rehearing, and accepting preliminary report. Washington, D.C., April 26.

FERC (U.S. Federal Energy Regulatory Commission). 2001b. Ensuring sufficient capacity reserves in today's energy markets: Should we? And how do we? Study team discussion paper. Washington, D.C., September 26.

FTC (U.S. Federal Trade Commission). 2000. Creating or maintaining a monopoly. Washington, DC. Available at: http://www.ftc.gov/bc/compguide/maintain.htm (accessed 7 February 2002).

Green, R. and D. Newbery. 1992. Competition in the British electricity spot market. *Journal of Political Economy* **100**(5): 929–953.

Green, R. 1996. Increasing competition in the British electricity spot market. *Journal of Industrial Economics* **44**: 205–216.

Green, R. 1998. Draining the pool: The reform of electricity trading in England and Wales. Working paper, Department of Applied Economics, University of Cambridge.

Gribik, P. 1995. Learning from California's QF auction. *Public Utilities Fortnightly* April 15.

Hirst, E. 2001. Real-time balancing operations and markets: Key to competitive wholesale electricity markets. Report prepared for the Edison Electric Institute, Washington, D.C., and the Project for Sustainable FERC Energy Policy, Alexandria, Virginia.

Hirst, E. and S. Hadley. 1999. Maintaining generation adequacy in a restructuring U.S. electricity industry. Report prepared for the Environmental Protection Agency; ORNL/CON-472. Oak Ridge, Tennessee.

Hobbs, B., J. Iñón, and S. Stoft. Installed capacity requirements and price caps: oil on the water, or fuel on the fire? *Electricity Journal* July, **14**(6): 23–34.

Hogan, W. 1992. Contract networks for electric power transmission: Technical reference. Working paper, John F. Kennedy School of Government, Harvard University.

Hogan, W. 1995. Poolco: What's the trick? Slide presentation at the Summer Membership Meeting, Electric Generation Association, Greenbrier, West Virginia.

Hogan, W. 1997. A concurrent auction model for transmission congestion contracts. Working paper, Center for Business and Government, John F. Kennedy School of Government, Harvard University.

Hogan, W. 1998. Competitive electricity market design: A wholesale primer. Working paper, Center for Business and Government, John F. Kennedy School of Government, Harvard University.

Hogan, W. 1999. Getting the prices right in PJM: Analysis and summary: April 1998 through March 1999 The First Anniversary of Full Locational Pricing. April 2.

Hogan, W. 2001b. Coordination for competition: Electricity market design principles. Lecture presented at the Public Utility Commission of Texas Workshop on ERCOT Proposals, Austin, Texas.

Hogan, W. 2001c. Designing market institutions for electric network systems: Reforming the reforms in New Zealand and the U.S. Paper presented at The Utility Convention, Auckland, New Zealand.

Jaffe, A. and F. Felder. 1996. Should electricity markets have a capacity requirement? If so, how should it be priced? *Electricity Journal* December, **9**: 52–60.

Joskow, P. 2000a. Why do we need electricity retailers? Mimeo, Department of Economics, Massachusetts Institute of Technology, Boston, February 13.

Joskow, P. 2000b. Deregulation and regulatory reform in the U.S. electric power sector. Paper prepared for the Brookings-AEI Conference on Deregulation in Network Industries, Washington, D.C.

Joskow, P. 2001a. California's electricity market meltdown. OpEd, *New York Times*, January 13.

Joskow, P. 2001c. Statement before the Governmental Affairs Committee, U.S. Senate, June 13.

Joskow, P. and E. Kahn. 2001b. Identifying the exercise of market power: Refining the estimates. Working paper, Department of Economics, Massachusetts Institute of Technology and Analysis Group/Economics, San Francisco.

Joskow, P. and E. Kahn. 2001a. A quantitative analysis of pricing behavior in California's wholesale electricity market during Summer 2000. Working paper, Department of Economics, Massachusetts Institute of Technology and Analysis Group/Economics, San Francisco.

Joskow, P. and R. Schmalensee. 1983. *Markets for power: An analysis of electric utility deregulation.* Cambridge, MA: The MIT Press.

Kahn, A., P. Cramton, R. Porter, and R. Tabors. 2001. Pricing in the California exchange power electricity market: Should California switch from universal pricing to pay-as-bid pricing? Blue Ribbon Panel report commissioned by the California Power Exchange.

Klemperer, P. 1999. Auction theory: A guide to the literature. *Journal of Economic Surveys* 13(3): 227–286.

Klemperer, P. 2001a. Why every economist should learn some auction theory. Paper presented at the 8th World Congress of the Econometric Society, Seattle.

Klemperer, P. 2001b. What really matters in auction design. Working paper, Department of Economics, Oxford University.

Klemperer, P. and M. Meyer. 1989. Supply function equilibria in oligopoly under uncertainty. *Econometrica* 57(6): 1243–1277.

Knittel, C. 1999. The origins of state electricity regulation: Revisiting an unsettled topic. PWP-048, University of California Energy Institute, University of California, Berkeley.

Littlechild. 2000. Why we need electricity retailers: A reply to Joskow on wholesale spot price pass-through. WP 21/2000, Research Papers in Management Studies, Judge Institute of Management Studies, and DAE Working paper 0008, Department of Applied Economics, University of Cambridge, England.

Mas-Collel, A., A. Whinston, and J. Green. 1995. *Microeconomic Theory*. New York: Oxford University Press.

NECA (National Electricity Code Administrator Limited). 1999a. Review of VoLL in the national electricity market: Issues paper. Report of the Reliability Panel. Adelaide, Australia.

NECA (National Electricity Code Administrator Limited). 1999b. Review of VoLL in the national electricity market: Report and recommendations. Final report of the Reliability Panel. Adelaide, Australia.

NECA (National Electricity Code Administrator Limited). 1999c. A plain English guide to VoLL. Report of the Reliability Panel. Adelaide, Australia.

NERC (North American Electric Reliability Council). 1996. Glossary of terms. Report prepared by the Glossary of Terms Task Force. Princeton, New Jersey.

NERC (North American Electric Reliability Council). 2000. Reliability assessment 2000–2009. Princeton, New Jersey.

PJM (PJM Interconnection, L.L.C). 1999. PJM manual for dispatching operations, Manual M-12. Valley Forge, Pennsylvania.

PJM (PJM Interconnection, L.L.C). 2001. State of the market report 2000. Report of the Market Monitoring Unit. Valley Forge, Pennsylvania.

Platt, H. 1991. *The electric city: Energy and growth of the Chicago area, 1880–1930*. Chicago: University of Chicago Press.

Reiss, P. and M. White. 2001. Household electricity demand, revisited. Working paper, Graduate School of Business, Stanford University.

Rosenberg, A. 2000. Congestion pricing or monopoly pricing. *Electricity Journal* **13**(3): 33–41.

Rothwell, G. and T. Gomez. 2002. *Electricity Economics: Regulation and Deregulation*. New York: Wiley-IEEE Press.

Rudolph, R and S. Ridley. 1986. *Power struggle: The hundred-year war over electricity*. New York: Harper & Row, Publishers.

Ruff, L. 1999. Competitive electricity markets: Why they are working and how to improve them. Working paper, National Economic Research Associates, Cambridge, MA.

Ruff, L. 2001. Origins of the Original UK pool. Private memorandum to Stephen Littlechild. Available at www.stoft.com.

Samuelson, P. 1973. *Economics*. New York: McGraw-Hill.

Smith, V. 1995. Regulatory reform in the electric power industry. Working paper, Department of Economics, University of Arizona.

Schweppe, F., M. Caraminis, R. Bohn, and R. Tabors. 1988. *Spot pricing of electricity*. Kluwer Academic Publishers, New York.

Tirole, J. 1997. *The theory of industrial organization*. Cambridge, MA: The MIT Press.

Varaiya, P. 1996. Coordinated multilateral trades. Slides from a paper presented at the POWER Conference, University of California Energy Institute, University of California, Berkeley.

Wilson, R. 1997. Implementation of priority insurance in power exchange markets. *Energy Journal* **18**(1): 111–123.

Wilson, R. 1999. Market architecture. Working paper, Graduate School of Business, Stanford University, Available at: http://faculty-gsb.stanford.edu/wilson/E542/classmaterial.htm (accessed 7 February 2002).

Wolak, F. 1999. Report on the redesign of California real-time energy and ancillary services markets. Report to the California Independent System Operator, Folsom, California.

Wolak, F. 2000. An empirical analysis of the impact of hedge contracts on bidding behavior in a competitive electricity market. Working paper, Department of Economics, Stanford University.

Wolak, F. 2001. Want 10,000 megawatts now? Use variable power pricing. OpEd, *San Jose Mercury News*, May 4.

Wolak, F., Nordhaus, C. Shapiro. 2000. An analysis of the June 2000 price spikes in the California iso's energy and ancillary services markets. Report to the California Independent System Operator, Folsom, California.

Wolfram, C. 1999. Measuring duopoly power in the british electricity spot market. *American Economic Review* **89** (September): 805–826.

Wood, A. and B. Wollenberg. 1996. *Power generation, operation, and control*. NY: John Wiley & Sons, Inc.

Wu, F. and P. Varaiya. 1995. Coordinated multilateral trades for electric power networks: Theory and application. PWP-031, University of California Energy Institute, University of California, Berkeley.

# Index